Unless Recalled Earlier

Bacterial Artificial Chromosomes

METHODS IN MOLECULAR BIOLOGY™

John M. Walker, SERIES EDITOR

Bacterial Artificial Chromosomes

Volume 2
Functional Studies

Edited by

Shaying Zhao

The Institute for Genomic Research, Rockville, MD

and

Marvin Stodolsky

Office of Biological and Environmental Research,
US Department of Energy, Germantown, MD

HUMANA PRESS ✳ TOTOWA, NEW JERSEY

© 2004 Humana Press Inc.
999 Riverview Drive, Suite 208
Totowa, New Jersey 07512

www.humanapress.com

All papers, comments, opinions, conclusions, or recommendations are those of the author(s), and do not necessarily reflect the views of the publisher. Methods in Molecular Biology™ is a trademark of The Humana Press, Inc.

This publication is printed on acid-free paper. ∞
ANSI Z39.48-1984 (American Standards Institute)
Permanence of Paper for Printed Library Materials.

Production Editor: Tracy Catanese
Cover design by Patricia F. Cleary.

For additional copies, pricing for bulk purchases, and/or information about other Humana titles, contact Humana at the above address or at any of the following numbers: Tel.: 973-256-1699; Fax: 973-256-8341; E-mail: humana@humanapr.com; or visit our website: www.humanapress.com

Printed in the United States of America. 10 9 8 7 6 5 4 3 2 1

1-59259-753-X (e-book)
ISSN 1064-3745

Library of Congress Cataloging in Publication Data

Bacterial artificial chromosomes / edited by Shaying Zhao and Marvin Stodolsky.
    p. ; cm. -- (Methods in molecular biology ; v. 255-256)
Includes bibliographical references and index.
   ISBN 0-89603-988-9 (v. 1 : alk. paper) -- ISBN 0-89603-989-7 (v. 2 : alk. paper)
  1. Bacterial artificial chromosomes--Laboratory manuals.
    [DNLM: 1. Chromosomes, Artificial, Bacterial. 2. Chromosome Mapping. 3. Cloning, Molecular. 4. Genomic Library. QW 51 B129 2004] I. Zhao, Shaying. II. Stodolsky, Marvin. III. Methods in molecular biology (Clifton, N.J.) ; v. 255-256.
   QH600.15.B338 2004
   660.6'5--dc22

                                                                                    2003023009

Preface

Several developmental and historical threads are woven and displayed in these two volumes of *Bacterial Artificial Chromosomes*, the first on *Library Construction, Physical Mapping, and Sequencing*, and the second on *Functional Studies*. The use of large-insert clone libraries is the unifying feature, with many diverse contributions. The editors have had quite distinct roles. Shaying Zhao has managed several BAC end-sequencing projects. Marvin Stodolsky during 1970–1980 contributed to the elucidation of the natural bacteriophage/prophage P1 vector system. Later, he became a member of the Genome Task Group of the Department of Energy (DOE), through which support flowed for most clone library resources of the Human Genome Program (HGP). Some important historical contributions are not represented in this volume. This preface in part serves to mention these contributions and also briefly surveys historical developments.

Leon Rosner (deceased) contributed substantially in developing a PAC library for drosophila that utilized a PI virion-based encapsidation and transfection process. This library served prominently in the Drosophila Genome Project collaboration. PACs proved easy to purify so that they substantially replaced the YACs used earlier. Much of the early automation for massive clone picking and processing was developed at the collaborating Lawrence Berkeley National Laboratory. However, the P1 virion encapsidation system itself was too fastidious, and P1 virion-based methods did not gain popularity in other genome projects.

Improving clone libraries was an early core constituent of the DOE genome efforts. Cosmid-based libraries with progressively larger inserts were developed within the DOE National Laboratories Gene Library Program. But quality control tests by P. Youdarian indicated that perhaps 25% of human insert cosmids had some instability, possible owing to the multicopy property of the system. Both for this reason and to provide for larger inserts of cloned DNAs, DOE supported the investigation of several new cloning systems. Of the eukaryotic host systems, the Epstein-Barr virus-based system from Jean-M. Vos (deceased) was quite successful. But the added costs and care needed for use of eukaryotic cells precluded its wide adoption in HGP production efforts.

Among the bacterial host systems, two developed in the lab of Melvin Simon provided pivotal service. Ung-Jin Kim developed fosmids. They are maintained as single copy replicons and utilize the reliable encapsidation processes developed for cosmids. Fosmids proved to be highly stable. BACs were developed by Hiroaki Shizuya. They were introduced into *E. coli* by electroporation and stability was generally good, though there is an unstable BAC minority *(1)*. This BAC resource emerged after the chimeric properties of the large YACs was recognized. BACs were thus initially viewed with appropriate suspicion. But at the nearby Cedar-Sinai Medical Center, J. R. Korenberg and X.-N. Chen implemented a very efficient FISH analysis. They found that chimerism in any of the BACs was at worst around 5% and the BACs were well distributed across all the chromosomes. Overall human genome coverage was estimated in the 98–99% range, with even centromeric and near telomeric regions represented.

Two examples of this good coverage soon emerged. Isolation of the BRAC1 breast cancer gene had failed with all other clone resources. But when Simon's group was provided with a short cDNA probe, they soon returned a BAC clone carrying an intact BRAC1 gene. Pieter de Jong had acquired the technology of cloning long DNA inserts from the Simon lab, initially using a PAC vector and electroporation. After a first successful library, DOE advised de Jong to broadly distribute this new PAC resource. Shortly thereafter, he assembled a 900 kb contig for the candidate region of the BRAC2 gene. The subsequent DNA sequence generated at the Washington University then revealed the BRAC2 gene. These striking easy successes stimulated broad usage of the BAC and PAC resources.

End sequences of clonal inserts have been used to facilitate contig building since the 1980s in small-scale mapping and sequencing projects. Glen Evans for example was piloting with DOE support a "mapping plus sequencing" strategy on chromosome 11, before the BAC resources were available. Once a covering set of cloned DNAs with sequenced ends is generated, clones to efficiently extend existing sequence contigs can be chosen. As the need for high throughput genome sequencing to meet HGP timelines became imminent, only a few human chromosomes had adequate contig coverage. L. Hood, H. Smith, and C. Venter proposed a Sequence Tag Connector (STC) strategy to alleviate this bottleneck. With application to the entire human genome, concurrent BAC contig building and sequencing would be implemented.

The DOE instituted a fast track review of two STC applications in the spring of 1996 *(2)*. One was from a team comprised of L. Hood, H. Smith, and C. Venter, and the second from a team comprised of G. Evans, P. de Jong, and J. R. Korenberg. A panel with broad international representation reviewed applications from two teams. Interested colleagues from the NIH and NSF were observers. Although the overall STC concept was reviewed favorably, initial pilot implementations to better define the economics were recommended. A year later, progress was reviewed and a DOE commitment to a full scale implementation was made. At the request of the NIH, the DOE later increased support to accelerate a 20-fold coverage of the genome.

The STC data set has had multiple beneficial roles. Sequence Tag Sites (STSs) were defined within the STC sequences and used to enrich the Radiation Hybrid (RH) maps of the genome, thus providing for an early correspondence of the RH maps and the maturing contig maps. Validity constraints on sequence contigs were provided by the spanning BACs. Most broadly, the STC resource had an indispensable role for both the strategies of Celera Genomics Inc., and the international public sector collaboration, in the rapid generation of draft sequences of the human genome. The STC strategy is now implemented in many current genomic projects, including the NIH sponsored mouse and rat genome programs.

Bacterial Artificial Chromosomes in its two volumes provides a comprehensive collection of the protocols and resources developed for BACs in recent years. These two volumes collectively cover four topics about BACs: (1) library construction, (2) physical mapping, (3) sequencing, and (4) functional studies. The laboratory protocols follow the successful *Methods in Molecular Biology*™ series format by containing a clear sequence of steps followed by extensive troubleshooting notes. The protocols cover simple techniques such as BAC DNA purification to such complex procedures as BAC transgenic mouse generation. Both routine and novel methodologies are presented. Besides protocols, chapter topics include scientific reviews, software tools, database resources, genome sequencing strategies, and case studies. The books should be useful to those with a wide range of expertise from starting graduate students to senior investigators. We hope our books will provide useful protocols and resources to a wide variety of researchers, including genome sequencers, geneticists, molecular biologists, and biochemists studying the structure and function of the genomes or specific genes.

We would like to thank all those involved in the preparation of this volume, our colleagues, and friends for helpful suggestions, and Professor John Walker, the series editor, for his advice, help and encouragement.

Shaying Zhao
Marvin Stodolsky

References

1. http://www.ornl.gov/meetings/ecr2/index.html
2. http://www.ornl.gov/meetings/bacpac/body.html

Contents of Volume 2

Contents of the Companion Volume

Volume 1: Library Construction, Physical Mapping, and Sequencing

Contributors

MATHIAS ACKERMANN • *Institute of Virology, University of Zurich, Zurich, Switzerland*

DONNA G. ALBERTSON • *Comprehensive Cancer Center, University of California, San Francisco, CA*

JANETTE ALLISON • *Department of Microbiology and Immunology, University of Melbourne, Melbourne, Victoria, Australia*

PHILIP AVNER • *Unité de Génétique Moléculaire Murine, Institut Pasteur, Paris, France*

JOSEPH C. BAKER, JR. • *North Carolina Central University, Durham, NC*

VLADMIR BENES • *Genomics Core Facility, European Molecular Biology Laboratory, Heidelberg, Germany*

CHRISTINE BIRD • *Wellcome Trust Genome Campus, The Sanger Institute, Cambridge, UK*

STEPHANIE BLACKWOOD • *Comprehensive Cancer Center, University of California, San Francisco, CA*

EVA-MARIA BORST • *Virus Cell Interaction Unit, Medical Faculty, Martin-Luther-University of Halle-Wittenberg, Halle, Germany*

SHANNON BRADY • *Stanford Human Genome Center, Department of Genetics, Stanford University School of Medicine, Palo Alto, CA*

PRADEEP K. CHATTERJEE • *Julius L. Chambers Biomedical/Biotechnology Research Institute, North Carolina Central University, Durham, NC*

XIAO-NING CHEN • *Department of Human Genetics and Pediatrics, Cedars-Sinai Medical Center, University of California, Los Angeles, CA*

SANGDUN CHOI • *Division of Biology, California Institute of Technology, Pasadena, CA*

DAVID R. COX • *Perlegen Sciences, Mountain View, CA*

IRENA CRNKOVIC-MERTENS • *Department of Medical Virology, University of Heidelberg, Heidelberg, Germany*

SALLY H. CROSS • *MRC Human Genetics Unit, Western General Hospital, Edinburgh, UK*

ELAHE ELAHI • *Stanford Genome Technology Center, Stanford University, Palo Alto, CA*

CHRIS ELKIN • *Production Sequencing Department, DOE Joint Genome Institute, Walnut Creek, CA*

PETER ENGLER • *Department of Molecular Genetics and Cell Biology, University of Chicago, Chicago, IL*

SIMON J. FOOTE • *The Walter and Eliza Hall Institute of Medical Research, Royal Melbourne Hospital, Victoria, Australia*

CORNEL FRAEFEL • *Institute of Virology, University of Zurich, Zurich, Switzerland*

TIJANA GLAVINA • *Production Sequencing Department, DOE Joint Genome Institute, Walnut Creek, CA*

CLAUDIA GÖSELE • *Max-Planck-Institute for Molecular Genetics, Berlin, Germany*

DARREN GRAFHAM • *Wellcome Trust Genome Campus, The Sanger Institute, Cambridge, UK*

JANE GRIMWOOD • *Stanford Human Genome Center, Department of Genetics, Stanford University School of Medicine, Palo Alto, CA*

TREVOR HAWKINS • *DOE Joint Genome Institute, Walnut Creek, CA*

HANS PETER HEFTI • *Institute of Virology, University of Zurich, Zurich, Switzerland*

IRMA HEID • *Institute of Virology, University of Zurich, Zurich, Switzerland*

THOMAS G. H. HEISTER • *Institute of Virology, University of Zurich, Zurich, Switzerland*

JAMIE JETT • *Production Sequencing Department, DOE Joint Genome Institute, Walnut Creek, CA*

HANS E. JOHNSEN • *The Research Laboratory, Department of Haematology L, Herlev Hospital, University of Copenhagen, Denmark*

PIETER J. DE JONG • *BACPAC Resource Center, Children's Hospital Oakland Research Institute, Oakland, CA*

HITESH KAPUR • *Production Sequencing Department, DOE Joint Genome Institute, Walnut Creek, CA*

UNG-JIN KIM • *Department of Biology, California Institute of Technology, Pasadena, CA*

MARGIT KNOBLAUCH • *Max-Planck-Institute for Molecular Genetics, Berlin, Germany*

JULIE R. KORENBERG • *Department of Human Genetics and Pediatrics, Cedars-Sinai Medical Center, University of California, Los Angeles, CA*

MAXIM KORIABINE • *Laboratory of Biosystems and Cancer, National Cancer Institute, Bethesda, MD*

DAN KOSACK • *The Institute for Genomic Research, Rockville, MD*

ULRICH H. KOSZINOWSKI • *Department of Virology, Max von Pettenkofer Institut, Munich, Germany*

NATALAY KOUPRINA • *Laboratory of Biosystems and Cancer, National Cancer Institute, Bethesda, MD*

SERGEI KOZYAVKIN • *Fidelity Systems Inc., Gaithersburg, MD*

THOMAS KREITLER • *Max-Planck-Institute for Molecular Genetics, Berlin, Germany*

TAMARA KUCABA • *Genome Sequencing Center, Washington University School of Medicine, St. Louis, MO*

VLADIMIR LARIONOV • *Laboratory of Biosystems and Cancer, National Cancer Institute, Bethesda, MD*

SUN-HEE LEEM • *Department of Biology, Dong-A University, Pusan, Korea*

CHANG-SU LIM • *Department of Microbiology and Molecular Genetics, University of Texas-Houston Medical School, Houston, TX*

ANGELIKA LONGACRE • *Department of Molecular Genetics and Cell Biology, University of Chicago, Chicago, IL*

SUSAN LUCAS • *Production Sequencing Department, DOE Joint Genome Institute, Walnut Creek, CA*

SOLIDA MAK • *Department of Microbiology and Molecular Genetics, University of Texas-Houston Medical School, Houston, TX*

ANDREI MALYKH • *Fidelity Systems Inc., Gaithersburg, MD*

OLGA MALYKH • *Fidelity Systems Inc., Gaithersburg, MD*

ANUP MADAN • *Institute for Systems Biology, Seattle, WA*

MARCO MARRA • *Genome Sciences Centre, BC Cancer Agency, Vancouver, BC, Canada*

VIKKI M. MARSHALL • *The Walter and Eliza Hall Institute of Medical Research, Royal Melbourne Hospital, Victoria, Australia*

ALISTAIR MCGREGOR • *Department of Molecular Genetics, University of Cincinnati School of Medicine, Cincinnati, OH*

MARTIN MESSERLE • *Virus-Cell Interaction Unit, Medical Faculty, Martin-Luther-University of Halle-Wittenberg, Halle, Germany*

NATHAN MISE • *Unité de Génétique Moléculaire Murine, Institut Pasteur, Paris, France*

JOHN J. MONACO • *Howard Hughes Medical Institute, University of Cincinnati, Cincinnati, OH*

LARS MÜLLER • *Institute of Virology, University of Zurich, Zurich, Switzerland*

JOEP P. P. MUYRERS • *Gene Expression Program, European Molecular Biology Laboratory, Heidelberg, Germany*

RICHARD M. MYERS • *Stanford Human Genome Center, Department of Genetics, Stanford University School of Medicine, Palo Alto, CA*

STEVEN J. NORRIS • *Department of Pathology and Laboratory Medicine, University of Texas-Houston Medical School, Houston, TX*

VLADIMIR N. NOSKOV • *Laboratory of Biosystems and Cancer, National Cancer Institute, Bethesda, MD*

MICHAEL OLIVIER • *Human and Molecular Genetics Center, Medical College of Wisconsin, Milwaukee, WI*

KAZUTOYO OSOEGAWA • *BACPAC Resource Center, Children's Hospital Oakland Research Institute, Oakland, CA*

SANGITA PAL • *Department of Microbiology and Molecular Genetics, University of Texas-Houston Medical School, and Human Genome Sequencing Center, Baylor College of Medicine, Houston, TX*

RASHMI PERSHAD • *Department of Molecular Genetics, The University of Texas M. D. Anderson Cancer Center, Houston, TX*

DANIEL PINKEL • *Cancer Research Institute, University of California, San Francisco, CA*

FRANK POGODA • *Max von Pettenkofer Institute, Department of Virology, Genzentrum of the Ludwig-Maximilians-University of Munich, Munich, Germany*

NIKOLAI POLUSHIN • *Fidelity Systems Inc., Gaithersburg, MD*

MIHAI POP • *The Institute for Genomic Research, Rockville, MD*

GYÖRGY PÓSFAI • *Institute of Biochemistry, Biological Research Center, Szeged, Hungary*

TIM S. POULSEN • *Probe Applications, DakoCytomation, Denmark*

PAUL F. PREDKI • *Production Sequencing Department, DOE Joint Genome Institute, Walnut Creek, CA*

DIANA RADUNE • *The Institute for Genomic Research, Rockville, MD*

RACHEL REEG • *Institute for Systems Biology, Seattle, WA*

JEANETTE M. J. RIENTJES • *Gene Bridges GmbH, Dresden, Germany*

BRUCE A. ROE • *Department of Chemistry and Biochemistry, University of Oklahoma, Norman, OK*

MOSTAFA RONAGHI • *Stanford Genome Technology Center, Stanford University, Palo Alto, CA*

JACQUELINE SCHEIN • *Genome Sciences Centre, BC Cancer Agency, Vancouver, BC, Canada*

MARK R. SCHLEISS • *Division of Infectious Disease, Children's Hospital Research Foundation, Cincinnati, OH*

JEREMY SCHMUTZ • *Stanford Human Genome Center, Department of Genetics, Stanford University School of Medicine, Palo Alto, CA*

RICHARD SEGRAVES • *Comprehensive Cancer Center, University of California, San Francisco, CA*

MANDEEP SEKHON • *Genome Sequencing Center, Washington University School of Medicine, St. Louis, MO*

SHYAM K. SHARAN • *Mouse Cancer Genetics Program, National Cancer Institute, Frederick, MD*

ALEXEI SLESAREV • *Fidelity Systems Inc., Gaithersburg, MD*

DUANE SMAILUS • *Genome Sciences Centre, BC Cancer Agency, Vancouver, British Columbia, Canada*

DAVID SMAJS • *Department of Microbiology and Molecular Genetics, University of Texas-Houston Medical School, Houston, TX*

ANTOINE M. SNIJDERS • *Comprehensive Cancer Center, University of California, San Francisco, CA*

A. FRANCIS STEWART • *BioInnovation Zentrum, University of Technology, Dresden, c/o Max-Planck-Institute for Cell Biology and Genetics, Dresden, Germany*

MARVIN STODOLSKY • *Office of Biological and Environmental Research, US Department of Energy, Germantown, MD*

MARK SUTER • *Institute of Virology, University of Zurich, Zurich, Switzerland*

SRIVIDYA SWAMINATHAN • *Center for Cancer Research, National Cancer Institute, Frederick, MD*

DEBORAH A. SWING • *Center for Cancer Research, National Cancer Institute, Frederick, MD*

TANYA TEMPLETON • *The Walter and Eliza Hall Institute of Medical Research, Royal Melbourne Hospital, Victoria, Australia*

GIUSEPPE TESTA • *BioInnovation Zentrum, University of Technology, Dresden, c/o Max-Planck Institut for Cell Biology and Genetics, Dresden, Germany*

HERVÉ TETTELIN • *The Institute for Genomic Research, Rockville, MD*

KRISTINA VINTERSTEN • *European Molecular Biology Laboratory, Heidelberg, Germany, and Samuel Lunenfeld Research Institute, Mount Sinai Hospital, Toronto, Canada*

ANDREA VÖGTLIN • *Institute of Virology, University of Zurich, Zurich, Switzerland*

MARKUS WAGNER • *Department of Pathology, Harvard Medical School, Boston, MA*

RUOPING WANG • *Leukemia Section, The University of Texas M. D. Anderson Cancer Center, Houston, TX*

ZUNDE WANG • *Genomics Division, EpiGenX Pharmaceuticals, Santa Barbara, CA*

ROBERT WATERSTON • *Genome Sequencing Center, Washington University School of Medicine, St. Louis, MO*

MICHAEL M. WEIL • *Environmental and Radiological Health Sciences, Colorado State University, Fort Collins, CO*

GEORGE M. WEINSTOCK • *Human Genome Sequencing Center, Baylor College of Medicine, Houston, TX*

SCOTT E. WENDERFER • *Department of Pediatrics, University of Texas Houston, Houston, TX*

YOUMING ZHANG • *Gene Bridges GmbH, Dresden, Germany*

SHAYING ZHAO • *The Institute for Genomic Research, Rockville, MD*

SHENG ZHAO • *Department of Biostatistics, M. D. Andersen Cancer Center, Houston, TX*

HEIKE ZIMDAHL • *Experimental Genetics of Cardiovascular Diseases, Max-Delbrueck Center for Molecular Medicine, Berlin, Germany*

1

Use of BAC End Sequences for SNP Discovery

Michael M. Weil, Rashmi Pershad, Ruoping Wang,
and Sheng Zhao

1. Introduction

Genetic markers have evolved over the years, increasing in their numbers and utility. Beginning with phenotypes such as smooth or wrinkled, the selection of genetic markers broadened to include blood group and histocompatibility antigens, and protein allotypes. Around 1980, DNA itself became the marker *(1)*, first with restriction fragment length polymorphisms (RFLPs) and then with amplification polymorphisms based on simple sequence lengths (SSLPs) *(2)*. Each advance in the availability and usefulness of genetic markers has contributed to advances in fundamental and applied genetics.

Single nucleotide polymorphisms (SNPs) are particularly powerful markers for genetic studies because they occur frequently in the genome, allowing the construction of dense genetic maps. Also, SNP-based genotyping should be more amenable to automation and multiplexing than genotyping based on other currently available markers.

A variety of strategies have been used for SNP discovery. These include resequencing approaches based on the standard dideoxy, cycle sequencing methodology, or DNA "chips." Recently, we undertook a search for SNPs between commonly used inbred strains of laboratory mice using a resequencing approach. We took advantage of bacterial artificial chromosome (BAC) end sequence data generated by others for the public mouse genome sequencing effort. These sequences allowed us to design polymerase chain reaction (PCR) primers for amplification of homologous sequences in different mouse strains. We then sequenced the PCR products and identified sequence variations between the strains. Whenever possible, we used publicly available software and

From: *Methods in Molecular Biology, vol. 256:*
Bacterial Artificial Chromosomes, Volume 2: Functional Studies
Edited by: S. Zhao and M. Stodolsky © Humana Press Inc., Totowa, NJ

commercially available reagents. The approach is suitable for any organism for which some sequence data are available.

2. Materials
2.1. Sequence Selection and Primer Design Programs

1. CLEAN N is publicly available at http://odin.mdacc.tmc.edu/anonftp/
2. INPUT PRIMER is available at http://odin.mdacc.tmc.edu/anonftp/
3. Primer3 is available from the Whitehead Institute/MIT Center for Genome Research at http://www-genome.wi.mit.edu/genome_software/other/primer3.html

2.2. Amplification

1. 10X buffer: 15 mM MgCl$_2$, 0.5 M KCl, 0.1 M Tris-HCl, pH 8.3 (Sigma).

2.3. Preparation for Sequencing

1. Exonuclease I (Amersham Life Science).
2. Shrimp Alkaline Phosphatase (USB).
3. Sybr Green Dye (Molecular Probes).

2.4. Sequencing

1. ABI Prism Big Dye Terminator Ready Reaction Kit v2.0 (Perkin-Elmer).
2. 5X Reaction Buffer (Perkin-Elmer).
3. Multiscreen plate (Millipore, MAHV4510).
4. Sephadex G50 Superfine (Amersham).
5. Deionized and distilled water (VWR).
6. 45 µL column loader (Millipore).

2.5. SNP Identification

1. PolyBayes Software is available from the University of Washington (http://genome:wustl.edu/gsc/Informatics/polybayes/).

3. Methods
3.1. Selection of BAC End Sequences and Primer Design

1. The initial step is to select sequences that are long enough to take advantage of the full accurate read length of the sequencer that will be used. Repetitive sequences are excluded to avoid designing PCR primers that will amplify more than one genomic region. In addition, some SNP genotyping assays are not well suited for discriminating SNPs within a repetitive sequence, so focusing SNP discovery on nonrepetitive sequences will avoid genotyping difficulties later. Sequence selection can be automated with CLEAN N, an in-house computer program that we developed. The input for this program is a flat sequence file in FASTA format in which repetitive sequences are masked with "N" symbols (*see* **Note 1**). CLEAN N

removes sequences shorter than 600 nucleotides and those containing one or more "N" symbols.

2. The remaining sequences are put into an input format for the primer design program by another in-house program, INPUT PRIMER. We then use the Primer3 program *(3)*, which is available from the Whitehead Institute/MIT Center for Genome Research, to design PCR primers. The basic conditions for designing the primer pairs are as follows:

   a. Exclude region from base 100 to base 500.
   b. Exclude primers with more than three identical bases in a row.
   c. Use default value for optimum T_m (60.0°C), minimal T_m (57.0°C), maximum T_m (63.0°C).
   d. Use default value for optimum size (20 bases), minimal size (18 bases), maximum size (27 bases).
   e. Use default value for optimum GC content (50%), minimal GC (20%), maximum GC (80%).

3.2. Amplification

The optimal annealing temperature for each primer set is determined empirically by amplifying DNA with the primers in a gradient thermocycler with annealing temperature covering a 12°C range at 2°C intervals centered on the Primer3 predicted annealing temperature. The PCR products are analyzed by agarose gel electrophoresis, and the annealing temperature that generates a single-band PCR product of the expected size is noted. Primer sets that do not generate a single amplification product are discarded.

Suitable amplification primer sets are then used to amplify DNA from the strains or individuals being surveyed for SNPs. The PCR conditions are as follows:

1. Template: 2 μL (200 ng).
2. Primer: 0.1 μ*M* each.
3. dNTPs: 200 μ*M* each.
4. 10X buffer: 2.5 μL.
5. Taq Polymerase: 0.02 U/μL.
6. Total volume: 25 μL.

PCR cycling conditions are as follows:

1. Presoak: 95°C for 4 min.
2. Denaturation: 95°C for 30 s.
3. Annealing: as determined above, 30 s.
4. Polymerization: 72°C for 30 s.
5. PCR Cycles: 36.
6. Final Extension: 72°C for 7 min.

3.3. Preparation for Sequencing

1. The amplification products are prepared for sequencing by treatment with Exonuclease I and Shrimp Alkaline Phosphatase. Each 25 µL reaction mixture receives 1 µL Exonuclease 1 and 1 µL Shrimp Alkaline Phosphatase.
2. The plate is returned to the thermocycler, and incubated at 37°C for 30 min and then at 80°C for 15 min.
3. The concentrations of the PCR products are determined by Sybr Green Dye fluorescence quantified on a Storm Fluorimager (Molecular Dynamics). 1 µL of each PCR is transferred to a microtiter plate well containing 4 µL of water and 5 µL of 5X Syber Green. The fluorescence intensity of each sample is compared to a standard curve encompassing 7.5–200 ng/µL.

3.4. Sequencing

1. The sequencing reactions are assembled in 96-well microtiter plates as follows:
   a. x µL PCR product (10 ng per 100 bases to be sequenced) (*see* **Note 2**).
   b. 3 µL Primer 1 pmol/mL (one of the PCR primers is used as the sequencing primer).
   c. 4 µL Big Dye Terminator Ready Reaction Mix (*see* **Note 3**).
   d. 4 µL 5X Reaction Buffer (*see* **Note 4**).
   e. dH$_2$0 to a total reaction volume of 20 µL.
2. The standard thermocycling protocol outlined in the ABI Prism Dye terminator Ready Reaction protocol is followed, except the 4 min extension at 60°C is reduced to 2 min because the PCR products are short (*see* **Note 5**):
   a. Presoak: 96°C for 5 min.
   b. Denaturation: 96°C for 30 s.
   c. Annealing: 50°C for 30 s.
   d. Polymerization: 60°C for 2 min.
   e. PCR cycles: 25.
3. Excess dye terminator molecules are removed by gel filtration on superfine Sephadex G50 spin columns made in the wells of a Millipore multiscreen plate (*see* Millipore Tech Note TN053 for detailed protocol) as follows.
   a. Dry Sephadex is added to the wells of the multiscreen plate with a 0.45-µL column loader. 300 µL of water is added to each well and the Sephadex allowed to swell for 2 h at room temperature (at this point, the plates can be stored in Ziplock bags at 4°C).
   b. In preparation for sample loading, the multiscreen plate is assembled with a 96-well collection plate using an alignment frame (Millipore) and centrifuged at 450 RCF for 2 min.
   c. The sequencing reactions are loaded onto the Sephadex and the multiscreen plate is reassembled with a collection plate.
   d. Following centrifugation at 450 RCF for 2 min, the purified sequencing reactions are in the collection plate. They are dried using in a vacuum centrifuge designed to accept 96-well microtiter plates, and then resuspended in 15 µL of deionised and distilled water.

4. The collection plates are loaded onto the deck of a 3700 DNA Analyzer (*see* **Note 6**). Samples are injected at 2500 V for 55 s and run under standard conditions.
    a. Cuvet temperature: 40°C.
    b. Run temperature: 50°C.
    c. Run voltage: 5250 V.
    d. Sheath flow volume: 5 mL.
    e. Run time: 4167 s.
    f. Sample volume: 2.5 μL.
    g. Polymer: POP6.
5. Chromatograms generated from the sequencing run are then electronically transferred to a DEC Alpha machine for downstream processing.

3.5. SNP Identification

The software program Phred/Phrap, which is part of the Phrap package, is provided by the University of Washington (http://www.phrap.org/). Phred/Phrap will run phred and phrap, which create quality information for each base and assemble the sequences from same primer into a contig or contigs. The output from the Phred/Phrap program is used by the SNP detection program PolyBayes *(4)*, also available from the University of Washington. We run PolyBayes using the default setting of $P = 0.003$ (1 polymorphic site in 333 bp) as the total *a priori* probability that a site is polymorphic and a SNP detection threshold of 0.4 (*see* **Note 7**).

4. Notes

1. If the available DNA sequences are not masked, masking can be done using RepeatMasker software from the University of Washington Genome Center (http://ftp.genome.washington.edu/cgi-bin/RepeatMasker). RepeatMasker screens a sequence in FASTA format and returns it with simple sequence repeats, low complexity DNA sequences, and interspersed repeats replaced with "N" symbols. Repetitive element libraries available for use with RepeatMasker are primates, rodents, other mammals, other vertebrates, Arabidopsis, grasses, and Drosophila.
2. The amount of DNA used in the sequencing reaction is based on the size of the PCR product, using 10 ng per 100 bases to be sequenced as a guide. In general, we have found that this approximation for calculating the amount of PCR product that goes into a sequencing reaction produces a balanced sequencing reaction for products up to 1 kb in size.
3. The version 2.0 Big Dye kit was used in preference to version 1.0 because it produces longer reads on the 3700 platform.
4. The 5X Reaction Buffer used in the cycle sequencing reaction contains 400 m*M* Tris-HCL at pH 9.0 and 10 m*M* magnesium chloride. Use of this buffer allows the use of 50% less Big Dye Ready Reaction Mix thus reducing sequence reaction costs.
5. In our cycle sequencing protocol, cutting the extension time from 4 min to 2 min per cycle reduces the overall cycling time by 50 min. This time saving can increase productivity in a high throughput environment.

6. Initially, problems were encountered with the electrokinetic injection of DNA when in house deionized water was used. Chemical impurities present in the water may have been preferentially injected into the capillary, resulting in low-quality sequence data. This problem was remedied by switching to a commercial water source.
7. The PolyBayes setting was not optimized for mouse SNP detection.

Acknowledgment

BAC end sequences were provided by Dr. Shaying Zhao at The Institute for Genomic Research. This work was supported by Grant CA-16672 from the National Cancer Institute (NIH) and HG02057 from the National Human Genome Research Institute (NIH).

References

1. Botstein, D., White, R. L., Skolnick, M., and Davis R. W. (1980) Construction of a genetic linkage map in man using restriction fragment length polymorphisms. *Amer. J. Hum. Genet.* **32,** 314–331.
2. Weber, J. L. and May, P. E. (1989) Abundant class of human DNA polymorphisms which can be typed using the polymerase chain reaction. *Amer. J. Hum. Genet.* **44,** 388–396.
3. Rozen, S. and Skaletsky, H. (2000) Primer3 on the WWW for general users and for biologist programmers. *Methods Mol. Biol.* **132,** 365–386.
4. Marth, G. T., Korf, I., Yandell, M. D., et al. (1999) A general approach to single-nucleotide polymorphism discovery. *Nat. Genet.* **23,** 452–456.

2

Exon Trapping for Positional Cloning and Fingerprinting

Scott E. Wenderfer and John J. Monaco

1. Introduction

Positional cloning involves the genetic, physical, and transcript mapping of specific parts of a genome *(1)*. Linkage analysis can map specific activities, or phenotypes, to a quantitative trait locus (QTL), a genomic region no smaller than 1 centiMorgan (cM) or megabase (Mb) in length. Physical mapping can then provide a map of higher resolution. Physical maps are constructed from clones identified by screening genomic libraries. Genomic clones can be characterized by fingerprinting and ordered to create a contig, a contiguous array of overlapping clones. Transcript identification from the clones in the contig results in a map of genes within the physical map. Finally, expressional and functional studies must be performed to verify gene content.

Bacterial artificial chromosomes (BACs) and P1 artificial chromosomes (PACs), both based on *Escherichia coli* (*E. coli*) and its single-copy plasmid F factor, can maintain inserts of 100–300 kilobases (kb). Their stability and relative ease of isolation have made them the vectors of choice for the development of physical maps. Once BAC clones are obtained, exon trapping can be performed as a method of transcript selection even before characterization of the contig is complete. Trapped exons are useful reagents for expressional and functional studies as well as physical mapping of BAC clones to form the completed contig.

Exon trapping was first used by Apel and Roth *(2)* and popularized by Buckler and Housman *(3)*. A commercially available vector, pSPL3 *(4)*, has been used in multiple positional cloning endeavors *(5–8)*. Exon trapping relies on the

From: *Methods in Molecular Biology, vol. 256:*
Bacterial Artificial Chromosomes, Volume 2: Functional Studies
Edited by: S. Zhao and M. Stodolsky © Humana Press Inc., Totowa, NJ

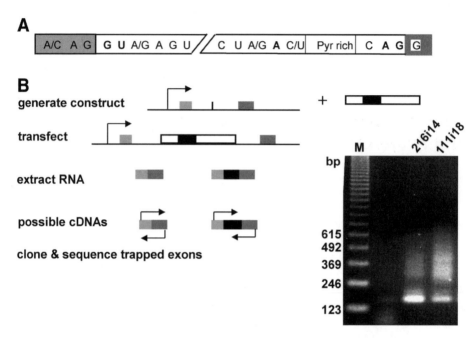

Fig. 1. **(A)** Exon splicing is conserved in eukaryotes. The sequences at the splice junctions are conserved. The gray box represents the 5′ exon and the checkered box represents the 3′ exon. The white box represents the intron. The bold bases indicate the 3′ splice acceptor, the branch point A, and the 5′ splice donor from left to right. **(B)** Because splicing is conserved, a genomic fragment (white bar) containing an exon (black box) from any species can be inserted within the intron of an expression construct for exon trapping. COS7 cells are transfected with the construct and 48 h later RNA is collected. The expressed recombinant mRNA can be isolated by RT-PCR using primers for the upstream and downstream exon of the expression construct. Genomic fragments lacking an exon would allow the upstream and downstream exons of the expression construct to splice together, resulting in a smaller RT-PCR product (the 177 bp band). We screened BAC clones by shotgun cloning small fragments into the intron of the HIV tat gene behind an SV40 early promoter. The RT-PCR products from two exon trapping experiments are shown.

conservation of sequence at intron–exon boundaries in all eukaryotic species (*see* **Note 1**). By cloning a genomic fragment into the intron of an expression vector, exons encoded in the genomic fragment will be spliced into the transcript encoded on the expression vector (*see* **Fig. 1**). Reverse transcriptase polymerase chain reaction (RT-PCR) using primers specific for the transcript on the expression vector will provide a product for analysis by electrophoresis and sequencing.

Because the expression vector utilizes its own exogenous promotor, exon trapping is independent of transcript abundance and tissue expression. Moreover, exon trapping provides rapid sequence availability. It has proven to be a very sensitive method for transcript identification *(9,10)* (*see* **Note 2**). By pooling subclones via shotgun cloning of cosmids, BACs, or yeast artificial chromosomes (YACs) into the pSPL3 vector, 30 kb–3 Mb can be screened in a single experiment.

Disadvantages include dependence on introns, splice donor and acceptor sites. False negatives are caused by missing genes with only one or two exons, interrupting exons by cloning into the expression vector, and possibly by not meeting unidentified splicing requirements. False positives are caused by cryptic splice sites *(11)*, exon skipping *(12)*, and pseudogenes.

No one method for transcript identification has become the stand-alone method for positional cloning. Genomic sequence analysis, when sequence is available, should be the primary tool for identification of genes within a genomic region of interest. Bulk sequencing provides a template for computer selection of gene candidates via long open reading frames (ORFs), sequence homology, or motif identification. Gene Recognition and Assembly Internet Link (GRAIL) analysis can be performed manually at a rate of 100,000 kb per person-hour *(13)*. PCR primer pairs can be made for each set of GRAIL exon clusters. Alternatively, predicted GRAIL exons may be represented in the expressed sequence tag (EST) database, a collection of sequences obtained from clones randomly selected from cDNA libraries encompassing a wide range of tissues or cell types. If an EST exists, corresponding cDNA clones can be purchased from the IMAGE consortium *(14)*. Motif and ORF searching does suffer from a lack of specificity and sensitivity and tend to be both time consuming and software/hardware dependent. Exon trapping is an excellent tool for verification of genes predicted in the sequence, as well as for identification of genes missed by computational techniques. A cluster of trapped exons likely encodes a functional gene product if several correspond to exons also predicted by GRAIL and together they encode a long ORF.

When no genomic sequence is available, exon trapping is the method of choice for initially identifying genes. Not only are new genes identified and known genes mapped, but also trapped exons, bona fide or false positives, become markers for the generation of a physical map. Southern or colony blots made from BAC clones can be hybridized with exon probes to map them to specific locations on individual BACs, or to BACs in a contig. Trapped exon probes an also be used to screen further genomic BAC libraries. In our experience, more than 100 markers were generated for every 1 Mb region, resulting in a marker density of one per 10 kb. Therefore, the number of markers generated during a completed exon trapping study will be sufficient for genome sequencing centers to begin obtaining and aligning sequence information in this contig *(15)*.

Most other strategies for positional cloning use "expression-dependent" techniques. Direct selection is the selection of transcribed sequences from a library of expressed cDNAs using solution hybridization with labeled genomic clones *(16,17)*. A similar technique, cDNA selection, selects transcribed sequences by hybridization screening of blotted genomic clones with labeled cDNA libraries *(18–20)*. Transcript selection techniques depend on the knowledge of mRNA distribution and abundance in different tissues. They are difficult to perform with BAC clones, as most will contain regions of repetitive sequence that must be blocked with competing unlabeled DNA. Performed together with exon trapping, they have been proven complimentary.

Exon trapping is not intended for extremely high-throughput gene identification or mapping. Whole genome sequencing and large-scale sequencing of cDNA library clones together have been the most efficient high-throughput gene identification methodology. EST databases contain a large number of gene markers that can be used for expressional profiling by RT-PCR or DNA chip technology. Radiation hybrid mapping of these EST clones has become a high-throughput technology for gene mapping *(21)*. However, EST databases tend to be overrepresented with genes expressed in high abundance. Researchers interested in a genomic region in a species that has been the subject of high-throughput analyses, such as *Homo sapiens*, may wish to obtain BAC clones and use exon trapping as a complimentary method.

Once trapped, exon clones can be used for expression analysis. Querying sequences of candidate exons against Genbank's EST dataset can be used to identify multiple tissues where the gene has been previously identified by sequencing of cDNA libraries. Hybridization to northern blots with total RNA from brain, heart, kidney, liver, lung, skeletal muscle, spleen, and thymus will give a general screen for expression appropriate for all candidate exons. Hybridization to blots with total RNA from cell lines can provide information on constitutive and inducible expression in different cell types. Alternatively, exon sequences can be used to generate a DNA chip for expressional profiling, allowing all exons to be tested in a single experiment.

2. Materials

2.1. Subclone BAC DNA into pSPL3 Exon Trapping Vector

1. Appropriate BAC or PAC clones may be purchased (Incyte Genomics, St. Louis, MO; Roswell Park Cancer Institute, Buffalo, NY).
2. BAC DNA should be isolated from 500 mL bacterial cultures by alkaline lysis. Lysates are passed through Nucleobond filters onto AX-500 columns (Clontech,

Palo Alto, CA), eluted, then precipitated with isopropanol, washed with ethanol, and reconstituted in 100 µL distilled H_2O. Aliquots of 5 µL of separate *EcoRI* and *NotI* digests can be analyzed by electrophoresis on agarose gels. Contamination of preps with bacterial DNA does not preclude their use, but may increase the false-positive rate.

3. *BamHI*, *BglII*, *DraI*, *EcoRV*, *EcoRI*, *NotI*, *HincI*, *NotI*, *PvuII*, and *T4* DNA ligase.
4. pSPL3 plasmid may be purchased as part of the exon amplification kit (Gibco-BRL, Gaithersburg, MD). Plasmid preps can be performed using alkaline lysis kits from Qiagen (Valencia, CA).
5. *E.coli* strain DH10b electromax cells can be purchased from Gibco BRL.
6. GenePulser bacterial cell electroporator and cuvets (Bio-Rad, Richmond, CA).
7. Luria Bertani broth with 100 µg/mL ampicillin (LB-amp).
8. Routine gels can be prepared from electrophoresis grade agarose (Bio-Rad).
9. DNA can be purified from low-melt agarose gel slices using the MP kit from U.S. Bioclean (Cleveland, OH).

2.2. Transient Transfections

1. COS-7 green monkey kidney cells may be obtained from ATCC (Rockville, MD) and maintained in 10 mL Dulbecco's modified Eagle's media (DMEM) with 10% fetal bovine serum (FBS) and 2 m*M* sodium pyruvate (GibcoBRL) at 37°C, 5–10% CO2. All manipulation should be performed in a hood under sterile conditions.
2. Phosphate buffered saline (GibcoBRL), stored at 4°C.
3. GenePulser mammalian cell electroporator and cuvets (Bio-Rad).

2.3. Exon Trapping

1. Superscript II RT, *Bst*XI, RNAse H, Taq DNA polymerase, Trizol reagent for total RNA isolation, uracil DNA glycosylase (UDG), prelinearized pAMP10 vector, and DH10b max efficiency competent cells.
2. Oligo SA2 sequence: ATC TCA GTG GTA TTT GTG AGC.
3. First strand buffer contains a final concentration of 50 m*M* Tris-HCl pH 8.3, 75 m*M* KCl, 3 m*M* MgCl$_2$, 10 m*M* dithiothreitol (DTT), and 0.5 m*M* dNTP mix.
4. PCR buffer contains a final concentration of 10 m*M* Tris pH 9.0, 50 m*M* KCl, 1.5 m*M* MgCl$_2$, and 0.2 m*M* dNTP mix.
5. Oligo SD6 sequence: TCT GAG TCA CCT GGA CAA CC.
6. Oligo dUSD2 sequence: ATA GAA TTC GTG AAC TGC ACT GTG ACA AGC TGC.
7. Oligo dUSA4 sequence: ATA GAA TTC CAC CTG AGG AGT GAA TTG GTC G.
8. RT reaction and PCR can be performed in a DNA thermocycler 480 (Perkin Elmer–Applied Biosystems, Norwalk, CT).
9. Water for manipulation and storage of RNA should be treated with 0.1% diethyl pyrocarbonate to remove RNAses and then autoclaved. When working with RNA, change gloves often and use only reagents prepared with RNAse-free water.

2.4. Screening Trapped Exons to Exclude False Positives and Previously Sequenced Exon Clones

1. LB-amp broth.
2. Sterile 96-well microtiter plates with lids (Fisher).
3. 96-pin replicator may be purchased from Fisher (Pittsburgh, PA), should be stored in 95% ethanol bath, and can be flame sterilized before and after each bacterial colony transfer.
4. Appropriately sized rectangular agar plates can be made by pouring molten LB agar into the lid of a standard 96-well microarray plate and solidifying overnight at 4°C.
5. Magnabond 0.45-μm nylon filters (Micron Separations Inc., Westborough, MA).
6. Prehyb solution contains a final concentration of 1 M NaCl, 1% sodium dodecyl sulfate (SDS), 10% dextran sulfate, and 100 μg/mL denatured salmon sperm DNA.
7. *Acc*I, *Ava*I, *Bgl*II, *Sal*I, T4 DNA kinase and exonuclease-free Klenow fragment.
8. T4 forward reaction buffer contains a final concentration of 70 mM Tris-HCL pH 7.6, 10 mM MgCl2, 100 mM KCl, and 1 mM 2-mercaptoethanol.
9. DNA replication buffer contains a final concentration of 0.2 M HEPES, 50 mM Tris-HCL pH 6.8, 5 mM MgCl$_2$, 10 mM 2-mercaptoethanol, 0.4 mg/mL bovine serum albumin (BSA), 10 μM dATP, 10 μM dGTP, 10 μM dTTP, and 5 OD$_{260}$ U/mL random hexamers mix.
10. [γ-^{32}P]dATP and [α-^{32}P]dATP. Proper shielding should be used when handling all solutions containing ^{32}P.
11. pSPL3$_{VV}$ oligo sequence: CGA CCC AGC A|AC CTG GAG AT.
12. pSPL3$_{1021}$ oligo sequence: AGC TCG AGC GGC CGC TGC AG.
13. pSPL3$_{1171}$ oligo sequence: AGA CCC CAA CCC ACA AGA AG.
14. pSPL3$_{1056}$ oligo sequence: GTG ATC CCG TAC CTG TGT GG.
15. pPSL3 intron probe can be prepared in bulk by double digest of pSPL3 vector with *Ava*I and *Sal*I. The 335 bp and 2086 bp bands can be isolated by agarose gel electrophoresis and purified using the U.S. Bioclean MP kit. It can be stored at –20°C, thawed on ice, and refrozen multiple times.
16. Previously sequenced exon clone (PSEC) probes can be prepared from double digests of trapped exons in pAMP10 using 5 U each of *Acc*I and *Bgl*II. Vector bands of 4 kb and either 50 or 109 bp (depending on direction in which trapped exon is cloned into pAMP10) should be avoided when probes are isolated from gel slices. PSEC probes can be stored at –20°C, thawed on ice, and refrozen multiple times.
17. Probe purification columns can be made by filling disposable chromatography columns with either Sephadex G-25 (for oligos) or G-50 (for longer single-stranded DNA probes) and spinning out buffer into a microfuge tube.
18. 2X SSC/SDS contains a final concentration of 0.3 M NaCl, 30 mM sodium citrate, and 0.5% SDS. 0.2X SSC/SDS contains 0.03 M NaCl, 3 mM sodium citrate, and 0.5% SDS.
19. X-OMAT AR film (Eastman Kodak Company, Rochester, NY).
20. Phosphor screen and phosphorimager (Molecular Dynamics (Amersham Pharmacia Biotech, Piscataway, NJ).

2.5. Size Selection of Trapped Exons for Sequencing of Unique Clones

1. LB-amp broth.
2. Sterile 96-well microtiter plates with lids.
3. PCR can be performed for sets of 96 samples using Gene Amp PCR system 9700. (Perkin Elmer–Applied Biosystems).
4. PCR buffer.
5. Individual bacterial clones may be transferred from 96-well plate via toothpicks, sterilized by autoclaving in tin foil, or by flame sterilized 96-pin replicator.
6. *Hind*III and *Pst*I.
7. Sequencing primers dUSA4, dUSD2.

3. Methods

3.1. Subclone BAC DNA into pSPL3 Exon Trapping Vector

1. Isolate genomic BAC clone (*see* **Note 3**).
2. Set up *Dra*I, *Eco*RV, and *Hinc*II digests for each BAC clone individually in three separate tubes (*see* **Note 4**). A total of 10 U restriction enzyme will digest 5 µg in 8 h.
3. Linearize pSPL3 exon trapping vector by digesting with the appropriate restriction enzyme and gel-purify.
4. Subclone each digest individually into linearized pSPL3 with 20,000 U T4 DNA ligase for 1 h at 42°C and transform DH10b bacterial cells by electroporation at 1.8 kV, 25 µF, 200 Ω (*see* **Note 5**).
5. Grow transformants overnight in 50 mL LB-amp broth, isolate DNA from shotgun subclones and test heterogeneity by running a *Pvu*II digest on a 1% agarose gel.

3.2. Transient Transfections

1. Plate 2×10^6 COS7 cells / 75 mm^2 dish and preincubate 24 h.
2. Harvest cells by centrifugation and wash twice in 5 mL ice cold PBS.
3. Resuspend to 4×10^6 cells/mL in ice-cold PBS and transfer 0.7 mL aliquots into labeled electroporation cuvets.
4. Add 15 µg supercoiled plasmid DNA, mix, and incubate on ice for 5 min.
5. Electroporate at a voltage of 350 V and a capacitance of 50 µF.
6. Incubate on ice 5–10 min then dilute cells 20-fold in 14 mL DMEM/FBS.
7. Plate transfected cells in T25 flasks and incubate 48 h (2 generation times).

3.3. Exon Trapping

1. Isolate total RNA using Chomczynski-based method. Resuspend total RNA yield from each T25 flask of cells in 100 µL RNAse-free H$_2$O and store RNA at –80°C. Run 3 µg RNA on a 1% agarose gel at 50 V to check purity (*see* **Note 6**).
2. Perform reverse transcription reaction on 3 µg total RNA (final concentration = 0.15 µg/mL) with 200 U Superscript II RT and 1 µ*M* SA2 oligo in 20 µL 1st strand buffer for 30 min at 42°C.

3. Preincubate cDNA 5 min at 55°C, then treat with 2 U RNAse H for 10 min, store at 4°C.

4. Perform PCR on 5 μL cDNA (approx 1.2μg) with 2.5 U Taq DNA polymerase and 1 μ*M* each oligos SA2 and SD6 in 40 μL PCR buffer for a total of six cycles (each cycle: 1 min denaturation at 94°C, 1 min annealing at 60°C, and 5 min extension at 72°C).

5. Continue final extension an additional 10 min at 72°C.

6. Treat PCR product with 20 U *Bst*XI restriction endonuclease at least 16 h at 55°C (*see* **Note 7**).

7. Add an additional 4 U *Bst*XI enzyme and treat for another 2 h at 55°C.

8. Perform secondary PCR on 5 μL *Bst*XI digest with 2.5 U Taq DNA polymerase and 0.8 μ*M* each oligo dUSA4 and dUSD2 in 40 μL PCR buffer for a total of 30 cycles (each cycle: 1 min denaturation at 94°C, 1 min annealing at 60°C, and 3 min extension at 72°C).

9. Run 9 μL secondary PCR product on >2% agarose gel to check heterogeneity. See **Fig. 1** for the appearance of a satisfactory exon trapping experiment.

10. Clone 2μL (approx 100 ng) heterogeneous exon mixture into pAMP10 vector using 1 U UDG in 10 μL.

11. Transform 3 μL UDG shotgun subclones into 50 μL DH10b max efficiency competent cells by heat shock, 42°C for 40 s, plate 20% of cells on each of two LB amp plates and grow >16h.

3.4. Screening Trapped Exons to Exclude False Positives and Previously Sequenced Exon Clones

1. Inoculate 200μL LB-amp broth per well with 286 CFU from each exon-trapping reaction in 96 well plates (three 96-well plates/BAC clone).

2. For each 96-well plate, inoculate one well with a bacterial clone transformed with pSPL3 vector alone (positive control) and a second well with a UDG clone from an exon trapping experiment where no genomic DNA was subcloned (negative control), and grow transformants >16 h.

3. Make three sets of colony dot blots by transferring 96 UDG clones *en mass* with 96-pin replicator to a nylon filter sterilely placed over a rectangular agar plate. Grow colonies >16 h, denature and wash away bacterial debris, and crosslink DNA to nylon at 120,000 μJ/cm^2.

4. Prehybridize for >1 h at 50°C in hybridization bottle.

5. Label 100 ng each of pSPL3$_{VV}$, pSPL3$_{1021}$, pSPL3$_{1171}$, and pSPL3$_{1056}$ oligos together with 75 μCi [γ-^{32}P]dATP and 10 U T4 kinase in 20 μL forward reaction buffer and purify with Sephadex G-25 column (*see* **Note 8**).

6. Add 1×10^7 CPM of labeled four pSPL3 oligo mixture for each milliliter prehybridized solution and hybridize 1 set of colony blots >8 h at 50°C.

7. Washing unbound oligos from blot with 2X SSC/SDS buffer twice at room temperature then four times at 60°C routinely results in appearance of specific signal on film within 16 h or on phosphor screen within 1 h.

8. Hybridize the second set of colony blots with pSPL3 intron, labeled with 75 μCi [α-^{32}P]dATP and 3 U exonuclease-free Klenow fragment in 50 μL DNA replication buffer and purify with Sephadex G-50 column.

9. Hybridize the third set of blots with previously sequenced exon clone (PSEC) mix, labeled with 75 μCi [α-^{32}P]dATP and 3 U exonuclease-free Klenow fragment in 25 μL DNA replication buffer and purify with Sephadex G-50 column (*see* **Note 9**).

10. Washing unbound single stranded DNA probe from blot twice with 2X SSC/SDS buffer, then twice with 0.2X SSC/SDS buffer at 65°C routinely results in appearance of specific signal on film within 16 h, or on phosphor screen within 1 h.

3.5. Size Selection of Trapped Exons for Analysis of Unique Clones

1. Grow bacterial clones transformed with "unsequenced, true positive" candidate exons in LB-amp broth in 96-well plates >16 h.

2. Using a 96-pin replicator, transfer bacterial clones to thin walled PCR tubes containing 40 μL PCR buffer. Colony PCR performed with 2.5 U Taq DNA polymerase and 0.8 μ*M* each of oligos dUSA4 and dUSD2 for a total of 30 cycles (each cycle: 1 min denaturation at 94°C, 1 min annealing at 60°C, and 3 min extension at 72°C).

3. Size select candidate exons by running on a 3% agarose gel (*see* **Note 10**).

4. Grow bacteria transformed with unique clones in LB-amp broth >16 h, and isolate DNA by alkaline lysis.

5. Test size selection by running *Hind*III/*Pst*I double digest on 3% agarose gel.

6. Sequence unique exons from plasmid preps using either oligo dUSA4 or dUSD2. If sequence obtained does not overlap, design additional primers from deduced sequence and repeat until full-length sequence is obtained (*see* **Note 11**).

4. Notes

1. Exon trapping detects exons encoded within the genome. The definition of an exon is well understood. Consensus sequences are present at both splice acceptor and splice donor sites *(22)*. Small nuclear RNA molecules hybridize to these consensus sequences in the messenger RNA, targeting the splicing machinery to excise the intervening sequence, or introns. Cryptic splice sites exist in the genome, defined as random sequence that mimics either a splice acceptor site or a splice donor site. The chance that a cryptic splice donor and a cryptic splice acceptor would be located close enough together in the genome to cause a false positive exon to be trapped is presumably rare, but the actual number is not known. Our data suggest that the specificity of exon trapping is high. At least 84% of clones have sequences with open reading frames and are expressed in vivo *(8)*. To help determine the specificity of exon trapping, one can analyze the flanking intron sequence to identify consensus splice sites. Because the sequences at the ends of exons are less conserved, we were unable to analyze the validity of

trapped exons by their sequence alone. Sequencing flanking intron sequence off the BAC clone for every trapped exon is a laborious task, not recommended routinely. However, one BAC clone used in our exon trapping experiments was also sequenced *(23)*. We did check for the presence of consensus splice sites in introns flanking 22 exons trapped from this BAC clone. Sixteen were exons from genes with published sequence. All 16 are flanked in the genome by consensus splice sites, but two used different splice sites from those published. Five trapped exon clones have open reading frames encoding previously unpublished sequence, and four of the five are flanked by consensus splice sites. The fifth is flanked only by a 5′ splice donor. Only one exon was trapped that lacked an open reading frame in any of the three reading frames, but it too is flanked by consensus splice sites. Therefore, the specificity of the splicing mechanism in our exon trapping experiments appears to be identical to the specificity of the endogenous splice machinery.

2. Our data suggest that exon trapping is 73% sensitive for transcript identification, when several hundred trapped exons are characterized per PAC or BAC clone *(8)*.

3. Sixfold redundant libraries will result in approximately 50 clones per one Mb. Up to six previously mapped genes or EST clones can be used as probes to screen a genomic BAC library in a single hybridization. A minimum contig of 10 clones should then be shotgun cloned into pSPL3 for exon trapping. With sequence information to aid in development of a contig, this can all be performed in less than a month. Screening 200 exons from each BAC or PAC clone tested should take two weeks, and up to 1000 additional clones can be characterized by PSEC screens in another two weeks.

4. Use of three separate restriction enzyme digests combined prior to ligation to vector minimizes the chance of missing an exon that happens to contain a restriction site within its sequence. An alternative method is to use a *Bam*HI and *Bgl*II double digest along with a *Sau*3AI partial digest in two separate tubes.

5. Transformation of competent cells by electroporation is much more effective than heat shock transformation for bacteria. In our experience, without electroporation of the BAC subclones, the sensitivity of identifying known genes using exon trapping decreased 10-fold.

6. Protocol for using Trizol reagent available from GibcoBRL. Yield of RNA prep is 5–7 µg per T25 flask (approx 10^6 cells). Using a spectrophotometer, the $A_{260/280}$ should be between 1.6–1.8 (less suggests phenol contamination or incomplete dissolution). Gel should show sharp ribosomal bands with the intensity of the 28S twice that of 18S. If the 5S band is as intense as the band at 18S, there is too much degradation to efficiently continue this protocol.

7. The success of the *Bst*XI digestion is critical for the elimination of false negatives. A short 177bp cDNA composed of only pSPS3 vector sequence will predominate unless *Bst*XI digestion is complete. Fresh GibcoBRL enzyme was the only formulation potent enough to approach 100% digestion using this protocol.

8. Cryptic splice sites within the pSPL3 intron were responsible for several false positives, from 10 to 50% of all products of an exon trapping experiment. Screening of trapped exons with four oligos and the entire pSPL3 intron removed 95%

of these false positives from further consideration. Three oligos are named by the location of the complimentary sequence on the pSPL3 vector. The pSPL3 intron sequence runs from 699 to 3094. The fourth oligo (pSPL3$_{v-v}$) contains sequence complimentary to the exons of the pSPL3 vector after being spliced together (splice junction indicated by a vertical bar in the sequence in the methods section). If the *Bst*XI digestion is incomplete and some pAMP10 clones without trapped exons remain, this fourth oligo will identify them.

9. A difficulty encountered with exon trapping was differential representation of trapped exons within the total pool. Some exons were present at proportions of 1:10 or even 1:4 when hundreds of exons were analyzed from a 100-kb BAC clone. Other exons required characterization of several hundred trapped exons from a particular BAC clone before a single copy was identified. The selection of smaller clones during PCR amplification or cloning does not explain the differences in abundance. Trapped exons from each BAC should be characterized hundreds at a time, first by size selection and sequencing, then by PSEC (spell out) screens. PSECs were isolated as probes, labeled individually and pooled in order to screen additional batches of cloned exons by hybridization. Hundreds of trapped exon clones could be easily screened with all PSECs after generating duplicate colony blots by transfer of bacterial clones from microtiter plates using a 96-pin replicator. Screening 200–300 exons from each exon trapping experiment is recommended. However, if known genes are not identified after characterizing 300, chances are very low that it will be identified in that experiment.

   Exon trapping yield varies between different species and between different regions on the same chromosomes, depending on the gene density. Yield is measured by the following equation:

$$\text{Yield} = \frac{\text{kb DNA screened}}{\text{exons trapped}}$$

   Each exon trapping experiment involves shotgun cloning multiple digests of the same BAC or PAC clone into the pSPL3 trapping vector. Additional experiments may be performed using different restriction endonucleases to generate inserts for shotgun cloning. Running a second experiment for the same BAC clone often doubles the number of exons trapped, but in our hands a third experiment does not result in many new exon clones. Exon trapping of a BAC was considered complete when >95% of trapped exons in a screen were positive for a PSEC. At that point, identification of missed genes by a complimentary "transcript identification" method (sequence analysis, zoo analysis, or expression analysis) would be warranted over screening more trapped exons.

10. Trappable exons have ranged in size from 49 to 465 bp, similar to the range observed for all exons in the genome. Electrophoresis of DNA in this size range is best visualized on 3% agarose gels. Estimating sizes then rerunning samples in order from smallest to largest can verify sizes and is often helpful. Isolation of DNA from 3% agarose gel slices to obtain PSEC probes is possible using the U.S. Bioclean MP kit.

11. Double-stranded sequence was not routinely obtained. Because neither 5′ nor 3′ exons can be trapped by this method, open reading frames are usually a property of true positives identified by exon trapping. An additional method for screening exon trapping products for true positives is zoo blotting. Zoo blotting involves the hybridization of DNA or cDNA from one species with genomic DNA or RNA from various related or divergent species. In one study, 85% of exon trapping products from human DNA demonstrated cross-hybridization to primate sequences, and 56% cross-hybridized to other mammalian sequences *(9)*. Finally, true positives can be verified by identifying transcripts by Northern blot or by screening cDNA libraries.

Unfortunately, one drawback of transcript identification is that not all transcripts encode functional gene products. EST databases exemplify this pitfall of transcript identification. An enormous number of cDNA clones represented in the EST database encode repetitive sequence. Sometimes this is owing to isolation of a pre-mRNA in which an intron containing a repeat element has not been spliced out. In other cases, the repetitive element is presumably expressed because of its own LTR, a *cis*-acting factor that drives transcription of the repeat sequence. The importance of repetitive transcripts in health and disease is debatable, but removal of EST sequences containing repeats is straightforward for transcript mapping. A simple algorithm called Repeatmasker is available over the Internet *(24)*. Entries in the EST database corresponding to novel single-copy sequences that lack ORFs present more of a problem during positional cloning. EST entries by definition are single pass single stranded sequences, and are therefore error-prone. However, there are some transcripts identified numerous times in several tissues, and multiple sequence alignments give a reliable sequence that still lacks an ORF. Moreover, as high-quality bulk genomic sequence becomes available, the presence of stop codons in all frames of EST sequences is often being confirmed. These transcripts have introns, and the resulting exons can be identified by exon trapping. Seeking the function of nontranslated RNAs has been laborious without the aid of sequence similarities. The continuing analysis of quantitative trait loci from spontaneous mutation and large scale induced mutagenesis projects will eventually result in the endorsement of transcribed sequences to convert transcript maps into gene maps.

Acknowledgments

This work was supported by the Howard Hughes Medical Institute and the John Wulsin foundation. The authors would like to thank Dr. Megan Hersh for critically reviewing this manuscript.

References

1. Menon, A. G., Klanke, C. A., and Su, Y. R. (1994) Identification of disease genes by positional cloning. *Trends Clin. Med.* **4,** 97–102.

2. Apel, T. W., Scherer, A., Adachi, T., Auch, D., Ayane, M., and Reth, M. (1995) The ribose 5-phosphate isomerase-encoding gene is located immediately downstream from that encoding murine immunoglobulin kappa. *Gene* **156,** 191–197.

3. Buckler, A. J., Chang, D. D., Graw, S. L., et al.: (1991) Exon amplification: a strategy to isolate mammalian genes based on RNA splicing. *Proc. Natl. Acad. Sci. USA* **88,** 4005–4009.

4. Church, D. M., Stotler, C. J., Rutter, J. L., Murrell, J. R., Trofatter, J. A., and Buckler, A. J. (1994) Isolation of genes from complex sources of mammalian genomic DNA using exon amplification. *Nat. Genet.* **6,** 98–105.

5. Haber, D. A., Sohn, R. L., Buckler, A. J., Pelletier, J., Call, K. M., and Housman, D. E. (1991) Alternative splicing and genomic structure of the Wilms tumor gene WT1. *Proc. Natl. Acad. Sci. USA* **88,** 9618–9622.

6. Taylor, S. A., Snell, R. G., Buckler, A., et al. (1992) Cloning of the alpha-adducin gene from the Huntington's disease candidate region of chromosome 4 by exon amplification. *Nat. Genet.* **2,** 223–227.

7. Lucente, D., Chen, H. M., Shea, D., et al. (1995) Localization of 102 exons to a 2.5 Mb region involved in Down syndrome. *Hum. Mol. Genet.* **4,** 1305–1311.

8. Wenderfer, S. E., Slack, J. P., McCluskey, T. S., and Monaco, J. J. (2000) Identification of 40 genes on a 1-Mb contig around the IL-4 cytokine family gene cluster on mouse chromosome 11. *Genomics* **63,** 354–373.

9. Church, D. M., Banks, L. T., Rogers, A. C., et al. (1993) Identification of human chromosome 9 specific genes using exon amplification. *Hum. Mol. Genet.* **2,** 1915–1920.

10. Trofatter, J. A., Long, K. R., Murrell, J. R., Stotler, C. J., Gusella, J. F., and Buckler, A. J. (1995) An expression-independent catalog of genes from human chromosome 22. *Genome Res.* **5,** 214–224.

11. Wieringa, B., Meyer, F., Reiser, J., and Weissmann, C. (1983) Unusual splice sites revealed by mutagenic inactivation of an authentic splice site of the rabbit beta-globin gene. *Nature* **301,** 38–43.

12. Andreadis, A., Gallego, M. E., and Nadal-Ginard, B. (1987) Generation of protein isoform diversity by alternative splicing: mechanistic and biological implications. *Annu. Rev. Cell Biol.* **3,** 207–242.

13. Xu, Y., Mural, R., Shah, M., and Uberbacher, E. (1994) Recognizing exons in genomic sequence using GRAIL II. *Genet. Eng.* **16,** 241–253.

14. http://image.llnl.gov/, webmaster@image.llnl.gov, Lawrence Livermore National Laboratory. The Image Consortium.

15. Collins, F. S., Patrinos, A., Jordan, E., Chakravarti, A., Gesteland, R., and Walters, L. (1998) New goals for the U.S. Human Genome Project: 1998–2003. *Science* **282,** 682–689.

16. Lovett, M. (1994) Fishing for complements: finding genes by direct selection. *Trends Genet.* **10,** 352–357.

17. Simmons, A. D., Goodart, S. A., Gallardo, T. D., Overhauser, J., and Lovett, M. (1995) Five novel genes from the cri-du-chat critical region isolated by direct selection. *Hum. Mol. Genet.* **4,** 295–302.

18. Parimoo, S., Patanjali, S. R., Shukla, H., Chaplin, D. D., and Weissman, S. M. (1991) cDna selection: efficient Pcr approach for the selection of cDnas encoded in large chromosomal Dna fragments. *Proc. Natl. Acad. Sci. USA* **88,** 9623–9627.
19. Fan, W. F., Wei, X., Shukla, H., et al. (1993) Application of cDNA selection techniques to regions of the human MHC. *Genomics* **17,** 575–581.
20. Goei, V. L., Parimoo, S., Capossela, A., Chu, T. W., and Gruen, J. R. (1994) Isolation of novel non-HLA gene fragments from the hemochromatosis region (6p21.3) by cDNA hybridization selection. *Amer. J. Hum. Genet.* **54,** 244–251.
21. Schuler, G. D., Boguski, M. S., Stewart, E. A., et al. (1996) A gene map of the human genome. *Science* **274,** 540–546.
22. Padgett, R. A., Grabowski, P. J., Konarska, M. M., Seiler, S., and Sharp, P. A. (1986) Splicing of messenger RNA precursors. *Annu. Rev. Biochem.* **55,** 1119–1150.
23. http://www-hgc.lbl.gov/human-p1s.html, Lawrence Berkeley National Laboratory, Human P1 sequence information.
24. http://ftp.genome.washington.edu/cgi-bin/RepeatMasker/, Smit, A. F. A. and Green, P., Univ. Washington Genome Center. (4/21/99) REPEATMASKER WEB SERVER.

3

Isolation of CpG Islands From BAC Clones Using a Methyl-CpG Binding Column

Sally H. Cross

1. Introduction

Vertebrate genomes are globally heavily methylated at the sequence CpG with the exception of short patches of GC-rich DNA, usually between 1–2 kb in size, which are free of methylation and these are known as CpG islands (*see* **refs.** *1* and *2* for reviews). In addition to distinctive DNA characteristics, CpG islands have an open chromatin structure in that they are hyperacetylated, lack histone H1, and have a nucleosome-free region *(3)*. The major reason for interest in CpG islands is that they colocalize with the 5′ end of genes. Both promoter sequences and the 5′ parts of transcription units are found within CpG islands. It has been estimated that 56% of human genes and 47% of mouse genes are associated with a CpG island *(4)* and these include all ubiquitously expressed genes as well as many genes with a tissue-restricted pattern of expression *(5,6)*. Before the draft human sequence became available the number of CpG islands in the human genome was estimated to be 34,200 (4 as modified by 7) and this figure is reasonably close to the 28,890 potential CpG islands that have been identified so far in the draft human genomic sequence *(8)*.

Usually CpG islands remain methylation-free in all tissues including the germline, regardless of the activity of their associated gene. There are three major exceptions to this: CpG islands on the inactive X chromosome *(9)*, CpG islands associated with some imprinted genes *(10)*, and CpG islands associated with nonessential genes in tissue culture cell-lines *(11)*. In both cancer and ageing aberrant methylation of CpG islands coupled with epigenetic silencing of their associated genes is found *(12,13, see 14* for a review). Why CpG

From: *Methods in Molecular Biology, vol. 256:*
Bacterial Artificial Chromosomes, Volume 2: Functional Studies
Edited by: S. Zhao and M. Stodolsky © Humana Press Inc., Totowa, NJ

islands are protected from methylation is not certain. However, the finding that deletion of functional Sp1 binding sites from either the mouse or hamster *Aprt* gene promoter leads, in both cases, to methylation of the CpG island suggests that the presence of functional transcription factor binding sites in CpG islands is involved *(15,16)*. Analysis of two of the rare CpG islands not located at the 5′ end of a gene *(17,18)* supports this idea because transcripts arising from the CpG island region were found in both cases *(19,20)*. Replication of CpG islands during early S phase has also been suggested to be involved in the protection of CpG island from methylation based on the finding that replication origins are often found at CpG islands *(21)*.

The unusual base composition and methylation-free status of CpG islands enables their detection by restriction enzymes whose sites are rare and, if present, usually blocked by methylation in the rest of the genome *(22)*. Here a method is described by which largely intact CpG islands can be isolated from BAC clones by exploiting the differential affinity of DNA fragments containing different numbers of methyl-CpGs for a methyl-CpG binding domain (MBD) column *(23,24)*. These columns consist of the MBD of the protein MeCP2 *(25,26)* coupled to a resin. MeCP2 is one of a family of proteins which bind symmetrically methylated CpGs in any sequence context and is involved in mediating methylation-dependent repression *(25,27–29)* and mutations in MeCP2 cause Rett syndrome, a neurodevelopmental disease *(30)*. DNA encoding the MBD was cloned into a bacterial expression vector to give plasmid pET6HMBD which, when expressed, yields a recombinant protein, HMBD, consisting of the MBD preceded by a tract of six histidines *(23)*. This histidine tag at the N terminal end enables the HMBD protein to be coupled to a nickel-agarose resin which can be packed into a column. DNA fragments containing many methylated CpGs bind strongly and unmethylated DNA fragments bind weakly to MBD columns *(23)*. On average, within CpG islands CpGs occur at a frequency of 1/10 bp and are unmethylated, whereas outside CpG islands CpGs are found at a frequency of 1/100 bp and are usually methylated. An average CpG island is between 1–2 kb in size and contains between 100 and 200 CpGs. When unmethylated, as is usually the case in the genome, they show little affinity for binding to MBD columns. However, when methylated they bind strongly and can be purified away from other genomic fragments which contain few methylated CpGs and, therefore, bind weakly.

Using MBD columns CpG island libraries have been made for several species *(23,31–33)*. Because CpG islands overlap the 5′ end of the transcription unit and are generally single-copy, they can be used to identify their associated full-length cDNA either by screening cDNA libraries or searching sequence databases. As they contain promoter sequences and therefore transcription factor binding sites, they can be screened for genes controlled by a particular transcription factor *(34)*. MBD columns have also been used to isolate CpG islands from large genomic clones *(24)*, which will be described in detail here, and sorted human chromosomes *(35)*. Finally, methylation of CpG islands appears to be is one route by which genes are epigenetically silenced in cancer (reviewed in *14*). Such methylated CpG islands have been identified both by screening the human CpG island library *(36)* or by directly isolating methylated CpG islands using the MBD column *(37)*.

The general protocol can be split into the following steps:

1. Production of HMBD and coupling to nickel-agarose to form the MBD column.
2. Calibration of the MBD column using plasmid DNAs containing known numbers of methyl-CpGs.
3. Restriction digestion of bacterial artificial chromosome (BAC) DNA so that CpG islands are left largely intact and other DNA is reduced to small fragments.
4. Methylation of the BAC DNA fragments at all CpGs.
5. Fractionation of the methylated DNA fragments over the MBD column. Elution at high salt yields a DNA fraction highly enriched for largely intact CpG islands.

2. Materials

2.1. Preparation of the MBD Column

1. LB broth: 1% bacto tryptone, 0.5% bacto yeast extract, and 1% NaCl (all w/v).
2. LB agar: As LB broth with the addition of 12 g/L Bacto agar.
3. 100 mM isopropyl β-D thiogalactopyranoside (IPTG) in water, filter-sterilized. Store at –20°C.
4. 2X SMASH buffer: 125 mM Tris-HCl (pH 6.8), 20% glycerol, 4% sodium dodecyl sulfate (SDS), 1 mg/mL bromophenol blue, 286 mM β-mercaptoethanol. Divide into aliquots, keep the one in use at room temperature and store the others at –20°C until required.
5. 100 mM phenylmethylsufonyl fluoride (PMSF) in isopropanol. Store at 4°C. Add to buffers A, B, C, D, and E to a final concentration of 0.5 mM just before use.
6. Stock solutions of the following protease inhibitors: leupeptin, antipain, chymostatin, pepstatin A and protinin prepared and stored as recommended by the manufacturer. Add to buffers A, B, C, D, and E to a final concentration of 5 µg/mL just before use.
7. 20% Triton X-100.
8. Buffer A: 5 M urea, 50 mM NaCl, 20 mM HEPES (pH 7.9), 1 mM ethylenediamine tetraacetic acid (EDTA) (pH 8.0), 10% glycerol.
9. Buffer B: 5 M urea, 50 mM NaCl, 20 mM HEPES (pH 7.9), 10% glycerol, 0.1% Triton X-100, 10 mM β-mercaptoethanol.
10. Buffer C: 2 M urea, 1 M NaCl, 20 mM HEPES (pH 7.9), 10% glycerol, 0.1% Triton X-100, 10 mM β-mercaptoethanol.
11. Buffer D: 50 mM NaCl, 20 mM HEPES (pH 7.9), 10% glycerol, 0.1% Triton X-100, 10 mM β-mercaptoethanol.

12. Buffer E: 50 m*M* NaCl, 20 m*M* HEPES (pH 7.9), 10% glycerol, 0.1% Triton X-100, 10 m*M* β-mercaptoethanol, 8 m*M* immidazole.
13. 1 *M* immidazole in water, filter-sterilized. Store at room temperature.

2.2. Basic Protocol for Running an MBD Column

1. MBD buffer: 20 m*M* HEPES (pH 7.9), 10% glycerol, 0.1% Triton X-100.
2. MBD buffer/x *M* NaCl: 20 m*M* HEPES (pH 7.9), x *M* NaCl, 10% glycerol, 0.1% Triton X-100.
3. 5 *M* NaCl.
4. 100 m*M* PMSF prepared and stored as in **item 2.1.** Add to MBD buffers to a final concentration of 0.5 m*M* just before use.

2.3. Calibrating the MBD Column and Preparation of BAC DNA

The reagents required for these protocols are generally available in molecular biology laboratories and an extensive list will not be included here. Specifically, reagents required for DNA isolation, purification, restriction enzyme treatment, and methylation will be needed. The reagents and the techniques are described in *(38)*.

3. Methods

In **Subheading 3.1.**, the preparation of an MBD column is described. **Subheadings 3.2.** and **3.3.** contain the basic protocol for running an MBD column and how to calibrate it. In **Subheading 3.4.** the preparation of the BAC DNA is described and in **Subheading 3.5.** the fractionation of the BAC DNA over the MBD column is described.

3.1. Preparation of the MBD Column

To prepare an MBD column the recombinant protein HMBD is expressed in the *Escherichia coli* (*E. Coli*) strain BL21 (DE3) pLysS, partially purified, coupled to nickel-agarose resin and packed into a column (*see* **Note 1**). The T7 RNA polymerase expression system is used to produce HMBD protein *(39)*. This protocol should produce sufficient HMBD protein to make a 1 mL column, and may be adjusted as required.

All steps after **step 6** of **Subheading 3.1.1.** are done on ice or in a cold room using ice-cold solutions (*see* **Note 2**).

3.1.1. Preparation of HMBD Protein

1. Streak BL21 (DE3) pLysS (pET6HMBD) from a –80°C stock onto an LB agar plate containing ampicillin (50 µg /mL) and chloramphenicol (30 µg /mL) and grow overnight at 37°C to obtain single colonies.

2. Innoculate 100 mL LB broth containing ampicillin (50 µg/mL) and chloramphenicol (30 µg/mL) with a single colony. At 37°C shake at about 300 rpm overnight in a 500 mL flask.

3. Inoculate 1.5 L of LB broth containing ampicillin (50 µg/mL) and chloramphenicol (30 µg/mL) with 45 mL of the overnight culture. Measure the OD_{600} (optical density at 600 nm). This should be approx 0.1. If not, adjust accordingly. Split the culture between two 2-L flasks and shake vigorously for 2–3 h at 37°C until the OD_{600} has reached between 0.3 and 0.5. Remove a 500-µL aliquot (sample 1).

4. To each flask, add IPTG to a final concentration of 0.4 m*M*. Grow the cultures for three hours at 37°C with vigorous shaking. Remove another 500 µL aliquot (sample 2).

5. Centrifuge samples 1 and 2 at 14K (full speed) in a microfuge for 5 min at room temperature. Resuspend the pellets in 100 µL sterile, distilled water plus 100 µL 2X SMASH buffer and store at –20°C until required for the analysis gel.

6. Centrifuge the rest of the cells at 2000*g* for 20 min at 4°C in two 1-L centrifuge bottles.

7. Discard the supernatants and resuspend each pellet in 12.5 mL buffer A. Transfer to a 50-mL tube, add Triton X-100 to 0.1% and mix by gentle swirling. The solution will become viscous as the cells begin to lyse on addition of the Triton X-100.

8. Disrupt the cells and shear the DNA by sonication. The extract will lose its viscosity and may darken in colour. Remove a 100-µL aliquot, add 100 µL of 2X SMASH buffer, mix and store at –20°C (sample 3).

9. Centrifuge the disrupted cells at 31,000*g* for 30 min at 4°C. Pour the supernatant into a 50-mL tube. Remove a 100-µL aliquot, add 100 µL of 2X SMASH buffer, mix and store at –20°C (sample 4). Store the remaining supernatant (approx 25 mL) at –80°C until required otherwise go on to **step 3** in **Subheading 3.1.2.**

3.1.2. Partial Purification of the HMBD Protein

To do this, the crude protein extract prepared in **Subheading 3.1.1.** is passed over a cation exchange resin to which most of the contaminating bacterial proteins bind weakly but the basic HMBD protein (predicted pI 9.75) binds tightly.

1. If the protein extract has been stored at –80°C, thaw in cold water or on ice. Add protease inhibitors (**Subheading 2.1., items 5** and **6**) and mix by swirling.

2. To remove insoluble material centrifuge at 31,000*g* for 30 min at 4°C. Pour the supernatant into a 50-mL tube and discard the pellet.

3. Prepare 12 mL of Fractogel EMD SO3^e-650(M) (Merck) resin as recommended by the manufacturer and pipet 5 mL into each of two plastic disposable chromatography columns, such as Econo-Pac columns (Bio-Rad 732-1010). Attach a syringe needle to each column. This increases the flow rate. Two 5-mL columns are used rather that one 10-mL column to reduce the time taken by this protocol. To equilibrate the columns wash each with 25 mL buffer B, followed by 25 mL buffer C and finally with 25 mL buffer B.

4. Arrange the two columns so that they can drip into the same tube. Simultaneously, load half of the supernatant on one column and the other half on the other column. Collect the flowthrough (FT) in a single 50-mL tube and keep on ice.

5. Next, elute the bound protein by washing the columns simultaneously in 12 elution steps. For each wash step collect the eluates from both columns into a single 15-mL tube. For washes 1–4, use 5 mL of buffer B/column, for washes 5–8, use 5 mL of buffer B+C (27.5 mL of buffer B + 12.5 mL of buffer C)/column, and for washes 9–12 use 5 mL of buffer C/column. Keep fractions 1–12 on ice.

6. To ascertain which fractions contain the HMBD protein, remove 10-μL aliquots from each fraction and the FT. Add 10 μL 2X SMASH buffer to each. Heat these samples and samples 1–4 (put aside in **Subheading 3.1.1.**) at 90°C for 90 s. Separate 20 μL of each on a 15% SDS-PAGE gel, along with molecular weight standards (for example Protein marker, Broad Range (2-212 kD), New England Biolabs 7701S) and stain with Coomassie Brilliant Blue R-250 using standard techniques *(38)*. The MW of the HMBD is 11.4 kD and should be present in samples 2–4 and fractions 9–12 (*see* **Note 3**). Pool all fractions enriched for the HMBD protein. The partially purified extract can be stored at –80°C until required, otherwise go to **Subheading 3.1.3.**

3.1.3. Coupling the HMBD Protein to Nickel-Agarose Resin

1. If the protein extract has been stored at –80°C thaw on ice or in cold water. Add protease inhibitors (**Subheading 2.1., items 5** and **6**) and mix by swirling. Remove a 10-μL aliquot and use it to measure the protein concentration by the Bradford assay *(40)* using, for example, the Protein Assay kit (Bio-Rad 500-0002). Typically, the total amount of protein will be about 20–50 mg (approx 1 mg/mL). Remove 50 μL of the protein extract, add 50 μL of 2X SMASH buffer, mix and keep on ice (sample 5).

2. Pipet 1 mL of nickel agarose resin (for example Ni-NTA Superflow, Qiagen 30410) into a 5-mL disposable plastic chromatography column (for example Poly-Prep chromatography column Bio-Rad 731-1550). Wash with 4 mL of buffer D to equilibrate (*see* **Note 4**).

3. Load the protein extract onto the column and collect the FT in a 50-mL tube.

4. Wash the column with 4 mL of buffer D, followed by 4 mL of buffer E, and finally with 4 mL of buffer D, collecting 12 1-mL fractions.

5. To ascertain if the coupling of the HMBD protein to the nickel-agarose resin has been successful remove 10-μL aliquots from the FT and each fraction. Add 10 μL 2X SMASH buffer to each. Heat these samples and sample 5 at 90°C for 90 s. Separate 20 μL of each on a 15% SDS-PAGE gel, along with molecular weight standards (for example, Protein marker, Broad Range (2–212 kD), New England Biolabs 7701S) and stain with Coomassie Brilliant Blue R-250 using standard techniques *(38)*. If the coupling reaction has been successful, the HMBD protein should be visible in sample 5, but absent or present in trace amounts in the FT and wash fractions (*see* **Note 5**).

6. Estimate the amount of HMBD protein coupled to the nickel agarose. Pool the FT and the 12 eluted fractions in a 50-mL tube and measure the protein concentration as in **step 1**. Subtract the amount of protein eluted from the amount of protein loaded to find the amount of HMBD coupled to the resin.

7. Pack the coupled resin into a column (*see* **Note 6**).

3.2. Basic Protocol for Running an MBD Column

When fractionating differently methylated DNAs using an MBD column the same basic procedure is followed and this is outlined here. DNAs are eluted from MBD columns by increasing the NaCl concentration in the wash buffer. Generally, a 1-mL column is suitable for most applications. MBD columns should be run in a cold room using ice-cold solutions. Do not allow the MBD column to dry out.

1. Prepare MBD buffer and MBD buffer/1 *M* NaCl. Mix these together to make MBD buffers containing the required NaCl concentrations. (*see* **Note 7**).

2. Equilibrate the MBD column by washing it with five column volumes of MBD buffer/0.1 *M* NaCl, followed by five column volumes of MBD buffer/1 *M* NaCl, followed by five column volumes of MBD buffer/0.1 *M* NaCl (*see* **Note 8**).

3. Load the DNA (in MBD buffer/0.1 *M* NaCl). Wash the column with 5 mL of MBD buffer/0.1 *M* NaCl (*see* **Note 10**).

4. To elute bound DNAs increase the NaCl concentration present in the wash buffer as either a linear gradient or in steps up to a maximum of 1 *M* NaCl. This is done by mixing MBD buffer and MBD buffer/1 *M* NaCl in the correct proportions (*see* **Note 10**).

5. During **steps 3** and **4**, collect fractions of the size required in the procedure being used. The usual size of the fractions collected is 1 or 2 mL, although in some cases larger volumes are collected.

6. Wash the MBD column with five column volumes of MBD buffer/1 *M* NaCl followed by five column volumes of MBD buffer/0.1 *M* NaCl after use and store at 4°C or in a cold room (*see* **Note 11**).

3.3. Calibrating the MBD Column

The amount of HMBD coupled on a MBD column determines the NaCl concentration at which DNAs methylated to different degrees elute. As this varies from column to column, each MBD column should be calibrated by determining the elution profile of artificially methylated plasmid DNAs that contain different numbers of methyl-CpGs. To do this a cloning vector such as pUC19 which contains 173 CpGs (accession number M77789) could be used, but any plasmid with a known sequence, and therefore a known number of CpGs, is suitable. Typically, heavily methylated DNA fragments (those containing greater than 100 methyl-CpGs) elute between 0.7 and 0.9 *M* NaCl (*see*

Note 12). Unmethylated DNA generally elutes at 0.5–0.6 *M* NaCl (but *see* **Note 10**) and DNAs containing intermediate numbers of methyl-CpGs (30–40) elute at 0.1–0.2 *M* less than heavily methylated fragments *(23)*.

3.3.1. Preparation of the Differentially Methylated Plasmid DNAs

1. Digest 5 μg of plasmid DNA using a restriction enzyme that has one site in the plasmid and leaves a convenient 5′ overhang for endlabeling. For example, if using pUC19 *Eco* RI is suitable.
2. Take two aliquots of 2 μg of the linearized plasmid DNA. One aliquot is "mock-methylated," i.e., treated in the same way as the other aliquot but with the omission of enzyme. Methylate the other aliquot using CpG methylase (New England Biolabs 226S) which methylates all CpGs as directed by the manufacturer.
3. Assay if the methylation reaction has been successful by testing if the methylated DNA is now resistant to digestion by methylation-sensitive restriction enzymes such as *Hha* I or *Hpa* II. Perform reactions with and without enzyme following the manufacturer's instructions using about 30 ng DNA/reaction and analyse on a 1% agarose gel stained with ethidium bromide *(38)*. The "mock-methylated" DNA should be digested to completion by both *Hha* I and *Hpa* II, and the methylated DNA should be resistant to digestion by both enzymes (*see* **Note 13**).
4. Purify the DNAs (*see* **Note 13**). Resuspend each DNA sample in 20 μL TE and measure the DNA concentration using standard procedures *(38)*.
5. Using standard procedures *(38)* endlabel 600 ng of both the unmethylated and methylated linearized plasmids using the Klenow enzyme and appropriate labelled and unlabelled nucleotides (*see* **Note 14**). For example, if the plasmid has been linearised with *Eco* RI, which leaves a 5′-AATT-3′ overhang, use [α]^{32}P dATP and dTTP.
6. To eliminate unincorporated radioactivity precipitate the DNAs using standard procedures *(38)*, washing the DNA pellets twice with 70% ethanol before drying the DNA pellets either by air-drying or under vacuum. Resuspend each in 600 μL of MBD buffer/0.1 *M* NaCl. Monitor each using a handheld Geiger counter to check for successful endlabeling. This amount is sufficient for six column runs and can be stored at 4°C for 2–4 wk.

3.3.2. Calibration of a MBD Column Using the Endlabeled Plasmid DNAs

1. Here only the modifications required for calibration are detailed. Refer to **Sub-heading 3.2.** for the basic procedure which should be followed when running an MBD column.
2. Mix together 100 ng (100 μL) of each of the endlabeled unmethylated and completely methylated plasmid DNAs (*see* **Note 15**).
3. Load the DNA mixture onto a 1-mL MBD column, wash with MBD buffer/0.1 *M* NaCl up to 5 mL. Then wash with 5 mL of MBD buffer/0.4 *M* NaCl followed by a 40-mL linear salt gradient to 1 *M* NaCl (i.e., increase the concentration of NaCl from 0.4 to 1 *M* over 40 mL). Finally, wash with 5 mL of MBD buffer/1 *M* NaCl. Collect 1-mL fractions in either 5 mL or 1.5 mL tubes as convenient (*see* **Note 16**).

4. Count the radioactivity in each fraction. The radioactivity should elute from the column in two peaks during the linear gradient part of the run. Typically, the first peak will elute at about 0.5 *M* NaCl and the second peak at about 0.8 *M* NaCl.
5. From peak fractions remove 400-μL aliquots. Ethanol precipitate and resuspend DNA from these in 10 μL of TE using standard procedures *(38)*.
6. Determine the methylation status of these DNA samples by restriction enzyme analysis as described in **Subheading 3.3.1., step 3** using 3 μL of the test DNA/reaction. After running the analytical gel, dry it down and expose it to X-ray film to visualize the endlabeled DNA fragments (*see* **Note 17**). The DNA in the first peak should be digested by both *Hpa* II and *Hha* I, showing that it is unmethylated. The DNA in the second peak should be resistant to digestion by both enzymes, showing that it is completely methylated.

3.3.3. Determination of the NaCl Concentration at Which Only Methylated DNA Binds to the MBD Column

To purify CpG island fragments from BAC clones, the DNA sample, prepared as described in **Subheading 3.4.**, is loaded onto the MBD column at an NaCl concentration at which unmethylated DNA fragments and those containing few methyl-CpGs do not bind whereas heavily methylated DNA fragments do. The NaCl concentration at which this happens varies between MBD columns and should be ascertained for each MBD column using endlabeled unmethylated and "partially methylated" plasmid DNAs (*see* **Note 15**). These are loaded on the MBD column, this time individually, in MBD buffer containing various test NaCl concentrations to identify the highest at which the unmethylated plasmid remains in the flow-through and the partially methylated plasmid binds to the column.

1. Here only the modifications required for calibration are detailed. Refer to **Subheading 3.2.** for the basic procedure which should be followed when running an MBD column.
2. Load 100 ng (100 μL) of the endlabeled, unmethylated test plasmid onto the MBD column in 500 μL of MBD buffer/0.5 *M* NaCl. Wash the column with 9.5 mL of the MBD buffer/0.5 *M* NaCl followed by 10 mL of MBD buffer/1 *M* NaCl. Collect 1-mL fractions in 1.5-mL microfuge tubes or 5-mL tubes as convenient.
3. Count all the collected fractions for radioactivity to determine where the DNA elutes.
4. Reequilibrate the MBD column by washing with 10 mL of MBD buffer/0.5 *M* NaCl.
5. Repeat **steps 1–3** with the partially methylated plasmid.
6. Repeat **steps 1–4** varying the NaCl concentration of the MBD buffer in which the DNA is loaded onto the column in increments of 0.05 *M* to determine the highest at which the unmethylated DNA elutes in the loading buffer and the partially methylated DNA binds and is eluted by MBD buffer/1 *M* NaCl. Between each round of testing, reequilibate the MBD column using MBD buffer containing the appropriate NaCl concentration.

3.4. Preparation of BAC DNA

1. Prepare BAC DNA using standard procedures *(38)* or using commercially available kits and as a final step purify using a CsCl-gradient (*see* **Note 18**).
2. Digest the DNA to completion using a restriction enzyme whose recognition sequence is found infrequently within CpG island DNA but frequently elsewhere in the genome, such as *Mse* I, as directed by the manufacturer (*see* **Note 19**).
3. Methylate the fragmented BAC DNA using CpG methylase (New England Biolabs 226S) which methylates all CpGs as directed by the manufacturer. To methylate 20–50 μg of BAC DNA, perform a 500-μL reaction. To monitor, remove two aliquots of 10 μL of the reaction mix before and after the addition of the enzyme. To these, add 1 μg of linearized plasmid DNA such as *Eco* RI digested pUC19. After incubation, analyze these as described in **Subheading 3.3.1., step 3** using 3 μL/ restriction digest. Successful methylation of the plasmid DNA, as indicated by resistance to digestion by *Hpa* II and *Hha* I shows that the methylation of the genomic DNA has also gone to completion. If not, do another round of methylation (*see* **Note 19**).
4. Purify the methylated DNA (*see* **Note 19**) and resuspend it in 250 μL of MBD buffer containing NaCl at the concentration at which unmethylated DNA does not bind to the MBD column as determined in **Subheading 3.3.3.** Store at –20°C until required.

3.5. Fractionation of the Methylated BAC DNA on the MBD Column

1. Here, only the modifications required are detailed. Refer to **Subheading 3.2.** for the basic procedure, which should be followed when running an MBD column.
2. Load the BAC DNA prepared as described in **Subheading 3.4.** onto the MBD column. Wash with 4.75 mL MBD buffer containing NaCl at the concentration determined in the calibration step (*see* **Subheading 3.3.3.**) at which unmethylated DNA does not bind to the column but methylated DNA does to remove fragments which bind weakly.
3. Elute the bound DNA either with a salt gradient or in steps and collect fractions. Use a 30-mL linear salt gradient to 1 *M* NaCl. Finally, wash with 5 mL of MBD buffer/1 *M* NaCl. Collect 1-mL fractions in either 5- or 1.5-mL tubes as convenient (*see* **Notes 16** and **20**).
4. Purify DNA from these fractions by precipitating with ethanol and resuspend the DNA in 20 μL of TE using standard procedures *(38)*. Include 20 μg of glycogen (1 μL of a 20 mg/mL solution Boehringher Mannheim 901 393) as a carrier to avoid losing the DNA. Alternatively use a DNA purification kit.
5. Clone this DNA into a suitable cloning vector (*see* **Notes 21** and **22**).
6. Analyze clones to check that they are derived from CpG islands (*see* **Notes 23–26**).

4. Notes

1. The plasmid pET6HMBD in the *E. coli* strain XL1-BLUE can be obtained by writing to Professor A. P. Bird, ICMB, Edinburgh University, King's Buildings, Mayfield Road, Edinburgh EH9 3JR. For expression of the recombinant protein HMBD pET6HMBD should be transformed into the *E.coli* strain BL21 (DE3) pLysS (F⁻ *ompT hsdS*β(rβ⁻mβ⁻) *gal dcm* (DE3) pLysS (Novagen 69388-1)). For some expression constructs, it has been found that expression levels tend to decrease if the same stock is used repeatedly. To avoid this, always use a freshly streaked plate from a frozen stock kept at –80°C. However, if expression problems persist retransform pET6HMBD into BL21 (DE3) pLysS.
2. Buffers A, B, C, D, and E should be freshly prepared just before use.
3. In the initial analysis gel the induced HMBD protein may not be visible in samples 2, 3, and 4 because of the excess of bacterial proteins. However, after purification by cation exchange chromatography, the HMBD protein should be clearly visible and the dominant band present in the fractions eluted at high NaCl concentrations as most bacterial proteins elute in the FT.
4. Alternative nickel-agarose resins to the Ni-NTA Superflow suggested here are available, but be aware that some of these have to be charged before use. Prepare the nickel-agarose resin to be used according to the manufacturer's directions. Failure of the HMBD protein to couple to nickel-agarose resin is most likely due to use of uncharged resin.
5. If a small amount of HMBD protein is present in the FT or wash fractions collected after the coupling, it is likely that the capacity of the resin has been exceeded. Generally, between 25 and 40 mg of HMBD is sufficient to saturate 1 mL of resin.
6. Ideally, differentially methylated DNAs are separated by running MBD columns in conjunction with automated chromatography and fractionation systems such as the FPLC System (Pharmacia 18-1035-00) or Gradifrac System (Pharmacia 18-1993-01) with the resin packed into an HR5/5 column (Pharmacia 18-0382-01) so that flow rates and elution gradients can easily be controlled. Generally a flow rate of 1 mL/min is used. However, if such a system is not available MBD columns can be made using a small disposable plastic chromatography column (for example Poly-Prep chromatography column Bio-Rad 731-1550) and run under gravity flow. In this case, I would suggest using Ni-NTA agarose (Qiagen 30210) rather than Ni-NTA Superflow as it is cheaper, has similar binding capacity and the superior mechanical stability and flow characteristics of the Superflow resin are not required for gravity flow applications.
7. MBD buffers should be freshly prepared just before use.
8. In cases where DNAs are loaded onto the MBD column at NaCl concentrations higher than 0.1 *M* use MBD buffer containing the appropriate NaCl concentration, instead of MBD buffer/0.1 *M* NaCl, when equilibrating the column.
9. MBD columns should be calibrated before use with test plasmid DNAs containing known numbers of methyl-CpGs (*see* **Subheading 3.3.**).

10. This is only the basic procedure and should be adjusted and modified according to requirements. First, the NaCl concentration of the MBD buffer in which DNAs are loaded onto the column can be adjusted. DNA binds to the MBD column, irrespective of methylation status, if loaded in MBD buffer/0.1 M NaCl. This is probably because the HMBD protein is very basic *(23)*. However, if DNAs are loaded in MBD buffer containing about 0.5 M NaCl, it has been found that unmethylated DNA does not bind and remains in the FT, but methylated DNA still does bind *(31,41)*. The highest molarity at which this happens will vary from column to column depending on the amount of coupled HMBD and should be determined as described in **Subheading 3.3.3.** Second, choose whether to elute bound DNAs by increasing the NaCl concentration of the wash buffer in steps, as a linear gradient or by a combination of the two. If using step-wise elutions wash the column with five columns volumes of buffer at each step. Generally, when eluting bound DNAs with linear gradients the more shallow a gradient chosen the better the resolution. When using a 1-mL column, a linear gradient of 0.5 to 1 M over 30 mL has been found to give good separation of methylated DNAs *(31)*. If using step-wise elution, increase the concentration of NaCl by 100 mM NaCl for each step, which also results in good separation (S. H. Cross, unpublished observations).

11. MBD columns are stable for at least 6 mo if kept at 4°C and can be reused many times. Do not allow MBD columns to dry out.

12. The NaCl concentration at which a fragment elutes from the MBD column is determined principally by the total number of methylated CpGs it contains, rather than the number of CpGs per unit length *(23)*. Therefore, it can be assumed that methylated CpG islands will elute at the same NaCl concentration as the heavily methylated plasmid DNA used to calibrate the column.

13. If the methylated plasmid is still susceptible to digestion by methylation-sensitive restriction enzymes repeat the methylation reaction. It is often necessary to do at least two rounds of methylation. Between each round purify the DNA. To purify the DNA either extract and precipitate the DNA using standard procedures *(38)* or use commercially available kits, for example Qiaquick (Qiagen).

14. Great care must be taken when using radioactivity. Use appropriate shielding and precautions to avoid exposure and follow the local radiation protection rules.

15. When calibrating an MBD column plasmids containing a range of different numbers of methyl-CpGs can be used to refine where DNA fragments containing different numbers of methyl-CpGs can be expected to elute from the column. They are used for assaying the highest NaCl concentration at which unmethylated DNA does not bind to the MBD column but methylated DNA still does and are used in **Subheading 3.3.3.** Such test plasmids can be prepared using methylase enzymes which modify CpGs within certain sequence contexts. For example, *Hha* I and *Hpa* II methylases (New England Biolabs 217S and 214S) methylate CpGs within the sequence contexts GCGC and CCGG respectively. In the case of pUC19 use of these enzymes together would yield a plasmid containing 30 methylated CpGs.

16. If it is not possible to increase the NaCl concentration using a linear gradient, increase the NaCl concentration in steps of 0.1 M NaCl from 0.4 to 1 M.

17. To avoid loss of small DNA fragments during drying of the analytical gel, place it on DE81 paper (Whatman 3658 915), which is then placed on two sheets of 3MM paper (Whatman 3030 917). Cover with clingfilm before drying down.

18. Generally, 20–50 μg of DNA is sufficient for fractionating CpG islands from BAC clones. DNA yield varies greatly, we have found that from 200-mL cultures, the amount of DNA obtained ranges between 4 and 28 μg for different BACs (Ruth Edgar, personal communication). Therefore, the amount of starting culture required has to determined for each BAC although for most BACs a 1-L culture will yield more than sufficient DNA. It is important to use CsCl-gradient purified DNA because any contaminating *E. coli* DNA present will copurify with the CpG island DNA, which it resembles in sequence composition.

19. *Mse* I recognizes the sequence TTAA which is predicted to occur, on average, every 1000 bp within CpG islands and every 150–200 bp elsewhere *(23)*. However, the dinucleotide TA is found less frequently than expected in the genome, for reasons that are not understood, so that *Mse* I sites occur less frequently than they are predicted to. This has the advantage that the chance of an *Mse* I site occurring within a CpG island is reduced. On the other hand, the size of other genomic fragments is larger than expected, but this does not matter because of the low frequency of CpG in the genome. Following *Mse* I digestion, up to two-thirds of CpG islands are left intact whereas other sequences are found on small fragments containing on average, 1 to 5 methylated CpGs *(23)*. Other restriction enzymes with a 4-bp recognition site containing only Ts and As, such as *Tsp*509 I, which recognizes the sequence AATT, could be used, although sites for such enzymes may be found more frequently within CpG islands. In addition, *Mse* I is a good enzyme to use because *Mse* I fragments can be cloned into the *Nde* I site of the pGEM®-5Zf(+/–) cloning vectors (Promega P2241 and P2351) (*see* **Note 22** for discussion of cloning of purified CpG island fragments).

20. Perform at least two rounds of binding so that the fraction containing heavily methylated fragments is purified away from unmethylated fragments efficiently. Between each round dilute with MBD buffer so that the NaCl concentration is reduced to that at which unmethylated DNA does not bind to the column and methylated DNA does as determined in the calibration **Subheading 3.3.3.**

21. As mentioned in **Note 19** *Mse* I fragments can be cloned into the *Nde* I site of plasmid vectors such as pGEM®-5Zf(–/+) (Promega P2241 and P2351). This is because *Mse* I and *Nde* I produce compatible cohesive ends, which are, therefore, compatible for ligation. As the cloning site is destroyed, the best way to examine clone inserts is to amplify them by PCR using primers flanking the cloning site. Dephosphlorylate the linearized vector before use to reduce background, using standard procedures *(38)*. Use standard techniques for both ligation of the CpG island fraction into the vector and transformation *(38)*.

22. The bacterial strain chosen for transformation should be one that does not restrict methylated DNA, such as SURE (Stratagene 200294).

23. Analysis of potential CpG island clones should be carried out to determine if they are derived from *bona fide* CpG islands. One major contaminant is likely to be

E. coli fragments which, because they have the same sequence characteristics as CpG islands, will copurify along with CpG islands (*see* **Note 18**). A class of genomic non-CpG island fragment, which will copurify along with CpG islands, is GC-rich repetitive DNA, which is normally methylated in the genome. This type of fragment is removed by the "stripping" step used when isolating CpG islands from genomic DNA *(23)*, but such a step cannot be carried here because when genomic DNA is cloned into BACs the native methylation pattern is erased. Ways on identifying these contaminants are described below.

24. Cloned inserts would be expected to be > 0.5 kb (the average size of inserts in the human CpG island was 0.76 kb *[23]*). The sequence of the clones would be expected to have a GC-content in excess of 50% and to contain close to the expected number of CpGs. The clones would also be expected to be derived from unmethylated genomic sequences. Suggested tests are: (a) Test clones for the presence of *Bst* UI sites. This restriction enzyme has the recognition sequence CGCG which occurs about 1/100 bp in CpG island DNA and about 1/10 kb in non-CpG island DNA. If a clone contains a *Bst* UI site, this is a good indication that it is derived from a CpG island. This is an easy and reliable way of quickly judging if clones are from CpG islands. (b) Sequence clones and search sequence databases. Discard any clones that match *E. coli* sequence. Examine other matches to see if the clones match to known genomic sequences, known CpG islands or repeats. Examine the sequences to determine if the clones have the sequence characteristics of CpG islands. Expect a G+C content of greater than 50% and close to the expected number of CpGs, as predicted from base composition. Mammalian genomic DNA has a G+C content of about 40% and contains only about 25% of the expected number of CpGs as predicted from base composition. An easy way to visualise this data is to plot a graph with base composition on the *x*-axis and CpG observed/expected values on the *y*-axis, *see* **ref. 23** for an example. (d) Determine whether the clones are derived from unmethylated DNA in the genome. Use clones that do not contain repeats (*see* **e**) to probe Southern blots of genomic DNA which has been digested with *Mse* I alone and *Mse* I and a methylation-sensitive restriction enzyme such as *Bst* UI or *Hpa* II, using standard procedures *(38)*. If the clone is derived from an unmethylated CpG island, the genomic *Mse* I fragments should be cleaved by the methylation-sensitive enzymes. However, bear in mind that in some cases CpG islands are methylated as discussed in the Introduction. (e) Another way to determine if clones contain repeated sequences is by hybridizing colonies with total genomic DNA, only repeat-containing clones will hybridize. Only about 10% of CpG islands contain highly repeated sequences *(23)*. Nonrepetitive clones can be used as probes as in (d).

25. It should be remembered that CpG islands are not found at the 5′ end of all genes, notable exceptions being many genes with a tissue-restricted pattern of expression *(5,6)*. Therefore, other methods such as exon trapping and cDNA selection should be used *(42–44)* when isolating such genes from a BAC. However, the method described here does have the advantage that it depends only on sequence composition and is unaffected by gene expression patterns.

26. CpG islands are useful gene markers because there is only one *CpG* island/gene, they colocalize with the 5′ end of the transcript, include promoter sequences, and as they are usually single copy, they can be used to map genes and to isolate full-length cDNAs.

References

1. Antequera, F. and Bird, A. (1993) CpG islands, in *DNA Methylation: Molecular Biology and Biological Significance* (Jost, J. P. and Saluz, H. P., eds.), Birkhauser Verlag, Basel, Switzerland, pp. 169–185.
2. Cross, S. H. and Bird, A. P. (1995) CpG islands and genes. *Curr. Opin. Genet. Dev.* **5,** 309–314.
3. Tazi, J. and Bird, A. (1990) Alternative chromatin structure at CpG islands. *Cell* **60,** 909–920.
4. Antequera, F. and Bird, A. (1993) Number of CpG islands and genes in human and mouse. *Proc. Natl. Acad. Sci. USA* **90,** 11,995–11,999.
5. Gardiner-Garden, M. and Frommer, M. (1987) CpG islands in vertebrate genomes. *J. Mol. Biol.* **196,** 261–282.
6. Larsen, F., Gunderson, G., Lopez, R., and Prydz, H. (1992) CpG islands as gene markers in the human genome. *Genomics* **13,** 1095–1107.
7. Ewing, B. and Green, P. (2000) Analysis of expressed sequence tags indicates 35,000 human genes. *Nat. Genet.* **25,** 232–234.
8. International Human Genome Sequencing Consortium (2001) Initial sequencing and analysis of the human genome. *Nature* **409,** 860–921.
9. Riggs, A. D. and Pfeifer, G. P. (1992) X-chromosome inactivation and cell memory. *Trends Genet.* **8,** 169–174.
10. Tilghman, S. M. (1999) The sins of the fathers and mothers: genomic imprinting in mammalian development. *Cell* **96,** 185–193.
11. Antequera, F., Boyes, J., and Bird A. (1990) High levels of *de novo* methylation and altered chromatin structure at CpG islands in cell-lines. *Cell* **62,** 503–514.
12. Greger, V., Passarge, E., Höpping, W., Messmer, E., and Horsthemke, B. (1989) Epigenetic changes may contribute to the formation and spontaneous regression of retinoblastoma. *Hum. Genet.* **83,** 155–158.
13. Issa, J-P. J., Ottaviano, Y. L., Celano, P., Hamilton, S. R., Davidson, N. E., and Baylin, S. B. (1994) Methylation of the oestrogen receptor CpG island links ageing and neoplasia in human colon. *Nat. Genet.* **7,** 536–540.
14. Baylin, S. B. and Herman, J. G. (2000) DNA hypermethylation in tumorigenesis: epigenetics joins genetics. *Trends Genet.* **16,** 168–174.
15. Macleod, D., Charlton, J., Mullins, J., and Bird, A. (1994) Sp1 sites in the mouse *Aprt* gene promoter are required to prevent methylation of the CpG island. *Genes Dev.* **8,** 2282–2292.
16. Brandeis, M., Frank, D., Keshet, I., et al. (1994) Sp1 elements protect a CpG island from *de novo* methylation. *Nature* **371,** 435–438.
17. Tykocinski, M. L. and Max, E. E. (1984) CG dinucleotide clusters in MHC genes and in 5′ demethylated genes. *Nucl. Acids Res.* **12,** 4385–4396.

18. Stöger, R., Kubicka, P., Liu, C. G., et al. (1993) Maternal-specific methylation of the imprinted mouse *Igf2r* locus identifies the expressed locus as carrying the imprinting signal. *Cell* **73**, 61–71.

19. Macleod, D., Ali, R. R., and Bird, A. (1998) An alternative promoter in the mouse major histocompatibility complex class II I-Aβ gene: implications for the origin of CpG islands. *Mol. Cell Biol.* **18**, 4433–4443.

20. Wutz, A., Smrzka, O. W., Schweifer, N., Schellander, K., Wagner, E. F., and Barlow, D. P. (1998) Imprinted expression of the Igf2r gene depends on an intronic CpG island. *Nature* **389**, 745–749.

21. Delgado, S., Gómez, M., Bird, A., and Antequera, F. (1998) Initiation of DNA replication at CpG islands in mammalian chromosomes. *EMBO J.* **17**, 2426–2435.

22. Bickmore, W. A. and Bird, A. P. (1992) Use of restriction endonucleases to detect and isolate genes from mammalian cells. *Meth. Enzy.* **216**, 224–245.

23. Cross, S. H., Charlton, J. A., Nan, X., and Bird, A. P. (1994) Purification of CpG islands using a methylated DNA binding column. *Nat. Genet.* **6**, 236–244.

24. Cross, S. H., Clark, V. H., and Bird, A. P. (1999) Isolation of CpG islands from large genomic clones. *Nucl. Acids Res.* **27**, 2099–2107.

25. Lewis, J. D., Meehan, R. R., Henzel, W. J., et al. (1992) Purification, sequence and cellular localisation of a novel chromosomal protein that binds to methylated DNA. *Cell* **69**, 905–914.

26. Nan, X., Meehan, R. R., and Bird, A. (1993) Dissection of the methyl-CpG binding domain from the chromosomal protein MeCP2. *Nucl. Acids Res.* **21**, 4886–4892.

27. Nan, X. Campoy, J., and Bird, A. (1997) MeCP2 is a transcriptional repressor with abundant binding sites in genomic chromatin. *Cell* **88**, 471–481.

28. Nan, X., Ng, H., Johnson, C. A., et al. (1998) Transcriptional repression by the methyl-CpG-binding protein MeCP2 involves a histone deacetylase complex. *Nature* **393**, 386–389.

29. Jones, P. L., Veenstra, G. J. C., Wade, P. a., et al. (1998) Methylated DNA and MeCP2 recruit histone deacetylase to repress transcription. *Nat. Genet.* **19**, 187–191.

30. Amir, R. E., Van den Veyver, I. B., Wan, M., Tran, C. Q., Francke, U., and Zoghbi, H. Y. (1999) Rett syndrome is caused by mutations in X-linked MECP2, encoding methyl-CpG-binding protein 2. *Nat. Genet.* **23**, 185–188.

31. Cross, S. H., Lee, M., Clark, V. H., Craig, J. M., Bird, A. P., and Bickmore, W. A. (1997) The chromosomal distribution of CpG islands in the mouse: evidence for genome scrambling in the rodent lineage. *Genomics* **40**, 454–461.

32. McQueen, H. A., Fantes, J., Cross, S. H., Clark, V. H., Archibald, A. L., and Bird, A. P. (1996) CpG islands of chicken are concentrated on microchromosomes. *Nat. Genet.* **12**, 321–324.

33. McQueen, H. A., Clark, V. H., Bird, A. P., Yerle, M., and Archibald, A. L. (1997) CpG islands of the pig. *Genome Res.* **7**, 924–931.

34. Watanabe, T., Inoue, S., Hiroi, H., Orimo, A., Kawashima, H., and Muramatsu, M. (1998) Isolation of estrogen-responsive genes with a CpG island library. *Mol. Cell Biol.* **18**, 442–449.

35. Cross, S. H., Clark, V. H., Simmen, M. W., et al. (2000) CpG islands libraries from human Chromosomes 18 and 22: landmarks for novel genes. *Mamm. Gen.* **11,** 373–383.
36. Huang, T. H., Perry, M. R., and Laux, D. E. (1999) Methylation profiling of CpG islands in human breast cancer cells. *Hum. Molec. Genet.* **8,** 459–470.
37. Shiraishi, M., Chuu, Y., and Sekiya, T. (1999) Isolation of DNA fragments associated with methylated CpG islands in human adenocarcinomas of the lung using a methylated DNA binding column and denaturing gradient gel electrophoresis. *Proc. Natl. Acad. Sci. USA* **96,** 2913–2918.
38. Sambrook, J., Fritsch, E. F., and Maniatis, T., eds. (1989) *Molecular Cloning. 2nd ed.* Cold Spring Harbor Laboratory Press, Cold Spring Harbor, NY.
39. Studier, F. W., Rosenberg, A. H., Dunn, J. J., and Dubendorff, J. W. (1990) Use of T7 RNA polymerase to direct expression of cloned genes. *Meth. Enzymol.* **185,** 60–89.
40. Bradford, M. (1976) A rapid and sensitive method for the quantitation of microgram quantities of protein utilising the principle of protein dye binding. *Anal. Biochem.* **72,** 248–254.
41. John, R. M. and Cross, S. H. (1997) Gene detection by the identification of CpG islands, in *Genome Analysis: A Laboratory Manual, Vol. 2 Detecting Genes* (Birren, B., Green, E. D., Klapholz, S., Myers, R. M., and Roskams, J., eds.), Cold Spring Harbor Laboratory Press, Cold Spring Harbor, NY, pp. 217–285.
42. Parimoo, S., Patanjali, S. R., Shukla, H., Chaplin, D. D., and Weissman, S. M. (1991) cDNA selection: Efficient PCR approach for the selection of cDNAs encoded in large chromosomal DNA fragments. *Proc. Natl. Acad. Sci. USA* **88,** 9623–9627.
43. Lovett, M., Kere, J., and Hinton, L. M. (1991) Direct selection: A method for the isolation of cDNAs encoded by large genomic regions. *Proc. Natl. Acad. Sci. USA* **88,** 9628–9632.
44. Buckler, A. J., Chang, D. D., Graw, S. L., et al. (1991) Exon amplification: A strategy to isolate mammalian genes based on RNA splicing. *Proc. Natl. Acad. Sci. USA* **88,** 4005–4009.

4

BAC Microarray-Based Comparative Genomic Hybridization

Antoine M. Snijders, Richard Segraves, Stephanie Blackwood, Daniel Pinkel, and Donna G. Albertson

1. Introduction

Alterations in DNA copy number underlie certain developmental abnormalities and are frequent in solid tumors. Microarray-based comparative genomic hybridization (array CGH) provides a high throughput means to measure and map copy number aberrations on a genome-wide scale and to link them directly to genome sequence. For applications involving the analysis of tumors, the technology must provide reliable detection of single copy gains and losses in mixed cell populations such as tumor and normal cells, accurate quantification of high level copy number gains, and confident interpretation of aberrations affecting only a single array element. Further, one would like to minimize the amount of specimen material required for an analysis. These requirements can be met if there are good signal-to-noise ratios in the hybridization.

A number of platforms for array CGH have been described and have used large genomic clones such as cosmids, P1s and BACs *(1–5)* or smaller clones such as cDNAs *(6,7)* as the array elements. The use of DNA from large insert clones (e.g., BACs) provides substantially more intense signals, than use of smaller clones such as cDNAs and, therefore, correspondingly better performance for detection of single copy gains and losses. However, preparation of sufficient DNA from BACs to use as array elements is problematic, because BACs are single copy vectors and the yield of DNA from these cultures is low compared to cultures carrying high copy number vectors such as plasmids. In addition, spotting high molecular weight DNA at sufficient concentration to obtain good signal-to-noise in the hybridizations may be difficult. These prob-

From: *Methods in Molecular Biology, vol. 256:*
Bacterial Artificial Chromosomes, Volume 2: Functional Studies
Edited by: S. Zhao and M. Stodolsky © Humana Press Inc., Totowa, NJ

lems can be overcome by generating a high complexity representation of the large genomic clone DNA that can subsequently be amplified using the polymerase chain reaction (PCR) *(8–10)*. The protocol described here uses a single tube ligation-mediated PCR method (*see* **Fig. 1**) that previously had been demonstrated to produce a high yield of DNA that was a good representation of the human genomic DNA from a single cell *(9)*. The BAC DNA is first digested with *Mse*I and then adapters are annealed to the digested BAC DNA. A first round of PCR amplification is carried out and then 1 µL of this material is used as template for a second round of amplification to generate the DNA used for spotting. The procedure produces sufficient spotting solution at a concentration of 0.8 µg/µL DNA in 20% dimethyl sulfoxide (DMSO) to make tens of thousands of arrays from 1 ng of BAC DNA, essentially an unlimited supply of reagent for array production. Using the hybridization protocols also described here, independently prepared representations yield highly reproducible data, as the average variation of the linear ratios on individual clones from two independent preparations is 6.6%. The ratios measured on arrays comprised of the BAC representations are essentially identical to ratios obtained by spotting BAC DNA directly isolated from the same BAC clones *(10)*.

2. Materials

2.1. Restriction Enzyme Digest of BAC DNA

1. 10X One-Phor-All Buffer Plus: 100 mM Tris-acetate (pH 7.5), 100 mM magnesium acetate, and 500 mM potassium acetate (available from Amersham Pharmacia Biotech Inc.). Stable at 4°C. (*See* **Note 1**)
2. *Mse*I restriction enzyme (20 U/µL) (available from New England Biolabs). Store at –20°C. Dilute tenfold in 10X One-Phor-All Buffer Plus prior to usage, store on ice. Leftover dilution should be discarded.
3. 20–500 ng of BAC DNA with minimal contamination from host DNA. A Qiagen Plasmid Mini Kit can be used to purify BAC DNA by following a modification of the manufacturer's protocol (*see* **Notes 2** and **3**).

Fig. 1. *(see facing page)* Schematic overview of the ligation mediated PCR procedure. BAC DNA is digested with the DNA endonuclease *Mse*I, leaving a 5′ phosphorylated TA overhang. Primer 2 guides Primer 1 to the overhang, where primer 1 is ligated at the 5′ overhang of the BAC DNA. Owing to the lack of a phosphate group at the 5′ end of Primer 2, this primer will not be ligated to the 3′ end of the BAC DNA. After melting off Primer 2, the 3′ end of the BAC DNA is extended by Taq/Pwo DNA polymerase (dashed line). This DNA is now used as template in the amplification reaction, which is primed by the excess Primer 1 remaining after ligation. A second round of amplification is carried out to generate DNA that will be made into spotting solution (not shown).

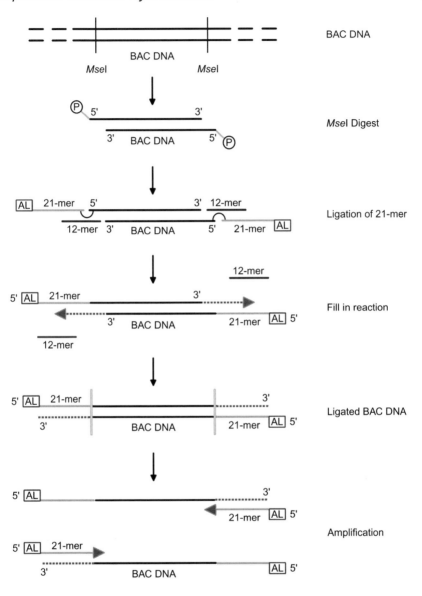

BAC DNA

*Mse*I Digest

Ligation of 21-mer

Fill in reaction

Ligated BAC DNA

Amplification

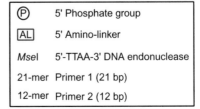

℗	5' Phosphate group
AL	5' Amino-linker
*Mse*I	5'-TTAA-3' DNA endonuclease
21-mer	Primer 1 (21 bp)
12-mer	Primer 2 (12 bp)

4. 0.2 mL Polypropylene PCR 8-tube strips with separate 8-cap strips. Strips and caps should be free of RNase, DNase, DNA, Pyrogens, and PCR inhibitors.
5. Sterile H_2O (e.g., autoclaved, deionized, and filtered water).
6. Ultrapure agarose (Invitrogen).
7. 1X TBE: 0.089 M Tris-borate, 0.089 M boric acid, 0.008 M ethylenediamine tetraacetic acid (EDTA).
8. 1% ethidium bromide solution. Ethidium bromide is a mutagen. Always wear protective clothes when handling any form of ethidium bromide (*see* **Note 4**).
9. 6X loading buffer: 0.25% bromophenol blue, 0.25% xylene cyanol, 30% glycerol in H_2O. Store at 4°C.
10. φ X174 RF DNA/Hae III Marker (Promega).
11. UV transilluminator.
12. 96-well plate seal (USA Scientific).
13. Multichannel pipettor (e.g., a Matrix Technologies 8- or 12-channel pipetor that dispenses 0.5–12.5 µL).

2.2. Ligation of Specific Primers to BAC DNA

1. 10X One-Phor-All Buffer Plus: 100 mM Tris-acetate (pH 7.5), 100 mM magnesium acetate, and 500 mM potassium acetate (available from Amersham Pharmacia Biotech Inc.). Stable at 4°C. (*See* **Note 1**.)
2. 1X TE buffer, pH 7.4: 10 mM Tris-HCl, pH 7.4, 1 mM EDTA, pH 8.0. Store at room temperature.
3. Primer 1: 5′-AGT GGG ATT CCG CAT GCT AGT-3′, 50 nmolar scale, containing a 5′ aminolinker. Dissolve in sterile TE, pH 7.4 to 500 µM, this is the primer stock solution. For the 100 µM working solution, dilute stock solution 1:5 in sterile H_2O. Store dry primer, stock solution, and working solution at –20°C.
4. Primer 2: 5′-TAA CTA GCA TGC-3′, 50 nmolar scale. Dissolve in sterile TE, pH 7.4 to 500 µM, this is the primer stock solution. For the 100 µM working solution, dilute stock solution 1:5 in sterile H_2O. Store dry primer, stock solution, and working solution at –20°C.
5. ATP 10 mM (available from Invitrogen). Store at –20°C.
6. T4 DNA ligase (5 U/µL) (available from Invitrogen). Store at –20°C, when in use keep the enzyme on ice. Return to –20°C as soon as possible.
7. PCR machine capable of ramping at approx 1.3°C/min.
8. 0.2 mL Polypropylene PCR 8-tube strips with separate 8-cap strips. Strips and caps should be free of RNase, DNase, DNA, pyrogens, and PCR inhibitors.
9. Sterile H_2O (e.g., autoclaved, deionized, and filtered water).
10. Multichannel pipetor (e.g., a Matrix Technologies 8- or 12-channel pipetor that dispenses 0.5–12.5 µL).

2.3. Ligation-Mediated PCR

1. 100 mM Stock solutions each of dATP, dCTP, dGTP, and dTTP. For the 10 mM dNTP working solution, dilute stock solutions 1:10 in sterile H_2O. Both stock solutions and working solution should be stored at –20°C.

2. 10X PCR buffer #1 (Expand Long Template PCR System available from Roche). Store at –20°C.

3. DNA polymerase mix containing Taq and Pwo DNA polymerase (3.5 U/µL) (Expand Long Template PCR System available from Roche). Store at –20°C, when in use, keep the enzyme mix on ice. Return to –20°C as soon as possible.

4. Sterile H_2O (e.g., autoclaved, deionized, and filtered water).

5. Ultrapure agarose (Invitrogen).

6. 1X TBE: 0.089 M Tris-borate, 0.089 M boric acid, 0.008 M EDTA.

7. 1% ethidium bromide solution. Ethidium bromide is a mutagen. Always wear protective clothes when handling any form of ethidium bromide (*see* **Note 4**).

8. 6X Loading buffer: 0.25% bromophenol blue, 0.25% xylene cyanol, 30% glycerol in sterile H_2O. Store at 4°C.

9. φ X174 RF DNA/Hae III Marker (Promega).

10. UV transilluminator.

11. Multichannel pipettor (e.g., a Matrix Technologies 8- or 12-channel pipetor that dispenses 5–250 µL).

12. Reagent reservoir with divider (25 mL, Matrix Technologies Corp.).

2.4. Re-PCR of Ligation-Mediated PCR

1. Primer 1: 5′-AGT GGG ATT CCG CAT GCT AGT-3′, 50 nmolar scale, containing a 5′ aminolinker. Dissolve in sterile TE, pH 7.4 to 500 µM, this is the primer stock solution. For the 100 µM working solution, dilute stock solution 1:5 in sterile H_2O. Store dry primer, stock solution, and working solution at –20°C.

2. 10X Taq buffer II (contains no $MgCl_2$; available from Perkin Elmer). Store at –20°C.

3. 100 mM Stock solutions each of dATP, dCTP, dGTP, and dTTP. For the 25 mM dNTP working solution, dilute stock solutions 1:4 in sterile H_2O. Both stock solutions and working solution should be stored at –20°C.

4. $MgCl_2$ (25 mM) (available from Perkin Elmer). Store at –20°C.

5. Amplitaq Gold DNA polymerase (5 U/µL) (available from Perkin Elmer). Store at –20°C, when in use keep the enzyme on ice. Return to –20°C as soon as possible.

6. 0.2 mL Polypropylene PCR 8-tube strips with separate 8-cap strips. Strips and caps should be free of RNase, DNase, DNA, pyrogens, and PCR inhibitors.

7. Sterile H_2O (e.g., autoclaved, deionized, and filtered water).

8. Multichannel pipettors (e.g., a Matrix Technologies 8- or 12-channel pipetor that dispenses 0.5–12.5 µL and a Matrix Technologies 8- or 12-channel pipetor that dispenses 5–250 µL).

9. Reagent reservoir with divider (25 mL, Matrix Technologies Corp.).

10. 96-well plate seal (USA Scientific).

11. Ultrapure agarose (Invitrogen).

12. 1X TBE: 0.089 M Tris-borate, 0.089 M boric acid, 0.008 M EDTA.

13. 1% ethidium bromide solution. Ethidium bromide is a mutagen. Always wear protective clothes when handling any form of ethidium bromide (*see* **Note 4**).

14. 6X loading buffer: 0.25% bromophenol blue, 0.25% xylene cyanol, 30% glycerol in sterile H₂O. Store at 4°C.
15. φ X174 RF DNA/Hae III Marker (Promega).
16. UV transilluminator.

2.5. Preparation of Spotting Solutions From Re-PCR Used for Array CGH

1. Fan oven (e.g., Techne Hybridiser HB-1D) capable of heating to 45°C, used to dry down the PCR products from **Subheading 2.4.**
2. 100% ethanol. Store at –20°C.
3. 70% ethanol.
4. 3 *M* sodium acetate pH 5.2: dissolve 40.8 g sodium acetate·3H₂O in 80 mL sterile H₂O, adjust pH to 5.2 with glacial acetic acid and adjust volume to 100 mL with sterile H₂O. Autoclave to sterilize. Store at room temperature.
5. 20% DMSO in sterile H₂O. Store at room temperature.
6. Multichannel pipetors (e.g., a Matrix Technologies 8- or 12-channel pipetor that dispenses 0.5–12.5 µL and a Matrix Technologies 8- or 12-channel pipetor that dispenses 5– 250 µL).

2.6. Random Primed Labeling of Genomic DNA for Array CGH Analysis

1. 2.5X Random Primers (BioPrime DNA labeling system, Invitrogen). Store at –20°C.
2. Sterile H₂O (e.g., autoclaved, deionized, and filtered water).
3. Genomic DNA (200–300 ng) (*see* **Note 3**). Stable at 4°C.
4. Klenow fragment (40 U/µL, BioPrime DNA labeling system, Invitrogen). Store at –20°C.
5. Cy3 and Cy5 labeled dCTP (1 m*M*, Amersham Pharmacia Biotech Inc.).
6. 0.5 *M* EDTA, pH 8.0.
7. 1 *M* Tris-HCl, pH 7.6.
8. 10X dNTP mixture in sterile water (store at –20°C):
   a. 2 m*M* each of dATP, dTTP, dGTP
   b. 0.5 m*M* dCTP
   c. 10 m*M* Tris-HCl base (pH 7.6)
   d. 1 m*M* EDTA
9. Sephadex G-50 spin column (Amersham Pharmacia Biotech Inc.).

2.7. Hybridization of Fluorescently Labeled Genomic DNA for Array CGH Analysis

1. Differentially labeled test and reference genomic DNA from **Subheading 2.6.**
2. Human Cot-1 DNA (1 mg/mL, Invitrogen) (*see* **Note 5**).
3. Salmon sperm DNA (10 mg/mL, Invitrogen).
4. Yeast tRNA (100 µg/µL, Invitrogen).
5. 20% SDS in sterile H₂O (heat to 68°C to dissolve). Store at RT.

6. 100% ethanol. Store at –20°C.
7. 3 *M* Sodium acetate pH 5.2.
8. Dextran sulfate (sodium salt, 500,000 MW).
9. Formamide (redistilled, ultrapure, Invitrogen). Store at –20°C.
10. 20X SSC: 3.0 *M* NaCl, 0.3 *M* sodium citrate, pH 7.0.
11. Hybridization mixture: dissolve 1 g dextran sulfate (sodium salt, 500,000 MW, available from Fisher Biotech) in 5 mL formamide (redistilled, ultrapure) and 1 mL 20X SSC (*see* **Note 6**). Hybridization mixture should be stored in aliquots at –20°C.
12. PN buffer: 0.1 *M* sodium phosphate, 0.1 % Nonidet P-40 (NP40), pH 8.0 (*see* **Note 7**).
13. Sterile H$_2$O (e.g., autoclaved, deionized, and filtered water).
14. UV Stratalinker 2400 (available from Stratagene) capable of producing 130,000 µJoules UV.
15. Very slowly rocking table (approx 1 rpm) inside a 37°C incubator (e.g., a VWR brand Rocker, Model 100).
16. Rubber cement (Ross, American Glue Corporation).
17. Silicone gasket (Press-to-Seal, PGC Scientific).
18. 100% glycerol.
19. 10X PBS: 1.4 *M* NaCl, 0.027 *M* KCl, 0.1 *M* Na$_2$HPO$_4$, 0.018 *M* KH$_2$PO$_4$, adjusted to pH 7.4 with HCl.
20. Stereomicroscope.
21. 10 mL syringe.
22. 200 µL disposable pipet tip without a filter.
23. Binder clips.

3. Methods

Subheadings 3.1.–3.5. describe the protocol for the preparation of 96 DNA spotting solutions from pure BAC DNA (*see* **Figs. 1** and **2**). The hybridization procedure is outlined in **Subheadings 3.6.–3.7.**

3.1. Restriction Enzyme Digest of BAC DNA

The restriction enzyme digest is carried out in a 5-µL reaction volume containing: 2.2X One-Phor-All Buffer Plus, 2 U *Mse*-I restriction enzyme and 20–500 ng BAC DNA. The reaction is set up as follows:

1. Dilute the 10X One-Phor-All Buffer Plus to a final concentration of 0.8X in a volume of 750 µL using sterile H$_2$O and dispense 93 µL into each tube of an 8-tube strip (*see* **Notes 1** and **8**).
2. Using a multichannel pipetor, dispense 2.5 µL of the 0.8X buffer solution into each tube of 12 8-tube PCR strips and seal the tubes using a 96-well plate seal to prevent evaporation and possible contamination (*see* **Note 9**).
3. Add 1.5 µL BAC DNA (*see* **Notes 2** and **10**) to each tube using a single channel pipettor. The BAC DNA concentration usually ranges from 20–400 ng/µL. During

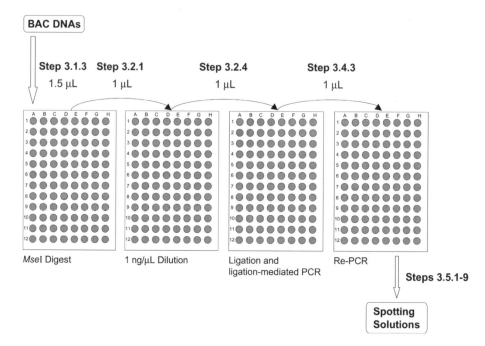

Fig. 2. Schematic overview of the ligation mediated PCR procedure for 96 BAC DNA clones. After digesting the BAC DNA, 1 μL of each digest reaction is used to make a dilution of 1 ng/μL. One μL from the dilution is used for the ligation and subsequent PCR. From this PCR (50 μL), only 1 μL is used as template for a Re-PCR. The Re-PCR yields approx 10–15 μg of DNA, which is precipitated and redissolved in 12 μL 20% DMSO. This solution is then loaded into the microtiter plate, for printing. Note that approx 20 ng of input BAC DNA can yield approx 1000 sets of printing solutions (input DNA amplification approx 5×10^5).

this step only one 8-tube PCR strip is handled at a time. The remaining 11 8-tube strips remain covered using a 96-well plate seal. After adding the BAC DNA to each 8-tube strip, place the strip in a different rack and seal with a 96-well plate seal (see **Note 11**).

4. Dilute the *Mse*I restriction enzyme to a final concentration of 2 U/μL in a volume of 120 μL using 10X One-Phor-All Buffer Plus (see **Note 1**). Keep the enzyme and the dilution on ice during this process.

5. Dispense 1 μL (2 U) of the *Mse*I enzyme dilution into each tube individually (see **Note 12**); the enzyme should stay on ice during this process. Cap each 8-tube strip using a 8-cap strip after adding the enzyme and place on ice.

6. The reaction is placed in a PCR machine for an overnight incubation at 37°C (12–16 h).

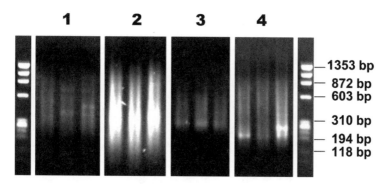

Fig. 3. Agarose gel analysis of ligation-mediated PCR process. Panel 1, typical size range and DNA distributions of three different *Mse*I digest reactions, panel 2, three ligation-mediated PCRs, panel 3, three failed ligation-mediated PCRs and panel 4, three Re-PCRs. All DNA samples were separated on a 1% agarose gel containing 0.5 μg/μL ethidium bromide together with a φ X174 RF DNA/*Hae* III DNA Marker, which has a size range from 72 to 1353 bp.

7. 1.75 μL of the digests can be run on a conventional 1% agarose gel containing 0.5 μg/mL ethidium bromide along with a φ X174 RF DNA/Hae III Marker to check fragment length. Restriction sizes should range from 100 to 1500 bp (*see* **Fig. 3**).

3.2. Ligation of Specific Primers to BAC DNA

The ligation reaction is carried out in a 10-μL reaction volume containing: 5 μ*M* Primer 1, 5 μ*M* Primer 2, 0.5X One-Phor-All Buffer Plus, 1 m*M* ATP, 5 U T4 DNA ligase, and 1 ng digested BAC DNA.

1. 1 μL of each of the restriction digests from **Subheading 3.1.** is diluted to 1 ng/μL using sterile H₂O. Determine the DNA concentration in each digest (*see* **Note 13**). Calculate the amount of sterile H₂O required to dilute 1 μL of each digest to a final DNA concentration of 1 ng/μL (for example, if the BAC DNA concentration in the digest reaction is X ng/μL, the appropriate amount of sterile H₂O to add is (X-1) μL). Set up 12 new 8-tube strips: individually add the calculated amount of sterile H₂O to each new tube at the locations corresponding to the locations of each digest. Transfer 1 μL of each digest using a multichannel pipettor. Cover using a 96-well plate seal (*see* **Fig. 2**).
2. The ligation reaction is set up in a 1.5-mL tube by combining:
   a. 56 μL Primer 1 (100 μ*M*)
   b. 56 μL Primer 2 (100 μ*M*)
   c. 56 μL One-Phor-All Buffer Plus (*see* **Note 1**)
   d. 616 μL sterile H₂O
   _____
   784 μL total

   Mix by pipeting and dispense 98 μL into each tube of an 8-tube strip.

3. Using a multichannel pipettor, dispense 7 µL of the above prepared primer solution (*see* **Subheading 3.2., step 2**) onto the bottom of 12 new 8-tube strips (*see* **Note 14**). Seal the 8-tube strips using a 96-well plate seal.
4. Add 1 µL of the 1 ng/µL BAC DNA digest prepared in **Subheading 3.2., step 1** to the 7 µL primer solution in each corresponding 8-tube strip (*see* **Fig. 2** and **Note 15**).
5. Put the 12 8-tube strips into a PCR machine. The annealing reaction is carried out at 65°C for 1 min, then the temperature is shifted down to 15°C, with a ramp of approx 1.3°C/min (in a Perkin-Elmer 9700 PCR machine this is a ramp rate of 5%).
6. During the annealing process, dispense approx 62.5 µL of 10 mM ATP solution into each tube of an 8-tube strip (*see* **Note 16**). Cap the 8-tube strip and place on ice.
7. As soon as the PCR machine reaches 15°C, promptly take out the tubes from the PCR machine and carefully open all 8-tube strips, including the 8-tube strip containing the ATP solution (**Subheading 3.2., step 5**). Using a multichannel pipettor on the repeat-pipeting-mode, pick up 12.5 µL of 10 mM ATP and dispense 1 µL on the inside wall of each of the 96 tubes containing DNA. Gently tap the PCR rack so that the ATP slides into the DNA/primer solution. Seal with a 96-well plate seal. This procedure for adding the ATP reduces the probability of carry over contamination.
8. Dispense 1 µL (5 U) of the T4 DNA ligase enzyme into each tube of the 96 tubes individually (*see* **Note 12**). Cap each 8-tube strip using an 8-cap strip after adding the enzyme and place on ice.
9. The reaction is placed in a PCR machine for an overnight incubation at 15°C (12–16 h).

3.3. Ligation-Mediated PCR

The ligation mediated PCR is carried out in a reaction volume of 50 µL containing: 0.6X 10X PCR buffer #1, 0.4 mM dNTP mixture, 3.5 U DNA polymerase mixture, and 10 µL ligation mixture (*see* **Subheading 3.2.**).

1. Combine in a 15-mL tube:
   a. 336 µL 10x PCR buffer #1
   b. 224 µL dNTP mixture (10 mM)
   c. 3920 µL sterile H_2O
       4480 µL total
   Mix briefly by vortexing and place on ice.
2. Remove the ligations from **Subheading 3.2., step 9** from the PCR machine. Pour the diluted dNTP mixture prepared in **Subheading 3.3., step 1** into a reservoir. Open all 8-tube strips carefully. Using a multichannel pipetor on the repeat-pipetting-mode, pick up 205 µL of the diluted dNTP mixture and dispense 40 µL on the inside wall of five subsequent 8-tube PCR strips. Purge the remaining 5 µL PCR mixture back into the reservoir. Discard the pipet tips for reasons discussed in **Note 14**. Repeat this procedure for the next five subsequent 8-tube PCR strips.

Again, purge the remaining 5 µL diluted dNTP mixture back into the reservoir. Adjust the multichannel pipettor fill volume to 85 µL, leaving the dispense volume at 40 µL. Pick up 85 µL diluted dNTP mixture from the reservoir and dispense 40 µL on the inside wall of the remaining two 8-tube PCR strips. Gently tap the PCR rack so that the diluted dNTP mixture slides into the DNA/primer solution. Seal with 8-tube strips.

3. To melt off Primer 2, place PCR tubes in the PCR machine at 68°C for 4 min. After 4 min, take out the tubes and add 1 µL of DNA polymerase mixture (3.5 U/µL) to each tube individually (*see* **Note 12**). Open one 8-tube strip at a time and close immediately after adding the enzyme mix and place on ice (*see* **Note 17**).

4. Place the 8-tube strips in the PCR machine: 68°C for 3 min; 94°C for 40 s, 57°C for 30 s, 68°C for 1 min 15 s for 14 cycles, 94°C for 40 s, 57°C for 30 s, 68°C for 1 min 45 s for 34 cycles, and 94 °C for 40 s, 57°C for 30 s and 68°C for 5 min for the last cycle, followed by incubation at 4°C.

5. 3.5 µL of each PCR should be run on a conventional 1% agarose gel containing 0.5 µg/mL ethidium bromide along with a φ X174 RF DNA/Hae III Marker to check fragment length. Size of the PCR product should range from 70 to 1500 bp, with the highest concentration of product around 200 to 800 bp (*see* **Note 18** and **Fig. 3**).

3.4. Re-PCR of Ligation Mediated PCR

The ligation-mediated PCR is used as a template in a Re-PCR reaction to generate DNA for spotting. Amplification is carried out in a 100-µL reaction containing: 1X Taq-buffer II, primer 1 (4 µ*M*), dNTP mixture (0.2 m*M*), MgCl$_2$ (5.5 m*M*), 2.5 U amplitaq Gold, and ligation-mediated PCR product (1 µL).

1. Combine in a 15-mL tube:
   a. 1000 µL 10X Taq-buffer II
   b. 400 µL primer 1 (100 µ*M*)
   c. 80 µL dNTP mixture (25 m*M*)
   d. 2200 µL MgCl$_2$ (25 m*M*)
   e. 50 µL amplitaq Gold (5 U/µL)
   f. 6170 µL sterile H$_2$O
      _____
      9900 µL total

2. Pour the above-prepared PCR mixture (*see* **Subheading 3.4., step 1**) into a reservoir and dispense 99 µL into each tube of twelve 8-tube strips using a multichannel pipetor. Seal using a 96-well plate seal.

3. Add 1 µL ligation-mediated PCR product, from **Subheading 3.3., step 4**, to each corresponding 8-tube strip using a multichannel pipettor. Do this by unsealing one 8-tube strip of PCR mixture and one 8-tube strip containing the ligation-mediated PCR. Cap the 8-tube strip after adding the template, then cap the 8-tube strip containing the remaining ligation-mediated PCR. Move on to the next 8-tube strip.

4. Place the 8-tube strips, now containing 99 µL PCR mixture and 1 µL ligation mediated PCR product, in a PCR machine: 95°C for 10 min; 45 cycles at 95°C for

30 s, 50°C for 30 s, 72°C for 2 min, and a final extension at 72°C for 7 min, followed by an incubation at 4°C.

5. 4 μL of each reaction should be run on a conventional 1% agarose gel containing 0.5 μg/mL ethidium bromide along with a φ X174 RF DNA/*Hae* III Marker to check fragment length. Size of the PCR product should range from 200 to 1500 bp (*see* **Note 19** and **Fig. 3**).

3.5. Preparation of Spotting Solutions From Re-PCR Used for Array CGH

1. Dry down the Re-PCR products to a volume of approx 50 μL by uncapping the 8-tube strips and placing the PCR rack face-up in a hybridization oven set at 45°C. This usually takes approx 75 min. If you are using an oven in which the heat inlet is biased toward one side of the PCR rack, the rack should be rotated every 15 min to allow even evaporation. Reducing the volume of the Re-PCR is necessary to accommodate the ethanol and sodium acetate for DNA precipitation (*see* **Subheading 3.5., step 2**).

2. Precipitate the PCR products by adding 150 μL prechilled 100% ethanol and 5 μL 3 *M* sodium acetate using a multichannel pipettor. Cap the 8-tube strips and invert the rack several times.

3. Chill the PCR rack at −20°C for 15 min.

4. Spin the plate at 1699*g* for 90 min at 4°C.

5. Remove the supernatant carefully, using a multichannel pipetor and add 150 μL 70% ethanol to the pellet using a multichannel pipetor. Cap the tubes after adding the 70% ethanol.

6. Vortex each 8-tube strip until the pellets come loose. After vortexing all 8-tube strips, centrifuge rack at 1699*g* for 45 min at 4°C.

7. Remove as much of the 70% ethanol as possible using a multichannel pipetor and dry the pellet at room temperature for approx 90 min. The time depends upon the amount of ethanol left on the pellet. Be careful not to over dry the pellets.

8. Add 12 μL of a 20% DMSO solution to the pellets using a multichannel pipetor and mix by pipeting. Only one 8-tube strip is handled at a time at this stage. Cap the tubes and resuspend by flicking the bottom of the PCR tubes to loosen the pellet.

9. Store the DNA solution at 4°C for an overnight (or longer) incubation. Use a multichannel pipettor to mix the solution until the pellet is completely in suspension. DNA solutions can be stored at 4°C (*see* **Note 20**).

3.6. Random Primed Labeling of Genomic DNA for Array CGH Analysis

A typical random primed labeling procedure is described. The random primed labeling is carried out in a 50-μL reaction volume containing: 200–300 ng genomic DNA, 1X random primers, 40 U Klenow DNA polymerase, Cy3 or Cy5 labeled dCTP (40 μ*M*) and 1X dNTP mixture. The random primed DNA product will typically range in size from 200 to 1500 bp, with the highest concentration around 400 bp.

1. Mix approx 200–300 ng of genomic DNA with 20 µL of 2.5X random primer solution and make up the volume to 42 µL with sterile H_2O.
2. Denature the DNA by heating the mixture at 99°C in a PCR machine for 10 min. Briefly centrifuge and place on ice.
3. Add:
   a. 5 µL of the 10X dNTP mixture (prepared in **Subheading 2.6., step 8**)
   b. 2 µL Cy3 or Cy5 labeled dCTP (1 m*M*)
   c. 1 µL Klenow DNA polymerase (40 U/µL).
   Mix well and place tube in a PCR machine at 37°C for an overnight incubation.
4. Place the Sephadex G-50 column in a 1.5-mL tube and prespin the column at 730*g* for 1 min.
5. Place the column in a clean 1.5-mL tube; apply the sample onto the column and spin at 730*g* for 2 min to remove unincorporated nucleotides from the DNA mixture. The color of the DNA mixture is an indicator of the amount of labeled nucleotide incorporated during the labeling reaction.

3.7. Hybridization of Fluorescently Labeled Genomic DNA for Array CGH Analysis

1. Preparation of array for hybridization:
   a. Expose a printed array to 260,000 µJoules of UV by using a Stratalinker (*see* **Note 21**).
   b. Fill a 10-mL syringe with rubber cement and fit a 200-µL pipet tip on the syringe outlet. Apply a rubber cement ring around the array using a stereomicroscope to observe the area of the array. Air-dry the rubber cement and apply a second thick layer of rubber cement on top of the first layer. Air-dry the rubber cement.
2. Preparation of samples for hybridization:
   a. Combine 50 µL labeled test genomic DNA, 50 µL labeled reference genomic DNA, and 25–50 µg of human Cot-1 DNA. Precipitate the DNA sample mixture with ethanol by adding 2.5 volumes of ice-cold 100% ethanol and 0.1 volumes of 3 *M* sodium acetate (pH 5.2). Mix the solution by inversion and collect the precipitate by centrifugation at 14,000 rpm for 30 min at 4°C.
   b. Carefully aspirate and discard the supernatant. Wipe the excess liquid from the tube with a paper tissue, being careful not to disturb the pellet. Air-dry the pellet for approx 5–10 min. Dissolve the pellet in 5 µL yeast tRNA (100 µg/µL), 10 µL 20% SDS, and 35 µL hybridization mixture (*see* **Note 22**) (for hybridization mixture preparation *see* **Subheading 2.7, step 11** and **Note 6**).
3. Preparation of array blocking solution:
   a. Precipitate 50 µL of salmon sperm DNA (10 mg/mL) by adding 2.5 volumes of ice-cold ethanol and 0.1 volumes of 3 *M* sodium acetate (pH 5.2). Mix the

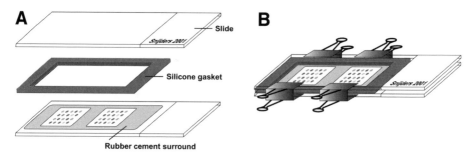

Fig. 4. Overview of hybridization slide assembly. After applying the hybridization probe solution inside the rubber cement ring on the array, a silicone gasket is placed around the rubber cement. A clean microscope slide is placed on top of the silicone gasket and the whole slide assembly is sealed using binder clips as shown. (**A**) Expanded view, showing the array slide with rubber cement around the array at the bottom, silicone gasket and slide used as a cover at the top. (**B**) Assembled hybridization chamber with binder clips.

        solution by inversion and collect the precipitate by centrifugation at 6400 rpm for 1–2 min.

b. Carefully aspirate and discard the supernatant. Wipe the excess liquid from the tube with a paper tissue, taking care not to disturb the pellet. Air-dry the pellet for approx 5–10 min. Dissolve the pellet in 5 μL sterile H_2O, 10 μL 20% SDS, and 35 μL hybridization mixture (*see* **Note 22**) (for hybridization mixture preparation, *see* **Subheading 2.7., step 11** and **Note 6**).

4. Denature the DNA sample solution from **Subheading 3.7., step 2** at 73°C for 10–15 min and incubate at 37°C for approx 60 min.

5. Place a silicone gasket around the rubber cement and apply 50 μL of the salmon sperm DNA blocking solution to each array inside the rubber cement ring. Place a clean microscope slide on top to prevent evaporation. Incubate at room temperature for 30 min. (*See* **Fig. 4**, binder clips are not necessary for this incubation.)

6. Carefully aspirate approx 30–40 μL of salmon sperm hybridization mixture from the array and quickly apply the hybridization solution onto the array (*see* **Note 23**). Place a microscope slide on top of the gasket and clamp the assembly together (*see* **Fig. 4**). Incubate the array for two nights at 37°C on a slowly rocking table.

7. Disassemble the slide assembly and rinse the hybridization solution from the slide under a stream of PN buffer.

8. Wash the slides twice in 50% formamide, 2X SSC, pH 7.0 for 15 min at 45°C followed by a 15-min wash in PN buffer at room temperature.

9. Carefully remove the rubber cement with forceps, while keeping the array moist with PN buffer.

10. If desired arrays may be mounted in a DAPI solution to stain the array spots (90% glycerol, 10% PBS, 1 μ*M* DAPI). Otherwise, rinse with sterile H_2O and allow to air-dry.

11. Arrays are ready for imaging (*see* **Note 24**).

4. Notes

1. The 10X One-Phor-All Buffer Plus is a universal buffer. In this protocol it is used at 2.2X and 2.5X final concentration for the *Mse*I digest (*see* **Subheading 3.1., step 1**) and ligation reaction (**Subheading 3.2., step 2**), respectively.

2. BAC DNA is purified from 25-mL cultures using a modification of the Qiagen Plasmid Mini Purification protocol. The volumes of buffers P1, P2, and P3 are increased from 0.3 to 1.5 mL. After adding buffer P3, incubate on ice for 10 min. Centrifuge at 4000 rpm for 45–60 min at 4°C. Filter the supernatant through a 35-μm nylon mesh (available from Small Parts Inc.) prior to loading onto Qiagen-tip 20 columns. Heat buffer QF to 65°C prior to adding it to the column. After elution, add 0.56 μL of isopropanol to the DNA, mix, and store at 4°C overnight. Mix and precipitate the DNA for 45 min at 14,000 rpm at 4°C. Air-dry the pellet for 15–30 min. Then dissolve in 50 μL sterile H_2O.

3. DNA concentrations should be determined using a fluorometer, rather than a spectrophotometer to obtain more accurate measurements.

4. Ethidium bromide is a mutagen and should be handled with care while wearing appropriate clothing. Ethidium bromide waste, i.e., agarose gels, should be discarded properly.

5. The DNA concentration of each new lot of Cot-1 DNA should be determined using a fluorometer and should measure 500 ng/μL or greater.

6. Distribute 1 g of dextran sulfate over the entire length of a 15-mL tube. While holding the tube horizontally, squirt in 5 mL formamide. Close the tube and shake vigorously for 30 s. Add 1 mL of 20X SSC and shake vigorously for 30 s. Dissolve overnight at room temperature. The final volume should be approx 7 mL.

7. Prepare 20 L of 0.1 M Na_2HPO_4, 0.1% NP40 pH9. While continuously measuring the pH, adjust the pH to 8.0 with 0.1 M NaH_2PO_4, 0.1% NP40 (approx 1 L). Make sure not to go below pH 8.0. Store at room temperature.

8. The restriction enzyme buffer dilution is enough for approx 3X 96 digest preparations; the leftover dilution can be stored at 4°C after capping with an 8-cap strip.

9. DNA/PCR contamination is possible when handling 96 different BAC DNAs at one time. Therefore, always wipe pipettors with 10 mM HCl (or other commercially available reagent). If the plate is sealed with a 96-well plate seal, unseal slowly. Keep 8-tube strips, 8-cap strips, tubes, and other labware clean. Amplification reactions should be carried out in a confined space, either a separate room or a PCR workstation. Do not open tubes with PCR products in these confined spaces. It is recommended to use filter pipet tips throughout the entire protocol.

10. Note here that 1.5 μL of BAC DNA is used for each restriction enzyme digest reaction regardless of the DNA concentration of each BAC preparation.

11. Roll back the plate seal of the 96-well plate (host plate), containing the digest buffer solution, until the first 8-tube strip is exposed. Take out the first 8-tube strip and add BAC DNA to each tube. Place this 8-tube strip in a clean 96-well plate (recipient plate) and place a 96-well plate seal on top of the plate so that it

covers the whole plate (one end of the plate seal will cover the 8-tube strip, the other end of the seal will stick to the 96-well plate). Go back to the host plate and roll back the plate seal until the next 8-tube strip becomes exposed. Take out the strip and add BAC DNA to each tube. Partially remove the 96-well plate seal of the recipient plate, whilst leaving the first 8-tube strip covered. While holding the 96-well plate seal in one hand, place the next 8-tube strip in the recipient plate. Reseal the recipient plate and repeat this procedure until all 8-tube strips are done.

12. The use of a multichannel pipettor for pipetting small amounts (~1 µL) of enzyme has not been reliable. We found that by pipeting the enzymes, *Mse*I, T4 DNA ligase, and the DNA polymerase mix separately using a single channel pipetor directly into the liquid in each tube has led to more reliable results. When the use of a multichannel pipettor is suggested, it is stated in the protocol. If not stated, we recommend not using a multichannel pipettor.

13. The DNA concentration in each restriction enzyme digest reaction is best determined by multiplying the DNA concentration of each original BAC DNA preparation by 0.3. For example, if the DNA concentration of the BAC DNA preparation used for the restriction enzyme digest reaction is 100 ng/µL then the amount of DNA in the digest reaction is (100 ng/µL) × (1.5 µL) = 150 ng of BAC DNA. The volume of each digest reaction is 5 µL, making the BAC DNA concentration in the digest reaction (150 ng) / (5 µL) = 30 ng/µL (or (100 ng/µL) × (0.3) = 30 ng/µL).

14. Use a 12.5-µL multichannel pipettor to pick up 8 µL, dispense 7 µL and throw away the remaining 1 µL and the pipet tips. The use of new pipet tips for each 8-tube strip is encouraged to ensure accuracy. Dispensing the remaining 1 µL and reusing the pipet tips might leave remnants of the primer solution, which might give rise to air bubbles during the next pick-up of 8 µL, which might inhibit the accurate dispensing of 7 µL in the next 8-tube strip.

15. First carefully unseal the BAC DNA dilution plate, and then using a multichannel pipetor pick up 1 µL of the first 8-tube strip. Unseal only the first 8-tube strip containing the primer solution, pick up the strip and add the DNA directly into the 7 µL primer solution and cap both 8-tube strips after adding the DNA.

16. The ATP solution is enough for approx 5X 96 ligation reactions; the leftover ATP solution can be stored at 4°C after capping with an 8-cap strip.

17. The 4-min incubation at 68°C will displace primer 2 (12 mer; Tm = approx 40°C) (*see* **Fig. 1**). It is essential to swiftly add the DNA polymerase mixture to each 8-tube strip, close the 8-tube strip with a 8-cap strip and place the strip-tube on ice. Acting swiftly during this step is essential to prevent significant re-annealing of primer 2. The subsequent 3-min incubation at 68°C will displace the reannealed primer 2 and extend the, now free, 3′ end of BAC DNA (*see* **Fig. 1**). The process of taking 12 8-tube strips out of the PCR machine, adding the enzyme mixture to each tube individually, placing the 8-tube strips on ice and finally placing the 12 8-tube strips back into the PCR machine for amplification should take no longer than 10–15 min.

18. If a banding pattern appears in the smear, the ligation and or ligation-mediated PCR has not been successful and should be repeated. All the products should have a smear ranging from 70 to 2000 bp. Any aberration from this smear: i.e., a small smear around 300–600 bp or a high molecular weight smear is unacceptable and should therefore be discarded (*see* **Fig. 3**).

19. The Re-PCR will most likely have the highest concentration of product around 200–400 bp. The average PCR concentration should be 150 ng/μL as determined with a fluorometer (*see* **Fig. 3**).

20. The final concentration of the spotting solution should be approx 0.8–1.3 μg/μL. After printing the spotting solutions on a glass surface, the spots will become nearly invisible due to the lack of salt in the spotting solutions. Breathing on the glass slide will make the arrays visible for a short period of time. Except for stratalinking the DNA to the surface (*see* **Note 21** and **Subheading 3.7., step 1a**), the slides will not need further treatment before hybridization.

21. Place the slide(s) in the Stratalinker, arrays facing up. The arrays should be given a fixed amount of *energy* (260,000 μJ) instead of other available options the Stratalinker might have, such as *autocrosslink* or *time*. Over-crosslinking the slide might result in a decrease in fluorescent hybridization signal.

22. Carefully pipet 5 μL sterile H_2O (for preparation of the array blocking solution) or 5 μL yeast tRNA (100 μg/μL) (for preparation of DNA sample solution) and 10 μL 20% SDS on the pellet. Incubate at RT for approx 15 min. Add 35 μL of hybridization mixture while stirring the solution with the pipet tip. Do not dissolve the pellet by pipeting to prevent foam formation due to the presence of SDS.

23. Aspirate the salmon sperm DNA solution by tilting the array at approx 45° allowing the solution to slide to one side of the array. Aspirate approx 30–40 μL of salmon sperm DNA solution using a 0.5–10 μL pipetor set at 10 μL by using approx 3–4 pipet tips. Make sure the array does not dry out by swiftly removing the salmon sperm DNA solution and applying the sample solution. If more than one array is being hybridized on one slide, tilt the array at approximately 45°, aspirate 20 μL of salmon sperm DNA solution from array 1., aspirate 10 μL salmon sperm DNA solution from array 2., then aspirate another 20 μL from array 1. and apply sample solution to array 1. Finally, aspirate 30 μL from array 2. and apply sample solution to array 2. Be careful not to generate bubbles when applying the sample solution to the array. This is to ensure that the microscope slide, which is placed on top of the silicone gasket (*see* **Fig. 4**), does not touch the sample solution.

24. Images can be acquired using commercially available CCD or laser scanning imaging systems (e.g., an Axon scanner 4000B). Image analysis and quantification can be done using commercially available image analysis programs (e.g., GenePix from Axon) or a program called UCSF Spot (11).

Acknowledgments

This work was supported by NIH Grants CA80314, CA83040, CA84118, HD17665, and CA58207; by California BCRP Grant 2RB-0225; and Vysis, Inc.

References

1. Solinas-Toldo, S., Lampel, S., Stilgenbauer, S., et al. (1997) Matrix-based comparative genomic hybridization: biochips to screen for genomic imbalances. *Genes, Chromosom. Cancer* **20,** 399–407.
2. Pinkel, D., Segraves, R., Sudar, D., et al. (1997) High resolution analysis of DNA copy number variation using comparative genomic hybridization to microarrays. *Nature Genet.* **20,** 207–211.
3. Geschwind, D. H., Gregg, J., Boon, K., et al. (1998) Klinefelter's syndrome as a model of anomalous cerebral laterality: testing gene dosage in the X chromosome pseudoautosomal region using a DNA microarray. *Dev. Genet.* **23,** 215–229.
4. Albertson, D. G., Ylstra, B., Segraves, R., et al. (2000) Quantitative mapping of amplicon structure by array CGH identifies *CYP24* as a candidate oncogene. *Nature Genet.* **25,** 144–146.
5. Bruder, C. E. G., et al. (2001) High resolution deletion analysis of constitutional DNA from neurofibromatosis type 2 (NF2) patients using microarray-CGH. *Hum. Mol. Genet.* **10,** 271–282.
6. Pollack, J. R., Perou, C. M., Alizadeh, A. A., et al. (1999) Genome-wide analysis of DNA copy-number changes using cDNA microarrays. *Nature Genet.* **23,** 41–46.
7. Heiskanen, M. A., Bittner, M. L., Chen, Y., et al. (2000) Detection of gene amplification by genomic hybridization to cDNA microarrays. *Cancer Res.* **60,** 799–802.
8. Lucito, R., West, J., Reiner, A., et al. (2000) Detecting gene copy number fluctuations in tumor cells by microarray analysis of genomic representations. *Genome Res.* **10,** 1726–1736.
9. Klein, C. A., Schmidt-Kittle, O., Schardt, J. A., Pantel, K., Speicher, M. R., and Riethmuller, G. (1999) Comparative genomic hybridization, loss of heterozygosity, and DNA sequence analysis of individual cells. *Proc. Natl. Acad. Sci. USA* **96,** 4494–4499.
10. Snijders, A. M., Nowak, N., Segraves, R., et al. (2001) Assembly of microarrays for genome-wide measurement of DNA copy number by CGH. *Nature Genet.* **29,** 263–264.
11. Jain, A. N., Tokuyasu, T. A., Snijders, A. M., et al. (2002) Fully automatic quantification of microarray image data. *Genome Res.* **12,** 325–332.

5

Large DNA Transformation in Plants

Sangdun Choi

1. Introduction

Large DNA (>100 kb) transformation methods have been reported in tomato using *Agrobacterium*-mediated transformation *(1)* and tobacco using biolistic bombardment *(2)*. Both methods used modified bacterial artificial chromosomes (BACs) such as BIBAC or pBACwich. BIBAC vector (used in tomato) is capable of transferring large DNA fragments from *Agrobacterium* into plants. pBACwich (used in tobacco) utilizes a Cre/*lox* site-specific recombination system to integrate large DNA fragments into plant chromosomes.

Biolistic bombardment *(3)* is a method by which foreign substances are introduced into intact cells and tissues via high-velocity microparticles. This technique uses tungsten or gold microparticles to deliver the DNA into the cells and is a convenient method for transforming intact cells, because very little pre- or post-bombardment manipulation is needed. Gold microparticles, driven by compressed helium, are employed most commonly due to their uniformity, spherical shape and chemically inert nature. Stable transformations of plants by the biolistic system have been reported in *Arabidopsis (4)*, rice *(5,6)*, wheat *(6,7)*, maize *(8)*, tobacco *(9,10)*, banana *(11)*, and others.

Cre/*lox* site-specific recombination is the Cre enzyme-mediated cleavage and ligation of two *lox* deoxynucleotide sequences. Cre, the product of the *cre* gene, is a 38.5-kDa recombinase that can reciprocally exchange the DNA at the 34 bp *lox* sites *(12)*. Cre has been shown in numerous instances to mediate *lox*-site specific recombination in animal and plant cells *(13–15)*. The products of the recombination event depend on the relative orientation of these asymmetric sequences *(16)*. By incorporating the Cre-*lox* system into a BAC vector, pBACwich (*see* **Fig. 1**) has been constructed and used for a more precise eval-

From: *Methods in Molecular Biology, vol. 256:*
Bacterial Artificial Chromosomes, Volume 2: Functional Studies
Edited by: S. Zhao and M. Stodolsky © Humana Press Inc., Totowa, NJ

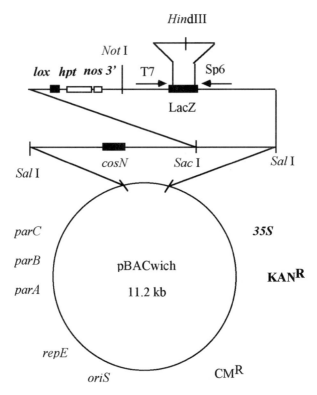

Fig. 1. Plant transformation BAC vector, pBACwich. CM^R: chloramphenicol resistance. *35S*: 35S dual enhancer promoter. KAN^R: kanamycin resistance. *hpt*: hygromycin resistance. *35S* and KAN^R can be used for a random integration and promoterless *hpt* can be used for a site-specific integration in plants.

uation of large DNA inserts. Transformation of large segments of DNA into plants would make it feasible to introduce a natural gene cluster or a series of interesting genes into a single locus in the chromosome. Such a group of genes can be constituted to create an entirely new metabolic pathway for production of the molecule in plants. This chapter outlines an approach for transforming large DNA into plants using two biological systems; biolistics and pBACwich.

2. Materials

1. Absolute ethanol (AAPER Alcohol and Chemical Co.).
2. Household bleach.
3. 100 mm × 20 mm Petri dish (Corning Inc.).
4. Tween-20 (Sigma).
5. Murashige Minimal Organics Medium (Life Technologies).

6. Kanamycin (Sigma).
7. Magenta GA7-3 Vessel (Life Technologies).
8. 1-μm gold particles (Bio-Rad).
9. 70% (v/v) ethanol (AAPER Alcohol and Chemical Co.).
10. 50% (v/v) glycerol (Life Technologies) solution.
11. TE buffer: 10 mM Tris-HCl, 1 mM ethylenediamine tetraacetic acid (EDTA), pH 8.0.
12. 2.5 M CaCl$_2$ (prepared fresh).
13. 0.1 M spermidine: spermidine free base (Sigma). Filter-sterilize and store as 25 μL aliquots at –20°C.
14. QIAGEN Large-Construct Kit (QIAGEN).
15. Biolistic PDS-1000/He unit (Bio-Rad).
16. Murashige and Skoog Complete Medium (Life Technologies).
17. Hygromycin B (Life Technologies).
18. PHYTAGAR (Life Technologies).
19. Extraction buffer: 0.35 M Sorbitol, 0.1 M Tris-base, 5 mM EDTA, 0.02 M NaBisulfite, adjust to pH 7.5.
20. Polytron homogenizer (Brinkmann, model PT10/35).
21. Nuclear lysis buffer: 0.2 M Tris-base, 0.05 M EDTA, 2 M NaCl, 2% (w/v) CTAB (ICN Pharmaceuticals, Inc.).
22. 5% Sarkosyl (Sigma).
23. Chloroform/isoamyl alcohol (24/1).
24. 0.5X TBE: 0.045 M Trizma base, 0.045 M boric acid, 1 mM EDTA, pH 8.3.
25. DOP1 mix: 2 μL of 5X Sequenase buffer (USB: 100 mM potassium phosphate, pH 7.4, 5 mM dithiothreitol (DTT), 0.5 mM EDTA), 1 μL of 2.5 mM dNTP, 1 μL of 10 mM DOP primer, and 3 μL of water.
26. DOP primer: 5′-CCG ACT CGA GNN NNN NAT GTG G-3′.
27. Sequenase (USB).
28. DOP2 mix: 5 μL of 10X Polymerase chain reaction (PCR) buffer (Perkin-Elmer: 500 mM KCl, 100 mM Tris-HCl, pH 8.3, 0.01% gelatin), 2 μL of 2.5 mM deoxynucleotide 5′-triphosphate (dNTP), 1 μL of 10 μM DOP primer, 1 μL of 10 μM pBACwich vector forward (5′-TGG GTA ACG CCA GGG TTT TC-3′) or backward (5′-CGG CTC GTA TGT TGT GTG GAA-3′) primer, 1 μL of 1 U/μL Ampli Taq (Perkin-Elmer), and 30 μL water.

3. Methods
3.1. pBACwich Vector

A new BAC vector, BACwich (*2*), was constructed, in which a promoterless hygromycin phosphotransferase gene (*hpt*) was designed to be activated and selected in a plant cell upon site-specific integration. As outlined in **Fig. 1**, pBACwich contains a *lox* site, a promoterless *hpt* gene, and a *nos 3′* terminator (the terminator sequence of the nopaline synthase gene). The pBACwich

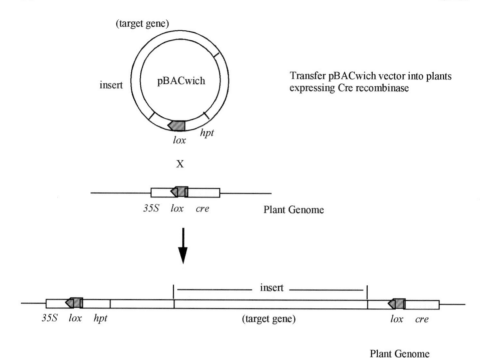

Fig. 2. A diagram of how pBACwich is integrated into a specific *lox* site in the plant genome. Plants containing the pBACwich DNA in a random position are sensitive to hygromycin due to the lack of *hpt* transcription. Integration of the pBACwich *via lox*-site recombination would produce a *35S-lox-hpt-lox-cre* linkage, resulting in a hygromycin-resistant (Hpt^R) phenotype. Termination of *cre* transcription should result in an insertion event that is stably maintained.

harboring target genes is inserted by site-specific recombination in vivo at the *lox* site already present on the plant chromosome using the existing Cre recombinase as diagrammed in **Fig. 2**. Integration of the pBACwich results in promoter displacement and termination of *cre* transcription.

3.2. pBACwich Genomic Library Construction

The general method for BAC library construction is well described in previous chapters and other literature *(17,18)*. pBACwich is derived from pBeloBAC11 *(19)*. A pBACwich library can be constructed in the same way using the *Hin*dIII cloning site and chloramphenicol selection. pBACwich is specially designed to integrate into a *lox* site located in the plant or fungal genome by Cre-*lox* site-specific recombination.

3.3. Cre/lox Site-Specific Transformation

3.3.1. Generation of a Promoter-lox-Cre Integration Site in Plants

Vectors used to create a *lox* site in the plant chromosome must have appropriate selective genes with appropriate promoters. In tobacco and *Arabidopsis*, several lines harboring a *p35S-lox-cre* were constructed *(20,21)*, in which *35S* is the 35S RNA promoter from cauliflower mosaic virus, *cre* is the coding region for Cre recombinase, and *lox* is the 34-bp recombination site introduced into the leading sequence of the transcript. Cre-mediated integration of pBACwich into the genomic *35S-lox-cre* target confers a hygromycin resistant (Hyg[R]) phenotype through a *35S-lox-hpt* linkage. pBACwich also contains the constitutively active promoter of the CaMV *35S* transcript fused with the *NPTII* coding region (KAN[R]), which can be used for non-specific random integration (*see* **Fig. 1**).

3.3.2. Plant Materials

Intact leaves of greenhouse or Petri dish grown tobacco (*Nicotiana tabacum*) or *Arabidopsis* can be used as the target material when they are between 15 and 40 d old. Both plants are grown at 23°C with a 16-h photoperiod in the vermiculite or Petri dish. Here, tobacco grown in a Petri dish was used as a target plant. Tobacco seeds were kindly supplied by Dr. David Ow (Plant Gene Expression Center, Albany, CA). A set of *35S-lox-cre* transgenic *Arabidopsis* lines, where *35S* is the 35S RNA promoter of the cauliflower mosaic virus (CaMV) and *lox* is placed in the leader sequence of the *cre* gene, have been produced *(21)*.

1. Sterilize the tobacco seeds (1999.5: harboring a single *35S-loxP-npt* construct) in 100% ethanol for 1 min and in bleach solution (1:2 dilution of household bleach in sterile ddH$_2$O with one drop of Tween-20 per 10 mL diluted solution) for 1 min.
2. Wash the seeds three to four times with sterile ddH$_2$O.
3. Germinate on hormone-free Murashige Minimal Organics Medium containing kanamycin at 150 µg/mL to select against segregates (*see* **Note 1**).
4. Transfer the resistant seedling into a Magenta GA7-3 Vessel on the same medium with kanamycin and grow until it has four to five nice expanded leaves.
5. Harvest the young green leaves, devein, and place on hormone-free Murashige Minimal Organics Medium at the center of the Petri dish (100 mm × 20 mm) a few hours prior to bombardment.

3.3.3. Particles Preparation

The microparticle (microcarrier or microprojectile) variables include the size and type of particles. The accelerator parameters include the helium pressure, the distance between the rupture disk and macrocarrier (macroprojectile), and the distance between the macrocarrier and the stopping screen. In order to obtain the maximum transformation efficiency, it is necessary to optimize the above parameters with the associated target plant tissue. Based on preliminary experiments biolistic transformation conditions of tobacco were selected to use 1 μm diameter gold microparticles, a helium pressure of 1100 psi, a chamber vacuum of 28 in Hg, and a target distance of 9 cm.

1. Place 60 mg gold microparticles (1 μm diameter) in a 1.5-mL microcentrifuge tube (*see* **Note 2**).
2. Add 1 mL 70% ethanol and vortex vigorously for 5 min.
3. Incubate for 15 min at room temperature and pellet by spinning for 15 s in a tabletop centrifuge.
4. Decant the supernatant, add 1 mL sterile distilled water, and vortex for 1 min.
5. Pellet the microparticles by spinning for 15 s in a tabletop centrifuge.
6. Discard the supernatant and wash two more times with sterile distilled water.
7. After removing the liquid, add sterile 50% (v/v) glycerol solution to bring the microparticle concentration to 60 mg/mL.
8. These particles can be stored at room temperature for up to 2 wk.

3.3.4. Coating Particles

1. Prepare pBACwich DNA using the QIAGEN Large-Construct Kit at a concentration of 1 μg/μL in TE buffer (*see* **Note 3**).
2. Vortex the microparticles in 50% glycerol for 5 min to disrupt agglomerated particles and resuspend.
3. Aliquot 50 μL of the gold suspension into 1.5 mL microcentrifuge tubes.
4. Add 5 μL of pBACwich DNA stock (1 μg/μL), 50 μL of $CaCl_2$ stock (2.5 *M*), and 20 μL of spermidine stock (0.1 *M*), in this order, while vortexing continuously (do not vortex vigorously) (*see* **Note 4**).
5. Allow the mixture to react and coat the particles with pBACwich plasmid DNA for 3 min while vortexing (do not vortex vigorously).
6. Allow the particles to settle for 1 min.
7. Pellet the coated particles by pulse centrifugation for 2 s.

3.3.5. Loading Particles

1. Remove all of the supernatant and wash the pellet in 140 μL of 70% (v/v) ethanol without disturbing the pellet.
2. Discard the liquid and wash the pellet in 140 μL of 100% ethanol without disturbing the pellet.
3. Discard the liquid and add 48 μL of 100% ethanol.
4. Mix the suspension by gentle tapping and aliquot 6 μL each onto macrocarriers (Kapton flying disks) (*see* **Note 5**).
5. Spread the suspension over an area 1 cm in diameter in the center of the disk using the pipette tip and desiccate immediately (*see* **Note 6**).

6. Set helium gun parameters at 1100 psi, target distance at 9 cm, vacuum at 28 mm Hg, and the distance between macrocarrier and stopping screen at 1 cm (*see* **Note 7**).

7. Bombard the plates with the selected pBACwich, and transfer the leaf explants ($\cong 1$ cm^2) to 100×20 mm Petri plates containing Murashige and Skoog Complete Medium, hygromycin B at 20 μg/mL and 0.25% (w/v) PHYTAGAR (*see* **Note 8**).

8. Transfer the bombarded tissues onto fresh selection medium (Murashige and Skoog Complete Medium, hygromycin B at 20 μg/mL and 0.25% (w/v) PHYTAGAR) 24 h after bombardment (*see* **Note 9**).

9. Cut the leaf tissues into smaller pieces as it expands on the selection medium and transfer onto fresh selection medium. Within two weeks after bombardment, the leaf tissues start to show regeneration of shoots at the cut edges.

10. Six weeks later, transfer shoots onto fresh medium containing 20 μg/mL hygromycin B. The shoots develop roots and grow into complete plants that reach maturity on soil and have set seeds. Plants from leaf explants, which give rise to shoots, are considered to have hygromycin resistant cells.

3.4. Molecular Analyses

3.4.1. PCR Analysis

Putative integrants can be analyzed by PCR for the presence of expected *35S-lox-hpt* junctions. The primer sets and PCR reaction conditions should be adjusted for each experimental design. For this case, PCR analysis incorporates two primer pairs (s & h; w & c); *35S* sequence (primer s: 5′-GTT CAT TTC ATT TGG AGA GG-3′), pBACwich *hpt* gene (primer h: 5′-GGT GTC GTC CAT CAC AGT TTG CCA G-3′), pBACwich backbone (primer w: 5′-GAT GGC CTC CAC GCA CGT TGT G-3′), and *cre* gene (primer c: 5′-CTA ATC GCC ATC TTC CAG CAG G-3′). The PCR conditions used for tobacco were 94°C for 1 min, 55°C for 1 min, 72°C for 1 min 10 s, and after 35 cycles, the products were extended at 72°C for 10 min.

3.4.2. Southern Analysis

Genomic DNAs are isolated from four to five week-old transgenic plants grown in Petri dishes containing Murashige and Skoog Complete Medium, hygromycin B at 20 μg/mL (without hygromycin for control groups) and 0.25% (w/v) PHYTAGAR.

1. Add 15 mL cold extraction buffer to a 50-mL Falcon tube containing 1 g fresh or frozen young tissue (*see* **Note 10**).

2. Grind the mixture with a Polytron homogenizer at speed 4 for 10 s and centrifuge at 3500*g* for 30 min.

3. Add extraction buffer (1.25 mL) to the pellet, vortex, and add 1.75 mL nuclear lysis buffer and 0.6 mL 5% Sarkosyl.

4. Incubate at 65°C for 20 min.

5. Add 7.5 mL chloroform/isoamyl alcohol (24/1), shake the mixture for 20 min, and centrifuge at 1800*g* for 10 min.
6. Transfer the aqueous supernatant to a new 50-mL Oakridge tube, add 0.6 vol cold isopropanol.
7. Mix well and spin the DNA in the Sorvall centrifuge (SS34 rotor) at 12,000 rpm for 25 min, 4°C.
8. Wash the DNA precipitate with 70% ethanol, dry and resuspend in 100 µL of TE.
9. Measure the genomic DNA concentration using spectrophotometer, and digest with various restriction enzymes such as *Hind*III, *Eco*RI, *Bam* HI, *Eco*RV, and *Xho*I (about 8 µg for each sample).
10. Separate on 0.9% agarose gels at a lower voltage (1.5–2.5 V/cm) in 0.5X TBE buffer and room temperature.
11. Transfer the DNAs from agarose gels to nylon membranes using Southern Blot.
12. Probe with various radiolabeled probes including internal fragments from the inserts.

3.4.3. DOP (Partially Degenerate Oligonucleotide Primer) PCR

To isolate the vector-insert junction sequences from integrated pBACwich clones, DOP *(22,23)* was used in combination with vector primers. The DOP primer (5′-CCG ACT CGA GNN NNN NAT GTG G-3′) has six bases specified at the 3′ end, six degenerate bases in the middle, and 10 arbitrary bases at the 5′ end.

1. Add 1 µL (5 ng/µL) of the pBACwich DNA to 7 µL DOP1 mix.
2. Use the following thermocycler program: 96°C for 3 min, 30°C pause (at this step, 2 µL 1 U/µL USB Sequenase was added), 30°C → 37°C ramping over 1 min, 37°C for 3 min, and 72°C for 10 min.
3. Add 40 µL of DOP2 mix to each tube.
4. Execute PCR for 35 cycles at 94°C for 30 s, 58°C for 30 s, and 72°C for 1 min.
5. In the secondary PCR for further enrichment of junction sequences, use 0.5 µL primary PCR product in a 50-µL reaction under the same conditions, except using the nested forward (5′-CCT GCA GGC ATG CAA GCT T-3′) or nested backward (5′-CAC TAT AGA ATA CTC AAG C-3′) vector primer together with DOP primer.

3.5. Random Integration Transformation

pBACwich has a *35S* promoter plus the *NPTII* (neomycin phosphotransferase coding region). The *NPTII* gene is used as the selectable marker for the pBACwich transgenic plants. For random integration transformation, Murashige and Skoog Complete Medium, kanamycin at 100 µg/mL and 0.25% (w/v) PHYTAGAR are used.

4. Notes

1. To maintain a single *lox* site in the plant chromosome, the wild type can be crossbred with a transgenic plant harboring a single *35S-loxP-npt* construct. The

tobacco seed harboring a single lox site has a kanamycin resistance gene and can be selected against other segregates.

2. Gold particles offer several advantages over tungsten such as size uniformity, spherical shape and chemically inert nature. An improved procedure has been reported *(9)*, in which the gold particles were heated in a dry oven at 180°C overnight and the ethanol was substituted with isopropanol. Heating in dry air or isopropanol treatment may be good to prevent agglomeration, thereby increasing the transformation efficiency.

3. Contaminating protein causes particle agglomeration during coating. The QIAGEN Large-Construct Kit procedure has yielded DNA of sufficient purity to avoid this problem.

4. Vigorous vortexing will shear the big pBACwich DNA (>100 kb) resulting in low transformation efficiency.

5. Kapton disks should be washed in 70% ethanol before use. The choice of rupture disk and microcarrier depends on the biological system being studied. Achieving high rates of pBACwith transformation for each plant requires optimization of these parameters in the preliminary experiments.

6. Exposure of Kapton flying disks to humidity after drying reduces transformation rates due to hygroscopic clumping and agglomeration.

7. The plates can be bombarded at several helium pressures (e.g., 1000, 1300, and 1600 psi) and several target distances (e.g., 6 and 12 cm) to determine optimal conditions and the best conditions can then be used. Helium pressure, target distance and microparticle size must be optimized for each plant.

8. The optimum cell age of each plant species must be determined empirically, but generally, actively dividing cells are transformed most efficiently.

9. A kill curve was done to determine the optimum concentration of hygromycin B to be used for transformation. Ten to 50 µg/mL of hygromycin B were tested on the leaf tissue. Within 2 wk leaf tissues showed different degrees of pigment discoloration with different concentrations of hygromycin applied. A concentration of 20 µg/mL hygromycin was found suitable for this experiment. The control leaves were completely bleached at this concentration.

10. DNeasy Plant System (QIAGEN) can be used to isolate genomic DNAs from plants. However, larger and more intact genomic DNAs can be obtained using the above method.

Acknowledgments

The author would like to thank Dr. Iain Fraser and Ms. Rebecca Hart for critically reading the manuscript and providing helpful discussions.

References

1. Hamilton, C. M., Frary, A., Lewis, C., and Tanksley, S. D. (1996) Stable transfer of intact high molecular weight DNA into plant chromosomes. *Proc. Natl. Acad. Sci. USA* **93,** 9975–9979.

2. Choi, S., Begum, D., Koshinsky, H., Ow, D. W., and Wing, R. A. (2000) A new approach for the identification and cloning of genes: the pBACwich system using Cre/*lox* site-specific recombination. *Nucl. Acids Res.* **28,** E19.
3. Sanford, J. C., Smith, F. D., and Russell, J. A. (1993) Optimizing the biolistic process for different biological applications. *Meth. Enzymol.* **217,** 483–509.
4. Seki, M., Iida, A., and Morikawa, H. (1999) Transient expression of the beta-glucuronidase gene in tissues of *Arabidopsis thaliana* by bombardment-mediated transformation. *Mol. Biotechnol.* **11,** 251–255.
5. Solis, R., Takumi, S., Mori, N., and Nakamura, C. (1999) *Ac*-mediated transactivation of the *Ds* element in rice (*Oryza sativa* L.) cells as revealed by GUS assay. *Hereditas* **131,** 23–31.
6. Drakakaki, G., Christou, P., and Stoger, E. (2000) Constitutive expression of soybean ferritin cDNA in transgenic wheat and rice results in increased iron levels in vegetative tissues but not in seeds. *Transgenic Res.* **9,** 445–452.
7. Sivamani, E., Bahieldin1, A., Wraith, J. M., et al. (2000) Improved biomass productivity and water use efficiency under water deficit conditions in transgenic wheat constitutively expressing the barley HVA1 gene. *Plant Sci.* **155,** 1–9.
8. Menossi, M., Puigdomenech, P., and Martinez-Izquierdo, J. A. (2000) Improved analysis of promoter activity in biolistically transformed plant cells. *Biotechniques* **28,** 54–58.
9. Sawant, S. V., Singh, P. K., and Tuli, R. (2000) Pretreatment of microprojectiles to improve the delivery of DNA in plant transformation. *Biotechniques* **29,** 246–248.
10. Drescher, A., Ruf, S., Calsa, T. Jr., Carrer, H., and Bock, R. (2000) The two largest chloroplast genome-encoded open reading frames of higher plants are essential genes. *Plant J.* **22,** 97–104.
11. Horser, C., Harding, R., and Dale, J. (2001) Banana bunchy top nanovirus DNA-1 encodes the 'master' replication initiation protein. *J. Gen. Virol.* **82,** 459–464.
12. Hoess, R., Abremski, K., Irwin, S., Kendall, M., and Mack, A. (1990) DNA specificity of the Cre recombinase resides in the 25 kDa carboxyl domain of the protein. *J. Mol. Biol.* **216,** 873–882.
13. Albert, H., Dale, E. C., Lee, E., and Ow, D. W. (1995) Site-specific integration of DNA into wild-type and mutant lox sites placed in the plant genome. *Plant J.* **7,** 649–659.
14. Medberry, S. L., Dale, E., Qin, M., and Ow, D. W. (1995) Intra-chromosomal rearrangements generated by Cre-*lox* site-specific recombination. *Nucl. Acids Res.* **23,** 485–490.
15. Stuurman, J., de Vroomen, M. J., Nijkamp, H. J., and van Haaren, M. J. (1996) Single-site manipulation of tomato chromosomes *in vitro* and *in vivo* using Cre-*lox* site-specific recombination. *Plant Mol. Biol.* **32,** 901–913.
16. Ow, D. W. and Medberry, S. L. (1995) Genome manipulation through site-specific recombination. *Crit. Rev. Plant Sci.* **14,** 239–261.
17. Choi, S. and Wing, R. A. (2000) The construction of bacterial artificial chromosome (BAC) libraries in *Plant Molecular Biology Manual* (Gelvin, S. B., and Schilperoort, R. A., eds.), H5, Kluwer Academic, The Netherlands, pp. 1–28.

18. Choi, S. and Kim, U.-J. (2001) Construction of a Bacterial Artificial Chromosome Library in *Methods in Molecular Biology: Genomics Protocols* (Starkey, M. P., and Elaswarapu, R., eds.), 175, Humana, Totowa, NJ, pp. 57–68.

19. Kim, U.-J., Birren, B. W., Slepak, T., et al. (1996) Construction and characterization of a human bacterial artificial chromosome library. *Genomics* **34,** 213–218.

20. Koshinsky, H. A., Lee, E., and Ow, D. W. (2000) Cre-*lox* site-specific recombination between *Arabidopsis* and tobacco chromosomes. *Plant J.* **23,** 715–722.

21. Chou, I., Kobayashi, J., Lee, E., Koshinsky, H., Medberry, S., and Ow, D. W. (1999) Chromosome rearrangements in *Arabidopsis thaliana* generated through Cre-*lox* site specific recombination. *Plant Animal Genome VII*, Jan. 17–21, San Diego, CA (Abs. P133).

22. Telenius, H., Carter, N. P., Bebb, C. E., Nordenskjold, M., Ponder, B. A., and Tunnacliffe, A. (1992) Degenerate oligonucleotide-primed PCR: general amplification of target DNA by a single degenerate primer. *Genomics* **13,** 718–725.

23. Wu, C., Zhu, S., Simpson, S., and de Jong, P. J. (1996) DOP-vector PCR: a method for rapid isolation and sequencing of insert termini from PAC clones. *Nucl. Acids Res.* **24,** 2614–2615.

6

Retrofitting BACs With a Selectable Marker for Transfection

Zunde Wang, Angelika Longacre, and Peter Engler

1. Introduction

Many functional studies require the transfection of bacterial artificial chromosome (BAC) clones into mammalian cells. Because most BAC vectors do not have a mammalian selection marker or do not have one suitable for all cell types, it is usually necessary to modify BAC clones to include a selection marker, a "retrofitting" process. Different BAC retrofitting strategies have been developed over the last 5 yr and are classified in the reference section (1–11).

In vitro retrofitting (2,4,6,11), in which the BAC DNA is modified through enzymatic reactions and transformed back into the *Escherichia coli* host, not only has the limitation of low transformation efficiency, but also will likely cause deletions to the genomic insert because smaller molecules have a selective advantage during transformation. In vivo retrofitting, on the other hand, modifies BAC molecules in bacteria, either through homologous recombination (3,5,8,9), transposition (1,7), or site-specific recombination (10), and does not require retransformation. Therefore, the efficiency of in vivo retrofitting will be independent of the BAC size and the fidelity is generally higher.

The BAC host, generally *E. coli* DH10B, lacks enzymatic machinery for homologous recombination, a property essential for the maintenance of BAC stability. To facilitate the modification through homologous recombination, the machinery has to be turned on temporarily, creating risks of undesirable recombination. Whether it will indeed increase insert instability remains to be investigated. Retrofitting via transposition has the disadvantage of marker integration either in targeted BAC or in bacterial genome, therefore, still requiring retransformation (7).

From: *Methods in Molecular Biology, vol. 256:*
Bacterial Artificial Chromosomes, Volume 2: Functional Studies
Edited by: S. Zhao and M. Stodolsky © Humana Press Inc., Totowa, NJ

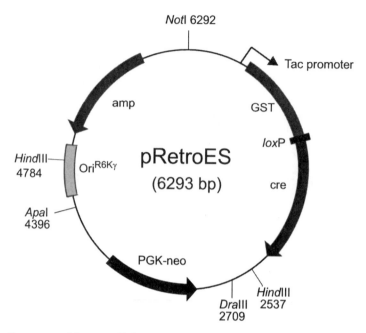

Fig. 1. Sturcture of the retrofitting vector, pRetroES. Essentially, it has a GST-*lox*P-cre fusion gene driven by a *tac* promoter that encodes a fusion cre recombinase for promoting cre/*lox*P recombination. The conditional replication origin, Ori$^{R6K\gamma}$, will not function in the usual BAC host. The PGKneo is a selectable marker useful for embryonic stem cell transfection. [Taken with permission from *Genome Research* **11(1)**: 139 with slight modification.

In vivo retrofitting through site-specific recombination has the advantage of not requiring BAC transformation or homologous recombination restoration. It does, however, require strictly regulated, transient expression of the recombinase to prevent the loop-out of the integrated selection marker. Recently, we constructed a plasmid, pRetroES, and described a strategy of using it for in vivo BAC retrofitting through cre/loxP site-specific recombination *(10)*. Since then, a few dozen researchers have requested the plasmid and many of them have successfully used it for BAC modification. Here we present the modified protocol based on responses that we received and include some alternatives and suggestions.

The main features of the plasmid are illustrated in **Fig. 1** and the retrofitting strategy is outlined in **Fig. 2**. The plasmid bears a fusion cre gene that, upon transformation into the BAC containing *E. coli* cells, produces cre recombinase to promote recombination between the loxP sites on the BAC and on the retrofitting vector, resulting in the integration of the plasmid into the BAC. After recombination, the cre coding region will be separated from its promoter

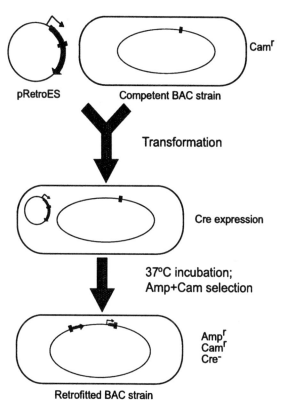

Fig. 2. Schematic representation of the retrofitting procedure. Briefly, competent BAC-bearing *E. coli* is prepared for each BAC and transformed with pRetroES by electroporation. During the posttransformation incubation, the fusion cre recombinase will be expressed and promote the recombination between the *lox*P sites on the retrofitting vector and the BAC vector, leading to the integration of pRetroES into the BAC at the *lox*P site. After integration, the Tac promoter is separated from the fusion gene and cre recombinase will not be produced. See text for detailed protocol. [Taken with permission from *Genome Research* **11(1):** 139.]

and recombinase will not be produced, preventing excision of the integrated plasmid. Other features include: the conditional replication origin, Ori^{R6K} that requires λpir protein for replication so the pRetroES vector cannot replicate on its own in the BAC-containing host; the *amp* gene to confer ampicillin resistance in bacteria; and the PGK*neo* gene which confers G418 resistance to mammalian cells. The PGK promoter has been shown to work particularly well in mouse embryonic stem (ES) cells as well as in other cell types. For applications involving other types of cells, it is possible to replace the PGK*neo* cassette with a suitable selection marker (*see* **Note 1**).

Briefly, the retrofitting procedure consists of three steps: preparation of competent BAC-containing *E. coli* cells; transformation of these cells with pRetroES; antibiotic selection and confirmation of the modified clones. The protocol described here works well for up to 16 retrofitting reactions at one time. For larger numbers of BAC clones, higher throughput can be achieved with some modifications of the competent cell preparation step (*see* **Note 3**).

2. Materials
2.1. Plasmid pRetroES

Can be obtained from the American Type Culture Collection (ATCC no. 87823). The deduced sequence is in GenBank, accession number AF397196 (*see* **Note 1**). We suggest preparing the plasmid DNA using Qiagen Spin Miniprep Kit: start from a 3-mL overnight culture of LB with 50 µg/mL ampicillin. After loading onto the column, wash twice with wash buffer, let dry for a few minutes, and elute with 50 µL elution buffer. Confirm the vector by *Hin*dIII digestion—one should see fragments of 4.1 and 2.2 kb. Determine the plasmid concentration and adjust to 0.05–0.2 µg/µL. Store the DNA at 4°C and use 1 µL for each retrofitting.

2.2. Reagents and Equipment

1. LB medium with chloramphenicol (20 µg/mL) or chloramphenicol plus ampicillin (or carbenicillin; 50 µg/mL).
2. Distilled water, autoclaved and stored at 4°C.
3. 10% glycerol, autoclaved and stored at 4°C.
4. LB agar plates with chloramphenicol (20 µg/mL) plus ampicillin (or carbenicillin; 50 µg/mL).
5. Refrigerated benchtop centrifuge.
6. Refrigerated microcentrifuge (or in a 4°C cold room).
7. Electroporator (Bio-Rad Gene Pulser) and 0.1 cm cuvets.
8. Solutions for plasmid miniprep *(12)*. Solution I is 50 m*M* Tris-HCl (pH 8.0), 10 m*M* ethylenediamine tetraacetic acid (EDTA) (pH 8.0), 100 µg/mL RNaseA. Solution II is 0.2 *N* NaOH, 1% sodium dodecyl sulfate (SDS). Solution III contains 3 moles potassium acetate and 115 mL glacial acetic acid per liter. Isopropanol and 70% ethanol.
9. DOTAP lipofection reagents (Roche/Boehringer).

3. Methods
3.1. Prepare Competent Cells for Each BAC Clone (up to 16 BACs. For higher throughput operations, see Note 3)

1. Inoculate 3 mL LB containing 20 µg/mL chloramphenicol with a single colony and incubate at 37°C overnight with shaking.

2. To a 250-mL flask with 50 mL LB containing 20 µg/mL chloramphenicol, add 1 mL of the overnight culture and incubate at 37°C with shaking until OD at 600 nm reaches 0.5 (usually about 2.5 h). Do not exceed this density. If desired, keep the rest of the overnight culture for DNA preparation later.

3. Transfer the culture into a 50-mL conical tube (e.g., Falcon 2070) and chill on ice for 10 min, then centrifuge at 3500 rpm at 4°C in a benchtop centrifuge for 15 min.

4. Remove supernatant very thoroughly using a Pasteur pipet attached to a vacuum line; resuspend in 40 mL distilled water at 0°C; centrifuge as in **step 3**; move to a cold room for **steps 5–7**.

5. Drain and resuspend in 1.5 mL cold distilled water; transfer to a chilled 1.7-mL microfuge tube.

6. Wash one more time with cold distilled water and twice with 10% glycerol, each time removing the fluid completely with a clean pipet tip attached to a vacuum line.

7. Resuspend the pellet in 60 µL 10% glycerol. Use 20 µL for transformation and freeze the rest for later use. Freeze by immersing the tube in a dry ice/ethanol bath, then store at –70°C (*see* **Note 2** for cautions).

3.2. Transform BAC-Containing Competent E. coli With pRetroES

1. Add 1 µL of a pRetroES miniprep (50–200 ng) to 20 µL of the competent cells prepared in **Subheading 3.1.**; transfer into a chilled 0.1 cm electroporation cuvet; electroporate with a Bio-Rad Gene Pulser set at 1800 V and 200 ohms.

2. Add 1 mL LB to the cuvet and transfer the contents to a 17×100 mm round bottom tube (e.g., Falcon 2059); incubate at 37°C with shaking for 90 min to allow cre expression and site-specific recombination.

3. Centrifuge at 3500 rpm for 15 min in a benchtop centrifuge; drain off the supernatant and resuspend in 100 µL LB; take 1 µL of the transformed cells and dilute in 100 µL LB.

4. Spread out both the undiluted and 1:100 diluted cells on separate chloramphenicol plus ampicillin agar plates; invert the plates and incubate at 37°C for about 20 h.

3.3. Confirmation of Retrofitted Clones

1. Select five well-separated colonies and transfer each into a 17×100 mm round bottom tubes (e.g., Falcon 2059) containing 5 mL LB with 20 µg/mL chloramphenicol and 50 µg/mL ampicillin. In another tube, dilute 5 µL of the overnight culture from section 3.2.2 in 5 mL LB with 20 µg/mL chloramphenicol. Incubate the tubes overnight at 37°C with shaking; in a separate tube, keep an aliquot of the overnight culture for future large-scale DNA preparation.

2. Prepare DNA by alkaline lysis and isopropanol precipitation. Centrifuge the cells in a benchtop centrifuge at 3500 rpm for 15 min; drain the supernatant and add 250 µL of solution I with RNase, resuspend and transfer to a clean 1.7 mL microfuge tube. Add 250 µL solution II, gently mix by inverting the tube, and incubate on ice for 5 min. Add 250 µL solution III and mix gently; centrifuge at 4°C for 10 min. Carefully transfer the supernatant to a clean tube and spin again

to remove trace precipitate. Transfer the supernatant to a clean 1.7 mL microfuge tube, add 0.7 vol cold isopropanol, mix, leave on ice for 5 min, then spin for 15 min at 4°C. Wash the pellet with 70% ethanol and air-dry; dissolve by adding 20 μL distilled water and incubating at room temperature for at least an hour.

3. Restriction digestion and pulsed-field gel electrophoresis (PFGE): carefully take 5 μL of the BAC DNA with a wide-bore pipet tip and digest with 2 U of *Not*I in a 10-μL reaction at 37°C for 1 h; in one lane, load 9 μL of the reaction; in the adjacent lane, load 1 μL so that both the insert and the vector fragments will be visible. The conditions for PFGE are as following: cast a 1% agarose gel, run the gel in 0.5X TBE buffer at 200 V with initial switch time of 5 s and final switch time of 15 s for 18 h. Stain the gel with ethidium bromide and photograph on an ultraviolet transilluminator.

4. Compare the PFGE patterns of the retrofitted clones with the original unmodified clones, choose and freeze retrofitted clones with intact insert. Sizes of the *Not*I vector fragments for the pBeloBAC11 vector: 6.7 kb if unmodified and 11.7 kb and 2.6 kb if retrofitted. For the pBACe3.6 vector: 8.6 kb if unmodified; 11.6 kb and 3.1 kb if the integration is at the loxP site; 13.4 kb and 1.4 kb if the integration is at the loxP511 site; 16.4 kb, 3.3 kb, and 1.4 kb if integrated at both sites (*see* **Notes 4–6**)

3.4. Transfection of Retrofitted BACs into Mouse ES Cells

1. Large-scale BAC DNA is prepared using Qiagen Large Construct Kit according to manufacturer's instructions.

2. For transfection of ES cells, 5 μg of supercoiled or linearized BAC DNA in 20 μL TE is mixed with 30 μL 0.1 *M* HEPES buffer (pH 7.4). Thirty microliters DOTAP is diluted with 70 μL 0.1 *M* HEPES in a 12 × 75 mm polystyrene tube (e.g., Falcon 2058) and the BAC DNA is gently mixed in. After incubation at room temperature for 30–45 min, 2×10^6 ES cells in 3 mL ES cell medium are slowly added and mixed by gently inverting the tube. After incubation at room temperature for 30 min with a few gentle mixings, half of the mixture is added to each of two 10 cm plates previously seeded with appropriate feeder cells. After another 24 h, selection media containing 255 μg/mL G418 is added. Change media every 24 h for the first 2 d and then every other day thereafter. Colonies start to appear after 10 d (*see* **Note 7**).

4. Notes

1. To modify the vector, one needs an *E. coli* strain expressing λpir protein. One such strain can be purchased from ATCC (ATCC number 47069). The deduced sequence for pRetroES can be found in GenBank, accession number AF397196. However, inaccuracies are possible due to uncertainty with its parental plasmids and caution needs to be taken when solely relying on it for further modification. One notable case is the presence of a second *Eco*RI site not shown in the sequence.

2. For highest efficiency of competent cells, it is important to keep all solutions, tubes, and other materials cold and to work relatively quickly. It is preferable to perform these steps in a 4°C cold room.

3. High throughput retrofitting can be achieved through reducing the volume of the culture for making competent BAC containing *E. coli* cells. We have tested this procedure with 5 mL cultures that allowed us to successfully scale up to 24 BACs a day. We believe with some optimization, a smaller culture volume in a 96-well format is also possible if large numbers of BACs are to be retrofitted.

4. We no longer suggest using PCR to confirm the recombination as we did in our earlier publication because virtually all colonies have desired recombinants (if no contaminating plasmids exist during transformation) and because colony PCR of BAC clones is not always consistent and cannot detect deletions or recombination to the genomic insert. In case this approach is needed (e.g., in a demonstration course), be aware that the sequence of the pBeloBAC primer was wrong in our earlier paper and the correct sequence should be: 5'-AGGAAACGACAGGT GCTGAA.

5. As we have discussed, in retrofitting pBACe3.6-derived BACs (many of the RPCI libraries), the integration may happen at either or both of the loxP sites, resulting in three types of recombinant vector fragment patterns. Our results show all of these types of recombinants are stable and work well in transfection.

6. Some researchers who used the plasmid noticed that some clones give an extra fragment after *Not*I digestion and PFGE. Whether they are recombinants or deleted inserts during site-specific recombination remains to be investigated.

7. Various lipofection reagents from several vendors may be used for BAC transfection and variations of the transfection procedure can be found in some of the references listed below *(4,6,10,11)*. We have not attempted to optimize the BAC transfection of ES cells but in our hands this procedure *(10)* generally yields ten to several hundred colonies.

Acknowledgments

Thanks to those who have tried our method and responded with helpful suggestions. We are especially grateful to Jose E. Mejia for many detailed comments on the procedure and to Ursula Storb, in whose laboratory this work was carried out.

References

1. Chatterjee, P. K. and Sternberg, N. L. (1996) Retrofitting high molecular weight DNA cloned in P1: introduction of reporter genes, markers selectable in mammalian cells and generation of nested deletions. *Genet. Anal.* **13,** 33–42.
2. Mejia, J. E. and Monaco, A. P. (1997) Retrofitting vectors for Escherichia coli-based artificial chromosomes (PACs and BACs) with markers for transfection studies. *Genome Res.* **7,** 179–186.

3. Yang, X. W., Model, P., and Heintz, N. (1997) Homologous recombination based modification in Escherichia coli and germline transmission in transgenic mice of a bacterial artificial chromosome. *Nat. Biotechnol.* **15,** 859–865.

4. Hejna, J. A., Johnstone, P. L., Kohler, S. L., et al. (1998) Functional complementation by electroporation of human BACs into mammalian fibroblast cells. *Nucl. Acids Res.* **26,** 1124–1125.

5. Jessen, J. R., Meng, A., McFarlane, R. J., Paw, B. H., Zon, L. I., Smith, G. R., and Lin, S. (1998) Modification of bacterial artificial chromosomes through chi-stimulated homologous recombination and its application in zebrafish transgenesis. *Proc. Natl. Acad. Sci. USA* **95,** 5121–5126.

6. Kim, S. Y., Horrigan, S. K., Altenhofen, J. L., Arbieva, Z. H., Hoffman, R., and Westbrook, C. A. (1998) Modification of bacterial artificial chromosome clones using Cre recombinase: introduction of selectable markers for expression in eukaryotic cells. *Genome Res.* **8,** 404–412.

7. Frengen, E., Weichenhan, D., Zhao, B., Osoegawa, K., van Geel, M., and de Jong. P. J. (1999) A modular, positive selection bacterial artificial chromosome vector with multiple cloning sites. *Genomics* **58,** 250–253.

8. Muyrers, J. P., Zhang, Y., Testa, G., and Stewart, A. F. (1999) Rapid modification of bacterial artificial chromosomes by ET- recombination. *Nucl. Acids Res.* **27,** 1555–1557.

9. Narayanan, K., Williamson, R., Zhang, Y., Stewart, A. F., and Ioannou, P. A. (1999) Efficient and precise engineering of a 200 kb beta-globin human/bacterial artificial chromosome in E. coli DH10B using an inducible homologous recombination system. *Gene Ther.* **6,** 442–447.

10. Wang, Z., Engler, P., Longacre, A., and Storb, U. (2001) An efficient method for high-fidelity BAC/PAC retrofitting with a selectable marker for mammalian cell transfection. *Genome Res.* **11,** 137–141.

11. Kaname, T. and Huxley, C. (2001) Simple and efficient vectors for retrofitting BACs and PACs with mammalian neo^r and EGFP marker genes. *Gene* **266,** 147–153.

12. Sambrook, J., Fritsch, E. F., and Maniatis, T. (1989) *Molecular Cloning. Second edition.* Cold Spring Harbor Laboratory Press, Cold Spring Harbor, NY.

7

BAC Modification Using a RecA Expressing Shuttle Vector System

Nathan Mise and Philip Avner

1. Introduction

Bacterial artificial chromosomes (BAC) are F-factor plasmid based vectors that are maintained as 1 or 2 copy plasmids in their bacterial host strain. BACs have the capacity for maintaining large, often more than 300 kb fragments of mammalian DNA as stable inserts and avoid much of the insert chimerism which characterizes yeast artificial chromosomes (YACs). Isolation of BAC DNA is easy because they exist as supercoiled circular plasmids that are resistant to shearing *(1)*. These advantages have made BACs into an important tool for genome research *(2)*. BACs are also being increasingly used as a tool for exploring the function of large genomic fragments either by generating transgenic animals by oocytes pronuclear injection *(3–5)* or by creating transgenic ES cell lines by liposome-mediated transfection *(6)*. The most widely used BAC vectors such as pBeloBAC11 or pBACe3.6 *(7,8)* lack the selection markers necessary for easeful transfection study in mammalian cells. To use such BAC clones for transfection analysis, it is usually necessary to retrofit selectable marker gene cassettes, which will both allow the positive selection of the transfectant and facilitate the stable maintenance of the transgene in mammalian cells. Functional analysis of the genomic elements lying within the BAC, may also require the targeted modification of the insert in order to create point mutations, deletions, and substitutions within the sequence. Development of BAC modification technology is therefore of critical importance and several different techniques to address this issue have been developed *(5,6,9–11)*. Some of these techniques are described in other chapters in this volume. Here we

From: *Methods in Molecular Biology, vol. 256:*
Bacterial Artificial Chromosomes, Volume 2: Functional Studies
Edited by: S. Zhao and M. Stodolsky © Humana Press Inc., Totowa, NJ

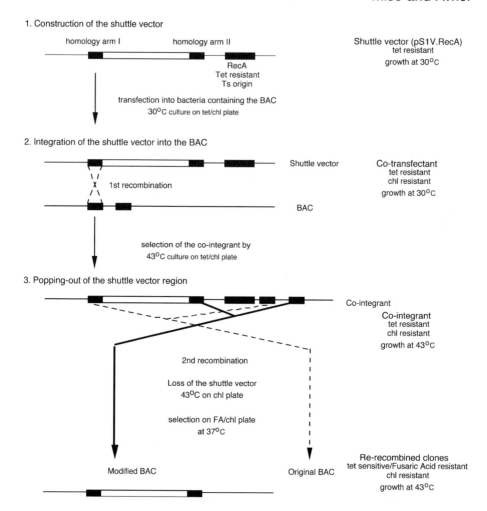

Fig. 1. Schematic illustration of the method. (1) The recombination cassette is first constructed in the building vector and is then transferred into the temperature-sensitive pSV1.RecA shuttle vector (*see* **Fig. 2**). (2) Co-integrants can be generated through homologous recombination on one of two homology arms. (3) Resolved BACs can be selected on plates containing fusaric acid and chloramphenicol.

describe a method for BAC modification based on the use of the RecA expressing shuttle vector system *(5)*.

The methodology is outlined in **Fig. 1**. The method can be divided into three parts. The first concerns the construction of a shuttle vector containing the insert flanked by two homology arms of more than 500 bp length. The shuttle vector, which is tetracycline resistant, can replicate at 30°C, but not at 43°C owing to the temperature sensitivity of the replication origin. The second step that concerns the integration into the BAC of the shuttle vector at a site corresponding to one of the two homology arms in the BAC depends on the action of the RecA enzyme expressed by the shuttle vector. The shuttle vector is transfected into the bacteria containing the BAC. Shuttle vector integration take place in a fraction of the cotransfectants which will grow in the presence of both tetracycline (tet) and chloramphenicol (chl) at 30°C. Such cointegrant are, however, the only cells able to grow at 43°C using the replication origin of the BAC in the presence of both tet and chl. The third step involves the popping-out of the shuttle vector region from the cointegrant mediated again by the RecA protein. The integrated shuttle vector sequence is excised from the BAC by homologous recombination within one of the two homology arms. To select the BAC clone which has lost the shuttle vector sequence, fusaric acid (FA) selection is used. In the presence of FA, while tet-resistant bacteria cannot grow, tet-sensitive bacteria grow slowly *(12)*. Only clones that have lost the shuttle vector sequence can grow on FA plates. Generally, as the frequency of recombination is the same for both homology arms, a modified BAC will be generated in half of the tet-sensitive/ FA-resistant clones.

Using this method, we have successfully modified our BAC by insertion both of a neomycin-resistant gene cassette in order to confer selection for stable maintenance in mammalian cells and a highly repeated *lac* operator cassette which tags the BAC and allows for in vivo visualization studies *(13)*. Though the *lac* operator repeat cassette is very large and has proved very instable in bacterial cells, the shuttle vector system worked effectively to allow both BAC modifications. We believe this protocol is suitable for insertion of fragments that may be difficult to amplify by polymerase chain reaction (PCR) alone such as highly repeated and/or very large sequences, because no PCR is required to put together the insert to be integrated into the BAC.

2. Materials

2.1. Construction of Shuttle Vectors

1. Building vectors: Use conventional high copy number plasmids having multiple cloning sites such as pBlueScript. The insert to be integrated into the BAC should first be subcloned into the building vector.
2. The temperature sensitive, recombination inducing shuttle vector: the vector can be obtained from Dr. Nathaniel Heintz at the Rockefeller University *(5)*. The vector and the method of use are protected by a patent. You can, therefore, use it only for academic purposes and should obtain the plasmid directly from Dr. N. Heintz.
3. Primers: Two sets of PCR primers amplifying 500 bp or longer homology arms are required. Primers should be designed to form restriction sites at each end of

the homology arms as outlined in **Fig. 2**. To ensure complete restriction digestion, an extra two or more additional nucleotides should be added to the 5′ end of each restriction recognition site (*see* **Note 1**).

4. High-fidelity PCR system: To prevent unwanted nucleotide exchanges or deletions, the homology arms should be amplified using a high-fidelity DNA polymerase having a proof-reading activity (*see* **Note 2**).
5. Qiaquick PCR purification kit (Qiagen) or equivalent.
6. Qiaquick Gel purification kit (Qiagen) or equivalent.
7. TE buffer: 10 mM Tris-HCl (pH 8.0), 1 mM ethylene diamine tetraacetic acid (EDTA).
8. Restriction enzymes and buffer system.
9. Calf intestinal Alkaline phosphatase (New England Biolabs).
10. T4 DNA ligase (New England Biolabs).
11. Phenol equilibrated with Tris-HCl buffer (pH 8.0).
12. Chloroform.
13. Isoamyl alcohol (IAA).
14. Phenol:Chloroform:IAA (25:24:1): A mixture consisting of equal parts of equilibrated phenol and chloroform:IAA (24:1).
15. Subcloning grade competent cells (*see* **Note 3**).
16. Plasmid minipreparation reagents. P1: 50 mM Tris-HCl (pH 8.0), 10 mM EDTA, 100 μg/mL RNaseA; P2: 200 mM NaOH, 1% sodium dodecyl sulfate (SDS); P3: 3.0M potassium acetate (pH5.5), and absolute ethanol. We use a simplified miniprep method to screen plasmid (*see* **Note 4**).
17. Stock solution of antibiotics: Ampicillin: dissolve ampicillin to a final concentration of 100 mg/mL in water, sterilize by filtration, and store at –20°C. Tetracycline: dissolve tetracycline to a final concentration of 50 mg/mL in absolute ethanol and store at –20°C. Avoid exposure of tetracycline solution to light by using black tubes or by wrapping the tube in aluminum foil. Chloramphenicol: dissolve chloramphenicol to a final concentration of 12.5 mg/mL in ethanol and store at –20°C.
18. LB medium: 10 g bacto-tryptone, 5 g bacto-yeast extract, and 10 g NaCl/1 L. Autoclave at 121°C for 20 min. Add appropriate antibiotics before using at a concentration of 100 μg/mL for ampicillin and 50 μg/mL for tetracycline.
19. LB plates: 10 g bacto-tryptone, 5 g bacto-yeast extract, 10 g NaCl, and 15 g bacto-agar/1 L. Autoclave at 121°C for 20 min. Add antibiotics/ drugs when the solution has cooled down to 65°C, and then pour the plates. Cool the plates down to room temperature, and store at 4°C for up to 2 mo. Wrap tetracycline containing plates in aluminum foil to prevent light exposure. LB/Amp, LB plate containing 100 μg/mL ampicillin: LB/Tet, LB plate containing 50 μg/mL tetracycline.
20. Bacterial incubator at 37°C.

2.2. Making Competent Cells Containing BAC to be Modified

1. LB medium containing 12.5 μg/mL of chloramphenicol.
2. Ice-cold distilled water.

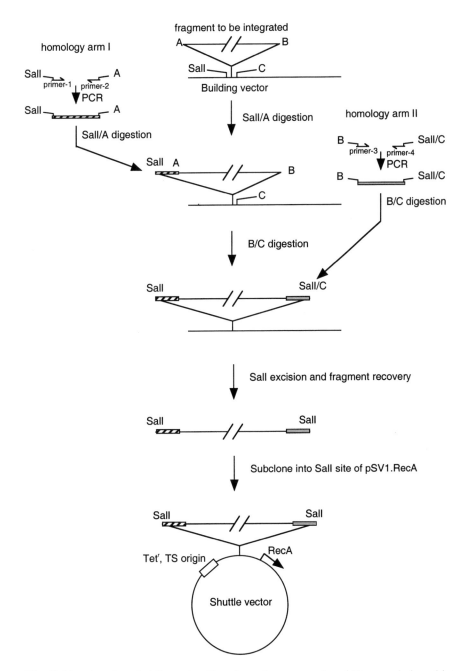

Fig. 2. Construction shuttle vector. Two homology arms, I and II, are subcloned into a building vector containing the insert to be integrated into the BAC. The fragment is then transferred to the pSV1.RecA shuttle vector by homology integration. PCR primers amplifying both homology arms are designed to have restriction sites as shown in the figure, which allow easy subcloning.

3. 10% glycerol in water: Sterilize by autoclaving.
4. Low temperature centrifuge.

2.3. Cointegration of the Shuttle Vector into the BAC

1. Electrotransformation apparatus such as Gene-pulser II (Bio-Rad).
2. LB/Tet plates and LB/Chl plates: LB/Chl, LB plate containing 12.5 μg/mL chloramphenicol.
3. SOC medium: dissolve 20 g of bacto-tryptone, 5 g of bacto-yeast extract, and 0.5 g of NaCl in 1 L of water. Autoclave as above. Add 10 mL of 1 M MgCl$_2$, 10 mL of 1 M MgSO$_4$, and 2 mL of 20% glucose. All solution including magnesium must be sterilized by filtration.
4. Bacterial incubators at 30 and 43°C.
5. Shaker at 30 and 43°C.

2.4. Resolving the Co-Integrant

1. LB/Chl plates.
2. TB/FA/Chl plates, TB plates containing fusaric acid and chloramphenicol: dissolve 10 g of bacto-tryptone, 1 g of bacto-yeast extract, 1 g of glucose, 8 g of NaCl, and 15 g of agar in 1 L of water, and add 0.5 mL of 0.1 M ZnCl$_2$ and 8 mL of 6.3 mg/mL chlorotetracycline. After autoclaving this solution, add 72 mL of 1 M NaH$_2$PO$_4$, 6 mL of 2 mg/mL FA, and 1 mL of 12.5 mg/mL chloramphenicol. Pour the plates and leave them on the bench overnight and then store at 4°C.
3. Bacterial incubator at 37°C.

3. Methods
3.1. Constructing Shuttle Vector (see Fig. 2)

1. Carry out PCR to make both homology arms using a high-fidelity amplification system. Diluted original BAC DNA should be used as template DNA. After PCR, run 5 to 10 μL of the reaction on an agarose gel to verify that the PCR products migrate as a single band.
2. Purify the PCR products using the Qiaquick PCR purification system or equivalent to remove free nucleotides and unincorporated primers.
3. Digest the purified PCR product of the homology arm I and the building vector having an insert to be integrated into the BAC with SalI and restriction enzyme A for 1 h at 37°C (*see* **Fig. 2**). The vector must be treated with DNA phosphatase to avoid self-ligation. Phosphatase treatment can be carried out by adding 1 U of phosphatase for 1 μg of the vector DNA to the restriction digestion mix and incubating one hour at 37°C.
4. Extract the digested PCR products and vector with Phenol/chloroform/IAA and chloroform and carry out an ethanol precipitation. Redissolve them in TE buffer.
5. Mix the digested vector and PCR products in 1X Ligation buffer (NEB). The vector: insert ratio should be 1:3. Add 1 μL (400 U) of T4 DNA ligase (NEB) and incubate the mixture at 16°C for overnight (*see* **Note 5**).

6. Transform competent bacterial cells with the ligation mixture and spread the cells on plates containing the appropriate antibiotics.
7. Screen several clones to find the plasmid carrying the PCR fragment.
8. When the clone(s) are identified, digest the plasmid DNA and homology arm II with restriction enzymes B and C (*see* **Fig. 2**). The vector must again be treated with DNA phosphatase.
9. Subclone the homology arm II as above (*see* **steps 4–7**).
10. Digest the building vector containing the two homology arms on either end of the insert with SalI and purify the fragment from an agarose gel using the Qiaquick Gel extraction kit.
11. Mix the insert and SalI digested and phosphatase treated shuttle vector in 1X ligation buffer, and add 1 µL of T4 ligase. The ligation reaction is carried out as described in **step 6**.
12. After transformation, the bacterial cells are selected on LB/Tet. The plate must be placed at 30°C overnight because the plasmid can replicate at 30°C but not at 37°C.
13. Inoculate several clones into 3 mL LB containing tetracycline and grow them at 30°C overnight.
14. Minipreparation of the shuttle vectors: Because of the low copy number of the shuttle vector in bacteria, only very low amounts of plasmid DNA are recovered from minipreps, and major bacterial genomic DNA contamination can sometimes be seen on the gel as background smear after restriction digestion. If the restriction fragment cannot be clearly seen on an agarose gel, the restriction pattern of the shuttle vector should be analyzed by conventional Southern hybridization. The PCR fragments or the insert to be integrated into BAC are useful probes to analyze the construct (*see* **Note 6**).

3.2. Making Competent BAC-Containing Cells

1. Inoculate a single colony of DH10B containing the BAC into 3 mL of LB containing chloramphenicol (12.5 µg/mL) and grow at 37°C for overnight.
2. Dilute the overnight culture 1/100 in 10 mL LB with chloramphenicol (12.5 µg/mL).
3. Culture them for exactly 3 h.
4. Chill on ice for 10 min (*see* **Note 7**).
5. Transfer 1.5 mL of culture to clean ice-cold 1.5 mL Eppendorf tubes.
6. Harvest the cells by centrifugation at 6000 rpm, 4°C for 1 min.
7. Remove as much medium as possible.
8. Resuspend the cells in 1.5 mL of ice-cold water.
9. Spin as above.
10. Repeat the washing step twice (**steps 7–9**).
11. Wash the cells once with 1.5 mL of ice-cold 10% glycerol as above (**steps 7–9**).
12. Resuspend the cells in 40 µL of ice-cold 10% glycerol.
13. Place the cells on ice when using them the same day, or freeze them in ethanol-dry-ice for storage. Frozen competent cells can be stored at −80°C at least several days.

3.3. Co-Integration of the Shuttle Vector into the BAC

1. Dilute the shuttle vector suitably for transformation with water: We usually use a 1/300 to 1/1000 dilution of plasmid DNA from a 1.5-mL culture.
2. Add plasmid DNA to the competent BAC containing cells and transfer cells to the electroporation cuvette.
3. Carry out electroporation in a 0.2-cm cuvet at 25 μF, 2.5 kV, and 200 ohms.
4. Add 1 mL SOC solution to the cuvet, transfer to a 15-mL Falcon tube, and culture it at 37°C with extensive shaking for 1 h.
5. Plate 100 μL of the bacterial culture onto one LB/Tet/Chl plate and spread the rest onto a second plate (*see* **Note 8**).
6. Place both plates at 30°C overnight.
7. Pick several (up to 6) colonies next day, and dilute them into 1 mL of LB. Spread the 1/10 of the dilution onto two LB/Tet/Chl plates. Place one plate at 43°C and the other plate at 30°C overnight (*see* **Note 9**).
8. Next day, only a dozen of clones will appear on the plate placed at 43°C. On the 30°C plates, you will note a thick bacterial layer.
9. Pick up several clones (up to 12 clones) from the 43°C cultured plates and dilute them into 1 mL of LB. Inoculate 100 μL of the dilution into 3 mL of LB containing tet and chl and streak 100 μL onto a LB/Tet/Chl plate to use as a master plate. Grow the miniculture at 43°C overnight. Incubate the master plate at 43°C overnight and store at 4°C for future use.
10. Carry out miniprep and analyze the BAC by conventional Southern hybridization. When one of the homology arms is used as the probe, two different bands, one from the BAC and the other from the integrated shuttle vector, may be detected. Restriction enzymes which reveal restriction fragments length difference between the original BAC insert and the integrated shuttle vector fragment should be chosen (*see* **Notes 10** and **11**).

3.4. Popping-Out the Shuttle Vector From the Co-Integrant

1. Once, the co-integrant is identified, spread the colonies from the master plates onto LB/Chl plates.
2. Place the plates at 43°C to select for the loss of the pop-out shuttle vector fragment.
3. Pick 8 to 16 colonies from the LB/Chl plates, and streak onto TB/FA/Chl plates (*see* **Note 12**).
4. Incubate the plates at 37°C until normal size colonies appear on the plates. Usually, it takes 2 to 3 d because of slow growth in the presence of FA.
5. Pick up 12 to 24 colonies and grow them in LB containing chloramphenicol (12.5 μg/mL).
6. Carry out miniprep of the candidates.
7. Analyze the targeting event having occurred in the modified BAC candidates by conventional Southern hybridization. The modified BAC DNA should be analyzed

by pulse field gel electrophoresis and PCR analysis for markers to confirm that no deletion occurred during modification.

4. Notes

1. The efficiency of restriction digestion within the primer depends on the nature of the restriction enzyme used. New England Biolabs' catalog provides useful information about restriction cleavage specificity close to the end of DNA fragments. Normally, we designed primers 33 to 40 base long consisting of a 25 to approx 30-mer conferring PCR specificity, a 6-mer as a restriction digest site, and 2 to 3-mer of additional nucleotides at the 3′ end of the primer. The annealing temperature should be calculated from the 25 to approx 30-mer PCR primer sequence. The restriction sites for subcloning must not be located within either the insert or the PCR product.

2. Various PCR systems are available. We have used both the Expand High Fidelity PCR system (Roche) and the PFU polymerase (Promega). Both kits worked very well.

3. We use home prepared electrocompetent cells for the various construction steps involved in the preparation of the shuttle vectors.

4. We prepare plasmid DNA using the following protocol. 1.5 mL of overnight culture of bacteria were harvested by centrifugation at 10,000 rpm for 1 min at 4°C. As much liquid was removed as possible. The bacteria were well suspended in 150 μL of P1, and 150 μL of P2 added to the suspension. After lysing the cells by inverting the tube, 150 μL of P3 were added to the viscous solution and the tube was inverted several times to mix it again. After centrifugation at 15,000 rpm for 10 min at 4°C, the supernatant was transferred to a fresh tube. Plasmid DNA can be precipitated by the addition of 1 mL of ethanol and centrifugation at 15,000 rpm for 10 min at 4°C. After a 70% ethanol wash, the plasmid DNA is redissolved in 50 μL of TE buffer. One microliter of high copy number plasmid and 10 μL of low copy number plasmid are sufficient for agarose gel detection.

5. The molar ratio between the PCR products and the vector must be determined on an agarose gel. Run the PCR product and the linearized vector on an agarose gel, stain the gel with ethidium bromide, and compare the amount of vector DNA and the PCR product by band intensity. If 500 bp homology arms and 3 kb vector are used, similar staining intensity for the bands indicates six times as many molecules of the PCR products as of the vector.

6. The orientation of the insert in the shuttle vector should also be verified.

7. After this step, all handling should be done in the cold room, and the bacteria must be kept on ice.

8. Because neither the precise quantity of plasmid nor the transformation frequency of the competent cells are known in advance, different amount of transformed cells should be plated on the two plates.

9. Each clone is picked using a disposable plastic inoculation loop and diluted in 1 mL of LB medium. The medium should be mixed by vortexing.

10. Sometimes, multiple bands are detected when probing with the PCR products because of the large amounts of DNA in each lane and/ or repeat nature of the genomic sequence. In such cases, purified insert can be used as probe instead.

11. If the homology arms are of the same length, the integration will occur at roughly the same frequency for both homology arms and two different integrant will be obtained.

12. Four to eight independent clones can be spread on the same plate. Divide each plate into four to eight areas with a marker pen. A hundred microliter of diluted bacteria is spotted onto the plate and spread using an inoculation loop. The medium will be quickly absorbed into the agar plate. As soon as this has occurred, put the plates in the incubator.

13. Recently, the Heintz lab has developed a modified shuttle vector system. We have received this new shuttle vector from them and are currently testing it for BAC modifications. This vector contains the *RecA* gene, the ampicillin resistant gene, the *SacB* gene and the pir protein dependent replication origin. Because the shuttle vector cannot replicate in normal BAC host bacteria such as DH10B which do not express the pir replication protein, bacteria resistant to both ampicillin and chloramphenicol should contain a BAC integrated the shuttle vector. The bacteria containing *SacB* gene grow very slowly on sucrose containing dishes. This gene allow positive selection of BAC clones which have lost the shuttle vector sequence.

Acknowledgments

We thank Dr. N. Heintz for giving us the pSV1.RecA shuttle vector and the original protocol for the BAC modification.

References

1. Shizuya, H., Birren, B., Kim, U.-J., et al. (1992) Cloning and stable maintenance of 300-kilobase-pair fragments of human DNA in *Escherichia coli* using an F-factor-based vector. *Proc. Natl. Acad. Sci. USA* **89,** 8794–8797.
2. Osoegawa, K., Tateno, M. Moon, P. Y., et al. (2000) Bacterial artificial chromosome libraries for mouse sequencing and functional analysis. *Genome Res.* **10,** 116–128.
3. Antoch, M. P., Song, E.-J., Chang, A.-M., et al. (1997) Functional identification of the mouse circadian clock gene by transgenic BAC rescue. *Cell* **89,** 655–667.
4. Nielsen, L. B., McCormick, S. P., Pierotti, V., et al. (1997) Human apolipoprotein B transgenic mice generated with 207- and 145-kilobase pair bacterial artificial chromosome. *J. Biol. Chem.* **272,** 29,752–29,758.
5. Yang, X. W., Model, P., and Heintz, N. (1997) Homologous recombination based modification in *Escherichia coli* and germ line transmission in transgenic mice of a bacterial artificial chromosome. *Nat. Biotechnol.* **15,** 859–865.
6. Wang, Z., Engler, P., Longacre, A., and Storb, U. An efficient method for high-fidelity BAC/PAC retrofitting with a selectable marker for mammalian cell transfection. *Genome Res.* **11,** 137–142.

7. Kim, U.-J., Birren, B. W., Slepak, T., et al. (1996) Construction and characterisation of a human bacterial artificial chromosome library. *Genomics* **34,** 213–218.

8. Frengen, E., Weichenhan, D., Zhao, B., Osoegawa, K., van Gel, M., and de Jong, P. J. (1999) A modular, positive selection bacterial artificial chromosome vector with multiple cloning sites. *Genomics* **58,** 250–253.

9. Muyrers, J. P., Zhang, Y., Testa, G., and Stewart, A. F. Rapid modification of bacterial artificial chromosomes by ET-recombination. *Nucl. Acid. Res.* **27,** 1555–1557

10. Jessen, J. R., Meng, A., McFarlane, R. J., et al. (1998) Modification of bacterial artificial chromosomes through chi-stimulated homologous recombination and its application in zebrafish transgenesis. *Proc. Natl. Acad. Sci. USA* **95,** 5121–5126.

11. Chatterjee, P. K. and Sternberg, N. L. (1996) Retrofitting high molecular weight DNA cloned in P1: introduction of reporter genes, markers selectable in mammalian cells and generation of nested deletions. *Genet. Anal.* **13,** 33–42.

12. Bochner, B. R., Huang, H. C., Schieven, G. L., and Ames, B. N. (1980) Positive selection for loss of tetracycline resistance. *J. Bacteriol.* **143,** 926–933.

13. Belmont, A. S., Li, G., Sudlow, G., and Robinett, C. (1999) Visualization of large-scale chromatin structure and dynamics using the *lac* operator/*lac* repressor reporter system. *Methods Cell Biol.* **58,** 203–222.

8

Bacterial Artificial Chromosome Engineering

Srividya Swaminathan and Shyam K. Sharan

1. Introduction

In the era of functional genomics, bacterial artificial chromosome (BAC) is an ideal choice of vector for cloning and manipulating large DNA fragments *(1)*. In addition to providing a system that is able to maintain large DNA fragments (average insert size approx 150–200 kb), BAC DNA can be purified by using standard plasmid purification method. Functional analysis of the BAC insert, which may represent coding or regulatory sequences or an uncharacterized DNA fragment, may require manipulation of the BAC DNA. For example, a drug resistance marker may have to be inserted to enable selection of the BAC DNA when transfected into eukaryotic cells, a gene present in the BAC insert may have to be mutated for functional studies or a transgene may have to be introduced for ectopic expression. Depending on the goal, the method of manipulation is selected. BACs are maintained in *Escherichia coli* (*E. coli*) cells that are *recA⁻* to ensure the stability of the insert. However, this hinders manipulation of BACs by homologous recombination in the bacterial cells. Consequently, the gene(s) required for recombination is transiently expressed to facilitate homologous recombination. Several different systems have been developed for BAC engineering, each utilizing a different protein(s) (such as recA, recE and recT, or exo and bet), to enable the BACs to undergo recombination *(2–5)*.

The BAC engineering technique that involves the bacteriophage genes *exo*, *bet*, and *gam*, and is independent of recA function is described here (*see* **ref. 5**, **Fig. 1**). The *exo* gene product has 5′–3′ exonuclease activity and the *bet* gene product is a single strand DNA binding protein that promotes annealing. This system also requires the *gam* gene function, which inhibits the recBCD nucle-

From: *Methods in Molecular Biology, vol. 256:*
Bacterial Artificial Chromosomes, Volume 2: Functional Studies
Edited by: S. Zhao and M. Stodolsky © Humana Press Inc., Totowa, NJ

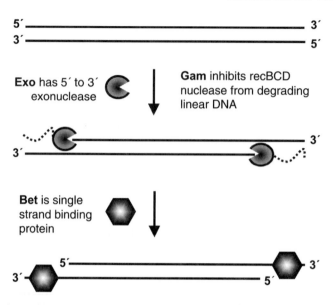

Fig. 1. Schematic representation of the bacteriophage λ recombination system. Exo has 5′ to 3′ exonuclease activity, which results in 3′ overhangs. Bet can bind single-stranded DNA and promote annealing while the *gam* gene product inhibits recBCD nuclease from degrading linear DNA fragment.

ase from degrading linear DNA fragments. An *E. coli* strain, Dy380, harboring a defective prophage to supply these functions has been generated *(6)*. The *exo*, *bet*, and *gam* genes are under the control of a temperature sensitive λ repressor, *cI857*, which is functional at 32°C but inactive at 42°C. Thus, the essential components for recombination can be provided by transiently shifting the bacterial culture from 32°C to 42°C. The length of homology required for efficient recombination is 30–50 bases *(5)*. These short homology arms can be easily added to the targeting vector by PCR, eliminating the use of conventional cloning methods. Here we describe methods to insert a selectable marker, to generate a subtle change without the use of selectable marker and to subclone DNA fragments from the BAC insert.

2. Materials

2.1. Bacterial Strains

The *E. coli* Dy380 strain harboring the defective λ prophage was generated from DH10B strain *(6)*. Two derivatives of Dy380 strain, EL250 and EL350, have been generated *(6)*. In addition to the prophage, these also have bacteriophage site specific recombinase gene *flpe* (EL250) or *cre* (EL350) under the control of an inducible promoter. These are all temperature sensitive strains and should be grown at 32°C. These stains can be obtained from Dr. Neal Copeland at NCI-Frederick (email: copeland@ncifcrf.gov).

2.2. Plasmids

Plasmids containing any antibiotic resistance gene (e.g., *amp*, *neo*) under the control of a prokaryotic promoter are used as template to generate targeting vectors by polymerase chain reaction (PCR). Bacterial cells containing the BAC that undergoes recombination can be selected for resistance to this antibiotic drug. In experiments where the modified BAC may have to be transfected into eukaryotic cells, antibiotic resistance genes that are selectable in both bacterial and eukaryotic cells are used (e.g. Neomycin, Zeocin, or Blasticidin resistance gene). These genes are under the control of prokaryotic as well as eukaryotic promoters *(3,4,6)*.

2.3. Equipment

Standard laboratory equipment are used, some of which are described below:

1. An incubator set at 32°C, a shaking incubator set at 32°C and shaking water bath (approx 150–200 rpm) set at 42°C.
2. Electroporator (Genepulser II with Pulse Controller II, Bio-Rad) and cuvets with a 0.1-cm gap (#870582, Bio-Rad).
3. A high-speed centrifuge and a refrigerated microcentrifuge.
4. PCR machine.
5. Spectrophotometer.
6. A –20°C freezer to chill cuvets and pipet tips for electroporation.
7. 2.2 mL deep-well plates with air permeable cover and plastic lid (Marsh Biomedical Products).
8. Multichannel pipets (for dispensing 1–10 and 10–50 µL vols).
9. Gel boxes for running agarose gels.

2.4. Reagents and Media

1. LB media without or with antibiotics (ampicillin 100µg/mL, chloramphenicol 20 µg/mL, kanamycin 25 µg/mL).
2. LB plates without or with antibiotics (Ampicillin 100 µg/mL, chloramphenicol 20 µg/mL or Kanamycin 25 µg/mL).
3. SOC medium *(7)*: 2% tryptone, 0.5% yeast extract, 10 mM NaCl, 2.5 mM KCl, 20 mM glucose, 20 mM MgSO$_4$.
4. Plasmid DNA purification reagents (Qiagen).
5. Gel extraction kit or reagents to purify DNA from Agarose gel (Qiagen).
6. Agarase enzyme and buffer (New England Biolabs).
7. Restriction enzymes and buffer (New England Biolabs or Invitrogen).

8. PCR reagents, *Taq* DNA Polymerase with proof reading activity (e.g., the Expand High Fidelity PCR system, Rosche Boehringer Mannheim or Deep Vent *Taq*, New England Biolabs) should be used for generating targeting vectors. For other purposes standard *Taq* DNA Polymerase (Perkin Elmer) can be used.

3. Methods

3.1. Electroporation of BAC DNA into Dy380 Cells

The first step of BAC manipulation is to introduce the BAC DNA into Dy380 cells (EL250, EL350 if flpe or cre proteins is required) that harbor the λ prophage, which provides the recombination function.

3.1.1. Preparation of Electrocompetent Dy380 Cells

1. Pick an isolated colony of Dy380 from LB plate and grow overnight in 3 mL of LB at 32°C.
2. Next morning, add 1 mL of the culture to 50 mL of LB in a 250-mL flask and grow at 32°C to an OD_{600} of 0.50–0.60.
3. Transfer 10 mL of the culture to a 50-mL Oak Ridge tube and spin at 6000g in prechilled rotor for 10 min at 1°C.
4. Wash the cell pellet with 10 mL of ice-cold water once and then resuspend in 1 mL water and transfer to a chilled 1.5 mL tube. Spin at 18,000g for 20–30 s at 1°C. Wash the cells two more times with 1 mL ice-cold water.
5. Resuspend cell pellet in water in a final volume of 50 μL and keep on ice.

3.1.2. Electroporation

1. Mix 100 ng of BAC DNA with 50 μL of electrocompetent Dy380 cells. Let it sit on ice for 5 min and then transfer into a 0.1-cm gap cuvet.
2. Set the Gene Pulser at 1.8 kV, 25 μF capacitance and 200 ohm resistance. Electroporate the BAC DNA into the cells and immediately add 1 mL SOC media.
3. Grow cells at 32°C for 1 h. Spin down cell for 20 s in a microcentrifuge. Resuspend the pellet in 200 μL LB.
4. Plate the cells on one 10-cm LB plate containing Chloramphenicol (20μg/mL) and incubate for 20–22 h at 32°C. Twenty to two hundred Chloramphenicol resistant colonies may be obtained.

3.1.3. Identification of Dy380 Cell Containing the BAC

1. Pick 5–10 individual colonies and grow in 3 mL LB containing Chloramphenicol.
2. Extract BAC DNA using an alkaline lysis method (*8*) and perform restriction digest (*see* **Note 1**). Compare the restriction pattern with that of the original BAC DNA. Most of the BAC DNA from Dy380 cells should be identical to the original DNA (*see* **Fig. 2**). However, occasionally some BACs show rearrangement and therefore it is important to confirm the integrity of the BAC clone after transforming them into Dy380 cells.

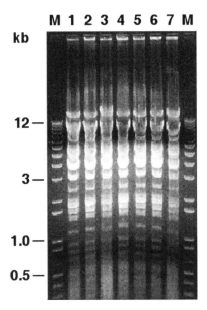

Fig. 2. Analysis of BAC DNA by restriction enzyme digest. Lane 1 shows BAC DNA in original host strain (DH10B) digested with *Eco*RI restriction enzyme. Lanes 2–7 are the *Eco*RI digest of the same BAC after being electroporated into Dy380 cells. Lanes marked "M" represent the 1 kb plus DNA molecular weight marker (Invitrogen).

3. Freeze aliquots of the Dy380 cells containing the BAC DNA in 15% glycerol at −70°C.

3.2. Insertion of a Selectable Marker into the BAC DNA

To insert a selectable marker into the BAC DNA, generate a targeting vector containing the selectable gene flanked by the homology arms (*see* **Fig. 3**). Electroporate this targeting vector into Dy380 cells containing the BAC and induced at 42°C to activate the expression of *exo*, *bet*, and *gam* genes. The correctly targeted cells are identified by the presence of the selectable marker.

3.2.1. Generating Targeting Vector by PCR

As the length of homology required for homologous recombination in bacterial cells is 30–50 bases, the targeting vectors can be generated rapidly by PCR to include the homology regions to a selectable marker. Optimal targeting efficiency is achieved when 50 bases of homology are used. For PCR, synthesize a pair of forward and reverse primers, each with 50 bases corresponding to the target region of the BAC DNA and 20 bases from the selectable marker to

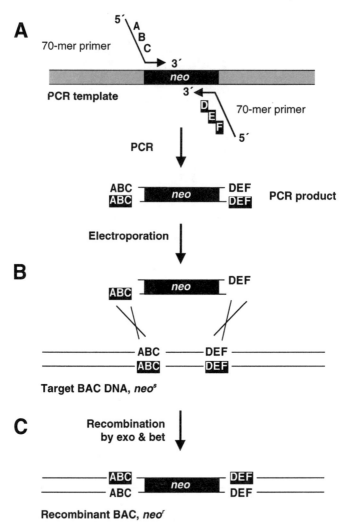

Fig. 3. Insertion of a selectable marker into BAC DNA. (**A**) Generation of targeting vector by PCR. The PCR primers used to generate the targeting vector are 70-mer oligonucleotides with 50 bases (ABC or DEF) from the target site to introduce the homology arm and 20 bases from the ends of the selectable marker. (**B**) The targeting vector is electroporated into the Dy380 cells that contain the BAC DNA and have been transiently induced to express the phage recombination genes. (**C**) After recombination the modified BAC DNA can be selected for the presence of the selectable marker. The alphabets in black represent the DNA sequence and those in white represent their complementary strand.

be included in the targeting vector as shown in **Fig. 2**. Use the two 70-mer primers without any purification other than the standard desalting procedure. Generally, 40 nmoles of oligonucleotides are resuspended in 500 μL 1X TE (pH 8.0).

3.2.2. Template for Generating Targeting Vector

The source of non-recombinant antibiotic resistant clones is the plasmid DNA that is used as template to generate the targeting vector. Zhang et al. *(3)* have described using *Dpn*I restriction enzyme to digest the template DNA (methylated site, recognized by the enzyme), whereas the unmethylated PCR product remains intact. In our hands, using linearized plasmid DNA for the PCR works very efficiently in eliminating the nonrecombinant clones.

1. Select a unique restriction enzyme site that is present in the vector sequence, but not within the sequence of the selectable marker to be included in the targeting vector.
2. Digest the plasmid DNA with the restriction enzyme and run the DNA on an agarose gel.
3. Cut the gel piece containing the linearized DNA and extract the DNA from the gel.
4. Resuspend the DNA in 1X TE (pH 8.0), check the concentration and dilute to 0.5–1.0 ng/μL.

3.2.3. PCR

Taq DNA Polymerase with proof reading ability is used to generate targeting vector using standard PCR conditions.

1. In general, use 1 μL of each oligo and 0.5–1.0 ng of the linear plasmid DNA in a 50 μL PCR.
2. Scaling up of the PCR may be required to obtain at least 600 ng of purified PCR product.
3. Conditions for the PCR cycle include an initial denaturation at 94°C for 4 min followed by 35 cycles of 94°C for 1 min, 55–60°C for 1 min and 72°C for 2 min and a final extension at 72°C for 7 min.
4. There are two ways to purify the PCR product. If a single correct size fragment is amplified then the PCR can be purified using commercially available PCR purification kits or by ethanol precipitation followed by a wash with 70% ethanol. The DNA should be resuspended in water and kept frozen at −20°.
5. Alternatively, if multiple fragments are amplified, the PCR is run on a low melting agarose gel. The correct size PCR fragment is cut out and the gel fragment is digested with Agarase enzyme to extract the DNA. This method is ideal for recovery of large DNA fragments (>3 kb) as it is gentle and the recovery rates are approx 80–90%.

3.2.4. Induction of Bacteriophage λ Recombination System in Dy380 cells

1. Pick an isolated colony of Dy380 containing the BAC and grow it overnight in 3 mL LB containing Chloramphenicol at 32°C.
2. Next morning add 1 mL of the culture to 50 mL LB containing Chloramphenicol in a 250-mL flask and grow at 32°C to an OD_{600} 0.50–0.60.
3. Transfer 10 mL of the culture to an Oak Ridge tube and place on ice, this will be used as uninduced sample.
4. To induce the Dy380 cells for recombination, transfer 10 mL of the culture into a 50-mL flask and shake vigorously (approx 200 rpm) at 42°C for 10–15 min. Stop the induction immediately by placing the flask into ice for 15 min with intermittent shaking.
5. Spin the induced and uninduced cells at $6000g$ in a prechilled rotor for 10 min at 1°C.
6. Wash with 10 mL ice-cold water once and then resuspend in 1 mL water and transfer to a chilled 1.5-mL tube.
7. Spin at $18,000g$ for 1 min at 1°C. Wash the cells two more times with 1 mL ice-cold water.
8. Resuspend cell pellet in water to a total volume of 50 μL.
9. Electroporate 100–300 ng of targeting vector into induced electrocompetent cells as well the uninduced control sample as described in **Subheading 3.1.**
10. Plate the cells on LB plates containing the appropriate antibiotics and incubate for 20–22 h at 32°C.
11. About 50–1000 colonies may be obtained on the plate containing the induced cells. The number of nontargeted clones (background) is estimated by counting the number of colonies on the control plate containing the uninduced electroporated cells. Generally, 0–5 colonies are obtained on the control plate suggesting that most of colonies on the "induced" plate represent recombinant clones (*see* **Note 2**).

3.2.5. Identification of Targeted Clones

Targeted clones can be identified by PCR, Southern blot analysis or by sequencing.

1. To identify by PCR, use primers flanking the targeted region (but not included in the targeting vector) to amplify the BAC DNA. The size of the PCR product should show the presence of the selectable marker into the desired site. Either the BAC DNA or this PCR product can be purified and sequenced using the same flanking primers in two separate reactions to confirm the insertion of the selectable marker and also to determine if any rearrangement occurred at the site of insertion.
2. The BAC DNA can be analyzed on a Southern blot by examining a change in the pattern of restriction fragments due to insertion of the selectable marker. It is also advisable to examine the restriction enzyme digestion pattern of the total BAC DNA on an agarose gel to see the integrity of the DNA as described in **Subheading 3.1.3.**

3.2.6. Insertion of a Nonselectable Gene

To insert a non-selectable transgene like *lacZ*, *Alkaline phosphatase*, *Green Fluorescent Protein* (*GFP*), *cre*, *flp*, into the BAC DNA, a similar approach can be taken. In such cases, the targeting vectors are selected by including a selectable marker, like any antibiotic resistance gene, 3′ of the poly A signal of the transgene. The selectable marker can be subsequently removed from the BAC by flanking it with two *lox*P or *FRT* sites. The selectable marker can be excised by expressing Cre or Flp protein. The two derivatives of Dy380, EL250, and EL350 are designed to express Cre and flpe proteins under the control of the arabinose inducible promoter, P_{BAD} (*6*). These cells also contain the bacteriophage λ prophage so the initial insertion can also be performed in these cells instead of the Dy380 cells.

3.2.7. Induction of Cre or Flpe Recombinase to Delete the Selectable Marker

1. Pick an isolated colony of EL250/EL350 containing the BAC that has the correct insertion of the transgene and a selectable marker (flanked by *lox*P or *FRT* sites).
2. Grow the colony at 32°C overnight in 3 mL LB containing Chloramphenicol.
3. Dilute 1 mL of the culture into 50 mL LB containing Chloramphenicol in a 250-mL flask and grow at 32°C.
4. After the culture reaches an OD_{600} of 0.3, add 500 µL 10% filter sterilized Arabinose to a final concentration of 0.1%. This induces the *flpe* gene in EL250 or the *cre* gene in EL350.
5. Grow the culture at 32°C until it reaches an OD_{600} of 0.5.
6. Plate 100 µL of 10^{-2}, 10^{-3}, 10^{-4} dilutions on LB plates containing Chloramphenicol and grow overnight at 32°C.
7. Clones that have lost the selectable marker by site specific recombination can be identified by selecting for sensitivity to the selectable drug. Streak approx 50–100 colonies in duplicate on LB plates with and without the selectable drug. Colonies that grow on LB plate, but not in the presence of the drug are the desirable clones that have lost the drug resistance gene and retained a single *lox*P or *FRT* site.
8. Confirm the loss of the selectable marker by sequencing the BAC DNA.

3.3. Modification of BACs Without Selection Using Oligonucleotides as Targeting Vector

Subtle modifications like single base changes or deletions and insertions in the BAC DNA can be generated by using a two step method (*3,6*). The first step involves targeting a positive selection marker along with a counterselectable marker (e.g., the *SacB* gene which is lethal to cells in the presence of sucrose) to the region of interest. In the second step, a DNA fragment containing the mutation is targeted to the same region and cells are selected for loss of the counterselectable marker. In addition, a rapid single-step method for generation of subtle changes using oligonucleotides as targeting vectors has been developed, which is described here *(9)*. In the absence of any selectable marker, a PCR-based selective amplification screen is used to identify the targeted clones. Because of the high-targeting efficiency, a small number of cells (1000–5000) are screened to obtain multiple targeted clones. The targeting efficiency depends on the length of homology and the strand of the DNA that is targeted by the oligonucleotide. About two-to-40-fold variation in recombination efficiency has been observed when different strands of the same region are targeted.

3.3.1. Targeting Vector

For generating a single base change, a 70-mer oligonucleotide with 34–35 bases of homology on either side of the mutation gives high targeting efficiency. However, for generating a deletion or an insertion, 70 bases of homology on each side of deletion or insertion site are required to achieve high efficiency. As the maximum length of an oligonucleotide that can be synthesized commercially is 100 bases, a PCR-based approach is used to generate longer oligonucleotides *(9)*. To generate a 140-mer oligonucleotide by PCR, use three oligonucleotides in a combination-PCR (*see* **Fig. 4A,B**).

1. A 100-mer with 50 bases of homology to each side of the deletion and two 40-mer oligonucleotides (a forward and a reverse primer). Each of the 40-mer primers has 20 bases of overlapping homology to either end of the 100-mer and an additional 20 bases that will be used to extend the homology to 70 bases on each side.
2. Set up combination-PCR by mixing 10 ng of the 100-mer and 300 ng of the 40-mers in a 50-μL PCR using *Taq* DNA Polymerase High Fidelity (Rosche Boehringer Mannheim). The resultant PCR product is a 140-base DNA fragment.
3. Similarly, to generate a targeting vector to insert up to 40 bases, use a forward and a reverse primer pair (*see* **Fig. 4A,C**). Each primer has 70 bases of homology at the 5′ end and the bases that need to be inserted at the 3′ end. The two oligonucleotides have 20 complementary bases at the 3′ end that help to anneal and extend the two by PCR to generate the targeting vector.
4. Purify the PCR product and denature at 95°C for 10 min to obtain single-stranded oligonucleotides. An additional advantage of using the denatured PCR products as targeting vectors is that both strands of the DNA are targeted resulting in optimal targeting efficiency (*see* **Note 3**). Similarly to create a single base change, use two complementary 70-mers to target both the strands simultaneously.

3.3.2. Electroporation of Targeting Vector

1. Prepare 50 μL induced and uninduced electrocompetent Dy380 cells containing the BAC as described in **Subheading 3.2.4.**

A

Target BAC DNA sequence

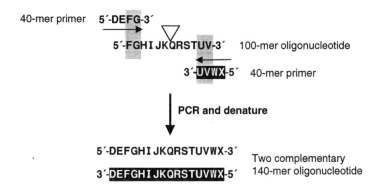

B

Generating targeting vector to delete "LMNOP" region

C

Generating targeting vector to insert "1 2 3 4" sequence

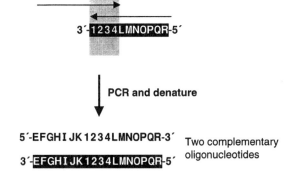

Fig. 4. Schematic representation of the PCR-based method to increase the length of homology. **(A)** The sequence from A to Z represents a double-stranded BAC DNA with the alphabets in white representing the bases complementary to those in black. **(B)** A 140-bp targeting vector is generated from a 100-mer oligonucleotide containing 50 bases (FGHIJK and QRSTUV) from either side of the deletion and two 40 mers each with 20 bases of homology to the 100-mer (FG and UV) and 20 bases to increase the length of homology (DE and WX). **(C)** A targeting vector to insert few bases (1234) can be generated from two oligonucleotides. Each oligonucleotide contains at the 5′end 70 bases of sequence from one side of the insertion site (EFGHIJK and LMNOPQR), and the sequence to be inserted (1234) at the 3′ end with 18–20 overlapping bases (required to anneal the two together during PCR).

2. Denature the complementary oligonucleotides or the PCR products by heating at 95°C for 10 min and then rapidly cooling on ice for 5 min prior to electroporation.
3. Electroporate 300 ng of the oligonucleotide under conditions described in **Subheading 3.1.2.** Grow the cells in 1 mL SOC medium (*see* **Subheading 2.4.** for composigion) for 1 h at 32°C.

3.3.3. Selective PCR Screen

Targeted clones can be detected by plating the electroporated cells at low dilution (approx 20–30,000 cfu/15 cm plate) on LB plates containing Chloramphenicol. The colonies can be hybridized with end labeled oligonucleotides that can specifically hybridize with the mutated clone and not with the wild-type BAC. In our hands, this procedure has resulted in high rates of nonspecific hybridization signal.

A PCR-based approach to identify a single base change, a small deletion or insertion from cultures of pooled bacterial cells is described here. This allele specific PCR amplification is called the mismatch amplification mutation assay-PCR (MAMA-PCR, **Fig. 5**, **ref. *10***). In the MAMA-PCR method, two PCR primers are used. One of the primers, which is not from the mutated region, has a perfectly matching sequence with respect to the wild-type and the mutant sequence. The second primer, called the detection primer, has two mismatches at the 3′ end sequence (ultimate and penultimate 3′ base) with respect to the wild-type sequence but only a single mismatch at the penultimate base with respect to the altered sequence. The two mismatched bases at the 3′ end of the primer when annealed to the wild-type template provide a conformation that is very inefficiently extended by DNA Polymerase. However, in the case of the mutant DNA, the ultimate 3′ base anneals to the complementary base allowing sufficient priming and amplification. Using a two-step PCR cycle consisting of a denaturation step and a common annealing/extension step enhances the specificity of the assay (*see* **Note 4**).

In case of a single base alteration, the 3′ base of the detection primer is the altered base. To detect deletions, the 3′ base of the detection primer should be the first base from the distal end of the deletion. Similarly, in case of small insertions, the 3′ base of the detection primer should be the first base of the inserted sequence. In case of insertions greater than 15–18 bases, primers specific to the inserted sequence can be used as detection primer and standard three step PCR conditions can be used.

3.3.4. Identification of Targeted Clone

1. After growing the cells at 32°C for 1 h in the SOC media, transfer them into a 1.5-mL tube and wash with 1 mL water a few times to remove the unelectroprated oligonucleotides.

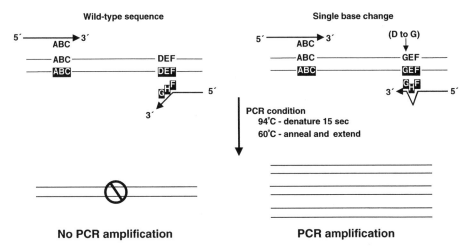

Fig. 5. Schematic representation of mismatch amplification mutation assay-PCR (MAMA-PCR). The sequence "ABC – DEF" represents a double-stranded DNA with the alphabet in white representing the base complementary to that in black. To detect a clone with a change from D to G, a detection primer with two mismatched bases (with respect to the wild-type sequence) at 3′end is used (D/G and E/H). It does not amplify the wild-type template under two-step PCR condition. The same detection primer has just one base mismatch at the penultimate base (E/H), with respect to the mutated sequence with D to G change. The primer amplifies a product under two-step PCR condition.

2. On an average, when 10 mL culture at OD_{600} of 0.55 are used to make the electro-competent cells and grown in SOC media for 1 h, there are approx $2 – 3 \times 10^8$ cells. Based on this estimate, dilute the culture to about 20 cells per milliliter of LB media containing chloramphenicol and add 500 µL to each well of a 2.2-mL deep 96-well plate (Marsh Biomedical Products, Inc.) using a multichannel pipet.

3. Grow the cells at 32°C for about 24 h. The exact number of cells per well should be determined by plating serial dilutions of the electroporated cells on agar plates.

4. After 24 h, transfer 10 µL of the culture from each well to a 96-well PCR plate using a multichannel pipet.

5. Make a cocktail of PCR reagents including the PCR buffer, deoxynucleotide 5′-triphosphates (dNTPs), forward and reverse primers (one of these should be the detection primer), and *Taq* DNA polymerase to set up a 50-µL PCR per well.

6. Add 40 µL of the PCR cocktail to each well containing 10 µL of the culture.

7. The PCR conditions include an initial denaturation for 4 min at 94°C followed by 40 cycles of 94°C for 15 s and 60°C for 1 min (a common annealing and extension temperature); and a final extension at 72°C for 7 min.

8. Run the PCR product on an agarose gel to identify the positive well(s).

9. Plate 100 µL of serially diluted (e.g., $1:10^{-5}$, $1:10^{-6}$, $1:10^{-7}$) cells from positive wells to obtain isolated colonies. Grow cells at 32°C overnight.
10. Pick 20–30 isolated colonies from each set and analyze them by PCR to identify few independent isolated clones.
11. Confirm the presence of the desired mutation by sequencing the BAC DNA and also examine the integrity of the BAC DNA by its restriction pattern.
12. Freeze the positive clones in 15% glycerol at –70°C.

3.4. Subclone DNA Fragment From BAC Insert

Subcloning a DNA fragment from the BAC insert by ligating desirable restriction enzyme digested fragments can be very time consuming. The λ recombination system can be rapidly and efficiently used to clone a DNA fragment from BACs. The only requirement is the sequence information of about 50 bases from the 5′ and 3′ end of the fragment that needs to be subcloned. The targeting vector consists of a plasmid sequence containing a selectable marker (e.g., Ampicillin or Kanamycin resistance gene), an origin of replication (for replication of the plasmid in the host cell) and 50 bases of homology to the 5′ and 3′ end of the fragment that needs to be subcloned. Any high copy plasmid (e.g., pBluescript) can be used for fragments up to 15 kb in size but a low copy plasmid (e.g., *pBR322*) is preferred for larger fragments. Lee et al. have used this technique to subclone DNA fragment up to 80 kb in size (*6*).

3.4.1. Retrieval Vector

To generate the retrieval vector by PCR, use a plasmid DNA as template. The PCR primers are two oligonucleotides consisting of 50 bases of homology to each end of the fragment that needs to be subcloned and 20 bases from the plasmid sequence flanking the region between the origin of replication and an antibiotic resistance marker. To excise the subcloned fragment from the retrieval vector, a restriction site should be included at each end. Care must be taken while designing the PCR primers, as unlike when a DNA fragment is inserted into the BAC DNA, here a circular molecule is generated after recombination and gap repair. This is achieved by having the homology arms in opposite orientation compared to the homology arms in the insertion vector as shown in **Fig. 6**.

1. To generate the targeting vector, use 0.5–1.0 ng of linearized pBR322 DNA as template and 300 ng of each PCR primer. Run the PCR product on low melting agaorse gel and purify the correct size fragment from the gel by digesting it with Agarase enzyme.
2. Generate 50 µL of induced and uninduced electrocompetent Dy380 cells containing the BAC DNA as described in **Subheading 3.2.4.**

A Target BAC DNA sequence

5′-ABCDEFGHIJKLMNOPQRSTUVWXYZ-3′
3′-ABCDEFGHIJKLMNOPQRSTUVWXYZ-5′

PCR template

B Primer sequences to generate targeting vector to retrieve "EFGHIJKLMNOPQRS"

Forward primer to include "EFGH" homology arm

5′-HGFE(50 bases) + 5'-neo (20 bases)-3′

Reverse primer to include "PQRS" homology arm

5′-PQRS(50 bases) + 3'-ori (20 bases)-3′

C Recombination of PCR product with BAC to retrieve "EFGHIJKLMNOPQRS"

Target site   5′-ABCDEFGHIJKLMNOPQRSTUVWXYZ-3′
3′-ABCDEFGHIJKLMNOPQRSTUVWXYZ-5′

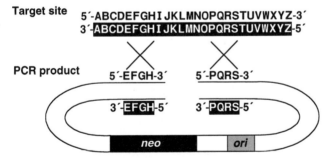

Fig. 6. Subcloning DNA fragments from the BAC insert. (**A**) The sequence from A to Z represents a double-stranded BAC DNA with alphabets in white representing the bases complementary to those in black. The PCR template is a linear plasmid DNA containing a selectable marker (*neo* gene) and origin of replication (*ori*) sequence for the propagation of the retrieved plasmid. (**B**) To generate a targeting vector by PCR, two primers are used. Each primer consists of 50 bases of homology to each end of the fragment that needs to be subcloned (EFGH and PQRS) and 20 bases from the plasmid sequence flanking the region between the origin of replication and an antibiotic resistance marker. (**C**) Recombination between the PCR product and the BAC DNA followed by a gap repair results in the formation of a circular plasmid that contains a fragment of the BAC DNA.

3. Electroporate the cells with 100 ng of the targeting vector as described in **Subheading 3.1.2.** Plate all the cells on LB plate containing appropriate antibiotic (depending on the selectable marker present in the targeting vector). Grow cells at 32°C overnight.
4. Pick 5–10 colonies and start culture in 3 mL LB containing antibiotics.
5. Extract the plasmid DNA and digest 200–300 ng of DNA with appropriate restriction enzyme (depending on the site present in the retrieval vector) to identify the positive clone (*see* **Note 5**).

4. Notes

1. To extract BAC DNA by alkaline lysis method, adding lysozyme (2.5 mg/mL) to the lysis buffer and replacing sodium dodecyl sulfate (SDS) with sarkosyl (Sodium N-Lauryl Sarcosine) in the denaturing solution significantly improves the yield and quality of BAC DNA (8).
2. When cells are plated after electroporation to select for insertion of a selectable marker, the colonies obtained contain a mixed population of cells containing the original BAC, as well as the recombinant BAC. Streak the colonies on LB plates containing the appropriate antibiotics to select for the recombinant BAC DNA. Pick and analyze a few isolated colonies to obtain pure recombinant clone.
3. A positive control targeting vector should be used to check the recombination efficiency of the Dy380 cells especially when oligonucleotides are used to manipulate the BAC DNA. This targeting vector can be designed to disrupt the Chloramphenicol resistance gene present in the BAC vector by inserting another antibiotic resistance gene, like *amp* or *neo*.
4. In MAMA-PCR, sometimes in spite of having a two-base mismatch at the 3′ end of the detection primer, a correct size fragment is amplified from the wild-type template. This can occur if the mutated base is same as one of the adjacent bases in the wild-type sequence. The ultimate base of the detection primer is able to pair with the template DNA by either skipping or looping out a base. Changing the orientation of the detection primer often solves this problem.
5. When cells are plated after electroporation to retrieve a DNA fragment from the BAC DNA, the colonies obtained may retain the BAC DNA. To obtain clones containing only the retrieved plasmid, transform any of the commonly used *E. coli* strains like *DH10B* or *DH5α* with 5–10 ng of the miniprep DNA and selected for resistance to the appropriate antibiotics. The transformants should be confirmed to contain the desirable plasmid by digesting the plasmid DNA with restriction enzyme.

Note Added in Proof
Mini-λ Recombination System

Although DY380 cells have been successfully used to modify exogeneous DNA cloned into a BAC vector, it has been difficult to introduce a few BAC clones into these cells. Recently, a nonreplicating, mini-λ circular DNA, which integrates into the bacterial chromosome and provides the Red recombination function has been developed *(12)*. This provides a recombination inducing system that can be introduced by electroporation into nearly any *E. coli* strain, including the *recA* mutant DH10B and its derivatives carrying BACs or PACs. The mini-λ provides a tractable system to express transiently the phage recombination genes at high levels.

Two-Step "Hit and Fix" Method

Although MAMA-PCR method can be used to identify recombinants containing a subtle alteration, it has some limitations. An example of this is when a unit of a small repetitive sequence (e.g., a mono-, di-, or tri-nucleotide repeat) has to be deleted or inserted, it is not possible to generate primers specific for the mutated sequence. To overcome these limitations, a two-step, "hit and fix" method has been developed recently *(13)*. In the first step of this method, a stretch of about 20 nucleotides is randomly changed around the base where the mutation is to be generated. In the second step, the modified bases generated in the first step are changed back to the original sequence except for the desired mutation. Since several nucleotides are changed in each of the two steps, the recombinant BACs can be identified by standard PCR methods using a primer specific for the altered bases or by oligo-specific hybridization.

Acknowledgments

The authors would like to thank E.-C. Lee, Daiguan Yu, Hilary Ellis, Nancy Jenkins, Neal Copeland, and Donald Court for the development of the phage recombination system in *E. coli* for BAC engineering. The authors would also like to thank Richard Frederickson of the Publication Department, NCI-Frederick, for help with the figures. Research sponsored by the National Cancer Institute, DHHS.

References

1. Shizuya, H., Birren B., Kim, U. J., et al. (1992) Cloning and stable maintenance of 300-kilobase-pair fragments of human DNA in *Escherichia coli* using an F-factor-based vector. *Proc. Natl. Acad. Sci. USA* **89,** 8794–8797.
2. Yang, X. W., Model, P., and Heintz, N. (1997) Homologous recombination based modification in *Escherichia coli* and germline transmission in transgenic mice of a bacterial artificial chromosome. *Nature Biotech.* **15,** 859–865.
3. Zhang, Y., Buchholz, F., Muyrers, J. P. P., and Stewart, A. F. (1998) A new logic for DNA engineering using recombination in *Escherichia coli. Nature Genet.* **20,** 123–128.
4. Muyrers, J. P. P., Zhang, Y., Testa, G., and Stewart, A. F. (1999) Rapid modification of bacterial artificial chromosomes by ET-recombination. *Nucl. Acids Res.* **27,** 1555–1557.

5. Yu, D., Ellis, H. M., Lee, E.-C., Jenkins, N. A., Copeland, N. G., and Court, D. L. (2000) An efficient recombination system for chromosome engineering in *Escherichia coli. Proc. Natl. Acad. Sci. USA* **97,** 5978–5983.
6. Lee, E.-C., Yu, D., Martinez de Velasco, J., et al. (2001) A Highly Efficient Escherichia coli-Based Chromosome Engineering System Adapted for Recombinogenic Targeting and Subcloning of BAC DNA. *Genomics* **73,** 56–65.
7. Sambrook, J., Fritch, E. F., and Maniatis, T. (1989) *Molecular Cloning: A laboratory Manual,* 2nd edition, Cold Spring Harbor Laboratory, Cold Spring Harbor, NY.
8. Sinnett, D., Richer, C., and Baccichet, A. (1998) Isolation of stable bacterial artificial chromosome DNA using a modified alkaline lysis method. *Biotech.* **24,** 752–754.
9. Blomfield, I. C., Vaughn, V., Rest, R. F., and Eisenstein, B. I. (1991) Allelic exchange in *Escherichia coli* using the *Bacillus subtilis* sacB gene and a temperature-sensitive pSC101 replicon. *Mol. Microbiol.* **5,** 1447–1457.
10. Swaminathan, S., Ellis, H. M., Waters, L. S., et al. (2001) Rapid engineering of bacterial artificial chromosomes using oligonucleotides. *Genesis* **29,** 14–21.
11. Cha, R. S., Zarbl, H., Keohavong, P., and Thilly, W. G. (1992) Mismatch amplification mutation assay (MAMA): Application to the c-H-ras gene. *PCR Meth. Appl.* **2,** 14–20.
12. Court, D. L., Swaminathan, S., Yu, D., et al. (2003) Mini-lambda: a tractable system for chromosome and BAC enginering. *Gene* **315,** 63–69.
13. Yang, Y., Sharan, S. K. (2003) A simple two-step, 'hit and fix' method to generate subtle mutations in BACs using short denatured PCR fragments. *Nucleic Acids Res.* **31,** e80.

9

ET Recombination

DNA Engineering Using Homologous Recombination in E. coli

Joep P. P. Muyrers, Youming Zhang, Vladmir Benes, Giuseppe Testa, Jeanette M. J. Rientjes, and A. Francis Stewart

1. Introduction

Recombinogenic engineering, or the modification and cloning of DNA molecules via homologous recombination, has opened up a new era. In contrast to conventional cloning strategies, recombinogenic engineering can be carried out at virtually any chosen position on a given target DNA molecule (which can be small or large). Any type of modification, including sequence deletions, substitutions, and insertions can be carried out via homologous recombination. Because homologous recombination is a precise process of high fidelity, recombinogenic engineering generates DNA clones efficiently and with high precision.

The possibilities of recombinogenic engineering have been reviewed recently *(1)*. The most straightforward and efficient type of recombinogenic engineering is termed ET recombination, or ET cloning. The term ET recombination is derived from the RecE and RecT proteins, which were first shown to be capable of mediating the required homologous recombination process *(2)*. However, the equivalent lambda proteins Redα and Redβ are also capable of mediating this process *(2–5)*. Therefore, this technology has also been called lambda (red) mediated DNA engineering in later descriptions *(5,6)*. In the last few years, ET recombination has been developed to allow a wide range of precise modifications of both small and large target molecules, including plasmids (such as mouse targeting constructs), bacterial artificial chromosomes (BACs), P1-vectors, PACs, and the *Escherichia coli* chromosome *(2–13*, for review, *see* **ref.** *1*). Recently, ET recombination has also been developed to allow one-step

From: *Methods in Molecular Biology, vol. 256:*
Bacterial Artificial Chromosomes, Volume 2: Functional Studies
Edited by: S. Zhao and M. Stodolsky © Humana Press Inc., Totowa, NJ

in vivo cloning of DNA sequences from plasmids and BACs *(6,14)*, or directly from purified genomic DNA *(14;* for review, *see* **ref. 1**). Because DNA cloning by ET recombination combines amplification and cloning into one step, and because the obtained clones are all proofread in vivo, this strategy provides an attractive alternative for traditional in vitro DNA cloning methods using polymerase chain reaction (PCR).

In the figures of this chapter, generic schemes for ET recombination-based DNA modification *(see* **Fig. 1**) and DNA cloning *(see* **Fig. 2**) are presented. ET recombination requires the generation of a linear targeting molecule, which is then contacted with the target DNA molecule in an ET-recombination-proficient *E. coli* cell. After homologous recombination has occurred, the obtained recombinant DNA product is isolated and analyzed. The purpose of this chapter is to provide a detailed experimental protocol and a trouble-shooting guide for those wanting to use the technology. The practical steps involved (PCR amplification, electroporation of ET-recombination proficient cells, and DNA analysis) are straightforward, and the described protocol aims to assist researchers in the establishment of the technology. A generic protocol for DNA engineering using ET recombination is presented. This protocol can be readily applied for various tasks, including BAC modification and in vivo DNA cloning.

2. Materials

All chemicals should be of molecular biology grade, and are purchased from Sigma unless indicated otherwise. All solutions should be made with double-distilled water (ddH$_2$O).

Fig. 1. *(see facing page)* DNA modification by ET recombination. **(A)** Design of the ET-oligonucleotides and generation of the linear targeting DNA. Each ET-oligonucleotide consists of parts **A**, **B**, and **C**: Part **A** and **C** are required, and part *B* is optional *(see* **Note 2**). For the downstream ET-oligonucleotide, these parts are **A′**, **B′**, and **C′**. The sequences of **A** and **A′**, the so-called homology regions, may be freely chosen and determine the exact position of modification on the target molecule. Any suitable template carrying the DNA of interest can be used for generation of the linear targeting molecule, which is usually done by PCR using a pair of ET-oligonucleotides. In many cases, the DNA of interest carries a selectable marker gene (for a review of DNA modification possibilities, *see* **ref. 1**). **(B)** The linear targeting DNA recombines with the target molecule (for example, a BAC) within an ET-proficient bacterial cell *(see* **Notes 7 and 8**). This results in the generation of the recombinant product, in which the linear targeting molecule has integrated into the target molecule at the desired position. In this example, ET recombination is used to substitute the sequence which was originally present between **A** and **A′** on the target molecule with the DNA of interest. However, the same principle is applied for ET recombination-based insertion and deletion of DNA sequences (for review, *see* **ref. 1**).

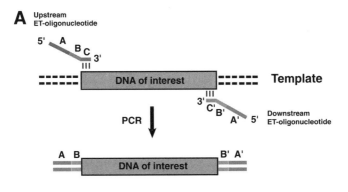

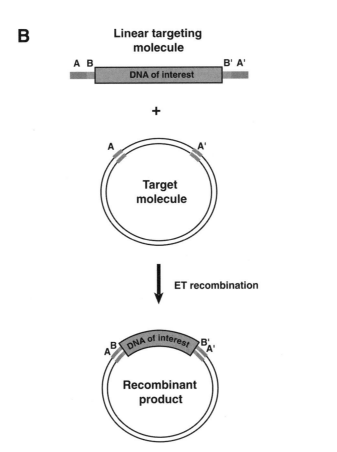

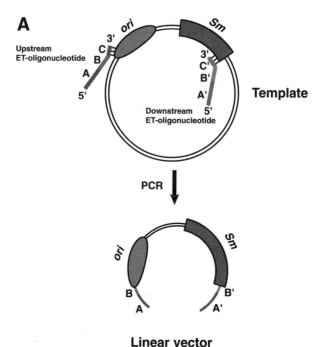

Linear vector

Fig. 2. DNA cloning and subcloning by ET recombination. **(A)** Design of the ET-oligonucleotides and generation of the linear vector. As in DNA modification, 3 parts of the ET-oligonucleotides are considered (again, part **B** and **B′** are optional, and parts **A** and **A′** determine the sequence to be cloned (the DNA of interest); *see* **Notes 2** and **3**). Choice of the template vector depends on the cloning task at hand. For example, in protein expression studies, a protein expression vector is chosen as the template. In this example, only the selectable marker gene (sm) and the origin of replication (ori) are amplified to generate a minimal linear vector. However, by the same principle, an entire vector backbone can be included in the linear vector, thereby generating a recombinant product containing a full vector backbone.

2.1. Generation of the Linear DNA by PCR Amplification

1. Oligonucleotides can be purchased from any reliable source (*see* **Note 1**); store at –20°C.
2. PCR amplification materials (Roche) and single nucleotide stocks can be bought as kits (Amersham Pharmacia Biotech). PCR amplification materials should be stored at –20°C.
3. 0.2-mL tubes (Stratagene) or 0.5-mL Eppendorf tubes, mineral oil (if required) and any functional PCR machine.

B

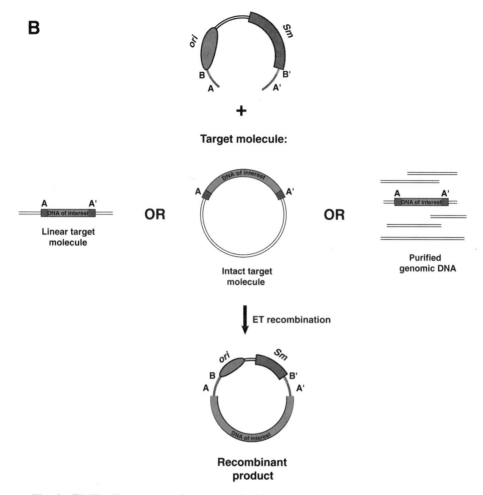

Fig. 2. **(B)** The linear vector is contacted with a target DNA molecule, which can be either a linear, or a circular target molecule (such as a BAC), or a mixture in which the target molecule is contained (such as purified genomic DNA from various organisms *[14]*). After homologous recombination in an ET-proficient strain (*see* **Note 7**), the recombinant product containing the DNA of interest is obtained. Amplification and cloning of the DNA of interest is thus carried out in a single in vivo step.

2.2. Gel Electrophoresis

1. 0.7% agarose (GibcoBRL) gels are used in combination with 1X TBE buffer *(16)*. The TBE buffer is stored as a 5X stock.
2. Ethidium bromide solution for visualization of DNA on agarose gels *(16)*.
3. DNA loading buffer for gel-electrophoresis *(16)*.
4. Power source and UV transilluminator.

2.3. DNA Enzymatic Modifications

1. All DNA digestions and other modifications are carried out as recommended by the manufacturer (New England Biolabs).
2. 1.5-mL Eppendorf tubes.

2.4. Purification and Concentration of DNA

1. After digestion or gel isolation, or as part of a DNA minipreparation, the obtained DNA products are purified using a phenol/chloroform/isoamylalcohol 25:24:1 extraction mixture (Amresco).
2. 100% ethanol (Merck).
3. 3 M NaAc (Merck), pH 5.2.
4. ddH$_2$O, or elution buffer EB (10 mM Tris-HCl, pH 8.5; Qiagen).

2.5. Preparation and Induction of Competent Cells

1. Ice-cold 10% glycerol solution. Made by mixing ddH$_2$O plus glycerol (Merck) to 10%, and prechilling on ice for at least 4 h (preferably overnight).
2. LB medium for growth of bacteria *(16)*.
3. 10% L-arabinose solution. The solution is prepared by dissolving 1 g of L-arabinose (Sigma) in LB-medium to a total volume of 10 mL. This solution is prepared freshly every time.
4. 35 mL centrifugation tube (Corning).

2.6. Electroporation

1. Electroporator (Eppendorf 2510); Electroporation settings are 1.35 kV, 10 µF, 600 Ohm.
2. 1-mm cuvets (Eppendorf).
3. SOC medium (Sambrook and Russell, Molecular Cloning: A Laboratory Manual, 3rd Edition (2001), Cold Spring Harbor Laboratory Press). Alternatively, LB medium can also be used.

2.7. Media for Antibiotic Selection

All antibiotics are purchased from Sigma, and stock solutions are stored at –20°C. For selective LB medium, the required antibiotic(s) is/are dissolved in LB medium to the indicated working concentration:

1. Chloramphenicol stock solution: A working concentration of 15 µg/mL is used for BACs and other large molecules, and a working concentration of 50 µg/mL is used for high-copy plasmids; stored as 30 mg/mL stock in EtOH.
2. Ampicillin stock solution: Working concentration 100 µg/mL, stored as 1000X stock in 50% EtOH.
3. Tetracycline stock solution: Working concentration 10 µg/mL, stored as 1000X stock in 75% EtOH.
4. Kanamycin stock solution: A working concentration of 15 µg/mL is used for BACs and other large molecules, and a working concentration of 50 µg/mL is used for high-copy plasmids; stored as 30 mg/mL stock in ddH$_2$O.

5. Streptomycin stock solution: Working concentration 50 µg/mL, stored as 1000X stock in ddH$_2$O.
6. Selective LB plates are made by adding 15 g of agar (Difco) to 1 L of LB medium. After boiling, allow to cool to approx 50°C, add the required antibiotics to the working concentration and pour into plates.

2.8. Isolation and Analysis of DNA Products From Bacteria

1. Selective LB medium, carrying the working concentration of the appropriate antibiotic(s) (*see* **Subheading 2.7.**).
2. P1, P2, P3 buffers (DNA purification kit; Qiagen).
3. 100% EtOH (Merck) and 70% EtOH (prepared by mixing 100% EtOH (Merck) with ddH$_2$O).

3. Methods
3.1. Generation of the Linear Targeting DNA by PCR Amplification

1. First, a pair of ET-oligonucleotides must be designed and ordered. As is indicated in **Figs. 1** and **2**, there are three parts to be considered in each ET-oligonucleotide. The sequences of these parts can be freely chosen according to the objective of the required cloning or modification step (*see* **Notes 2** and **3**).
2. The oligonucleotides are used in a PCR amplification reaction according to standard procedures (*see* **Note 4**).

3.2. Analysis of the Linear Targeting DNA: Separation From Template

1. After PCR amplification, an aliquot of the resulting mixture is analyzed by agarose gel electrophoresis, according to standard procedures *(16)*.
2. In many cases, the PCR-amplified linear targeting molecule needs to be separated from the template DNA (*see* **Note 5**). This can be done in two different ways:

3.2.1. DpnI Digestion

1. The entire PCR amplification reaction is purified using phenol/chloroform extraction, followed by DNA precipitation and dissolving in ddH$_2$O or buffer EB (*see* **Subheading 3.3.**). Alternatively, commercially available PCR-purification kits (Qiagen) can be used according to manufacturer's instructions. One column is used per 100 µL of PCR, and the final elution volume is 50 µL in elution buffer EB.
2. After either of these protocols, the obtained DNA (still consisting of the linear targeting DNA and the template) is digested with *Dpn*I enzyme in the provided buffer New England Biolabs; *see* **Note 6**). *Dpn*I digestions are carried out in a 60-µL total volume using at least 20 U of *Dpn*I per PCR. For complete digestion, *Dpn*I enzyme can be added twice. The digestion is carried out for a minimum of 2 h, and if convenient overnight.
3. After *Dpn*I digestion, the linear targeting molecule is phenol/chloroform purified, precipitated, and dissolved, as described in **Subheading 3.3.**

3.2.2. Gel Purification

The complete PCR is separated by agarose gel electrophoresis. The band containing the linear targeting molecule is excised with a scalpel, and purified using the following gel-purification protocol:

1. The excised band is transferred to an Eppendorf tube and incubated at −80°C for at least 20 min.
2. Then, the tube is transferred to 37°C for 5 min, followed by centrifugation at maximum speed using a table-top centrifuge for 3 min.
3. The resulting supernatant is transferred to a new Eppendorf tube.
4. To the resulting pellet, 300 μL of 1X TBE is added, followed by vortexing and incubation at −80°C for 20 min. Again, the tube is transferred to 37°C for 5 min, followed by centrifugation at maximum speed using a table-top centrifuge for 3 min. The resulting supernatant is transferred to the Eppendorf tube that already carries the supernatant of the first step.
5. The pellet is discarded.
6. The combined supernatants are further purified using phenol/chloroform extraction followed by DNA precipitation and dissolving in ddH₂O or buffer EB (*see* **Subheading 3.3**). Alternatively, commercially available gel-purification kits (Qiagen) can be used according to manufacturer's instructions. One column is used per 100 μL of PCR, and the final elution volume is 50 μL in elution buffer EB.

3.3. Purification and Precipitation of DNA

After enzyme digestion (such as **Subheading 3.2.1.**), gel purification (*see* **Subheading 3.2.2.**), or bacterial lysis (*see* **Subheading 3.6.**), the obtained solutions containing the DNA with a volume of V μL are purified as follows:

1. To V μL of DNA solution, add an equal volume V of phenol/chloroform/isoamylalcohol mixture and invert back and forth for at least 1 min.
2. Centrifuge for at least 5 min on a table-top centrifuge at maximum speed.
3. Carefully transfer the resulting top layer to a new eppendorf tube. Be sure not to transfer anything but the top layer, and discard the protein-containing middle layer and the bottom layer.
4. To the isolated top layer in the new tube, add 1/10 V of 3 *M* NaAc, pH 5.2, and 3 V of 100% ethanol. Mix by inverting and incubate at −80°C for at least 15 min.
5. Spin at 4°C or at RT for at least 10 min on a table-top centrifuge at maximum speed.
6. A pellet should be visible. Carefully remove the supernatant (thereby not disturbing the pellet), and add 1 mL of 70% ethanol. Vortex until pellet comes loose, and spin for 2 min on a table-top centrifuge at maximum speed.
7. Remove supernatant and allow pellet to dry completely.
8. To dry pellet, add 3 μL of ddH₂O or buffer Eb (Qiagen) per μL DNA product expected, and dissolve completely.

9. After dissolving, a small aliquot of the purified DNA solution should be checked by gel electrophoresis to allow an estimation of the concentration.

3.4. Preparation of Competent Cells

After amplification, separation from the template and purification, the linear targeting molecule is ready for use in ET recombination. Because ET recombination is carried out in vivo in *E. Coli*, competent bacterial cells need to be prepared. These cells need to be proficient for ET recombination (*see* **Notes 7–9**). The protocol for generation of electrocompetent ET proficient *E. coli* cells is as follows:

1. The host strain harbouring the target molecule (for example, the BAC host strain DH10B or HS996, Research Genetics) needs to be transformed with the ET-proficiency plasmid (*see* **Notes 7** and **8**). Because transformation of ET-proficiency plasmids is very efficient, any standard protocol for DNA transformation can be used (either chemically *[16]*, or by using electroporation, see protocol below). The transformed cells are obtained on a selective LB plate containing antibiotics for maintenance of the target molecule and the ET proficiency plasmid. If the target is a BAC, the used antibiotic is usually chloramphenicol. ET proficiency plasmids have been developed which carry the ampicillin resistance gene, or the tetracycline resistance gene; the appropriate antibiotic should be chosen for their maintenance.
2. From the plate containing the ET proficient strain (ET-proficient after **step 1**, or an endogenously proficient strain such as YZ2000, *see* **Notes 7** and **8**), grow at least two independent overnight cultures at 37°C in 5 mL LB selective medium.
3. Prepare ice-cold 10% glycerol. Per 1 volume of bacterial cell culture used to make competent cells (*see* **step 4** below), three volumes of 10% glycerol are needed.
4. Dilute the overnight culture (*see* **Subheading 3.4., step 2**) 100-fold in a total volume of 35 mL LB selective medium and place in a shaking incubator at 37°C, until OD_{600} equals 0.1.
5. Induce the expression of the required recombinases by adding L-arabinose to the culture, to a final concentration of 0.10 % (*see* **Notes 8–11**), and allow growth until OD_{600} of 0.3 to 0.45.
6. Transfer the cells to a centrifugation tube, and centrifuge at 7000 rpm in a Sorvall SLA-600 rotor for 5 min at −5°C (*see* **Note 12**).
7. Discard the supernatant, and carefully resuspend the cell pellet with 5 mL ice-cold 10% glycerol, using a precooled glass pipet. Add another 30 mL of ice-cold glycerol, and spin at 7000 rpm at −5°C for 6 min (*see* **Note 13**).
8. Discard the supernatant, and carefully resuspend the cell pellet with 5 mL ice-cold 10% glycerol, using a precooled glass pipette. Add another 30 mL of ice-cold glycerol, and spin at 7500 rpm at −5°C for 6 min (*see* **Note 13**).
9. Discard the supernatant, and carefully resuspend the cell pellet with 5 mL ice-cold 10% glycerol, using a precooled glass pipet. Add another 30 mL of ice-cold glycerol, and spin at 8000 rpm at −5°C for 6 min (*see* **Note 13**).

10. Discard the supernatant, invert the tube and quickly dry the sides with a clean tissue. Resuspend the cell pellet in the remaining liquid (usually approx 60 μL, *see* **Note 14**), and transfer 30 μL of cells to a precooled labeled Eppendorf tube (thus, two tubes can be prepared). These cells are now ready for electroporation.

3.5. Electroporation

1. To the Eppendorf tube containing the 30 μL electrocompetent cells, add 1 μL of the linear targeting molecule (or linear vector) solution (*see* **Note 15**).
2. Transfer the DNA/cell mixture to a prechilled 1-mm electroporation cuvet (Eppendorf).
3. Electroporate.
4. After electroporation, immediately add 1 mL of LB medium or SOC medium, transfer into an Eppendorf tube, invert the tube and incubate in a shaking incubator at 37°C for 75 min.
5. After incubation, the cells are spread on selective LB plates (*see* **Note 16**), and grow overnight at 37°C (*see* **Note 17**).

3.6. Isolation and Analysis of DNA Products From Bacteria

The vast majority of colonies obtained after electroporation contains the desired recombinant products (*see* **Note 18**). To analyze the colonies, the DNA has to be isolated using the following procedure:

1. Pick 4 (or more, if wanted) independent colonies from the plates obtained in **Subheading 3.5.**, and grow them overnight in 2 mL selective LB medium.
2. For each culture, perform the following steps:
   a. Transfer 1.5 mL of saturated culture into an Eppendorf tube, and spin for 1 min using a table-top centrifuge at maximum speed.
   b. Discard the supernatant, and resuspend the pellet in 150 μL P1 buffer.
   c. Add 150 μL of P2 lysis buffer, and invert the tube several times.
   d. Add 150 μL of P3 precipitation buffer, and vortex for 30 s at maximum output.
   e. Centrifuge at maximum speed in a table-top centrifuge for at least 5 min.
   f. Transfer the supernatant into a new, labeled Eppendorf tube.
3. From the thus obtained supernatants, purified DNA is isolated as described in **Subheading 3.3.** However, no NaAc needs to be added.
4. At least one-third of the obtained DNA solutions is analyzed by restriction analysis (*see* **Note 19**) and gel electrophoresis. The obtained clones can be further verified by Southern analysis (according to standard procedures; *[16]*) and/or by DNA sequencing (*see* **Note 20**).

4. Notes

1. Because the quality of the ET-oligonucleotides (*see* **Note 2**) is crucial to the technology, as a single mistake in the sequence of an ET-oligonucleotide may reduce the efficiency of ET recombination significantly, a reliable provider of oligonucleotides should be chosen. Purification of the oligonucleotides is usually not required.

2. For each ET-oligonucleotide, three parts must be considered (*see* **Figs. 1** and **2**). The first part, **A** (**A′** for the downstream ET-oligonucleotide) is the homology region. The sequences of the homology regions may be freely chosen, and determine the position on the target molecule that is modified by ET recombination (or, if ET recombination is used for DNA cloning, the homology region sequences determine the DNA fragment that is cloned, *see* **Note 3**). The homology regions are thus present on the target molecule as well. The length of **A** and **A′** is typically 40 to 50 nucleotides. Because ET recombination is mediated through homologous recombination, a process of extremely high precision and fidelity, it is essential to assure yourself that the sequences of the homology regions **A** and **A′** are accurate. We have experienced complications in DNA engineering when the sequence provided in a database was not completely accurate. If possible, it may be a good idea to sequence the homology regions on the target molecule to be sure they are accurate. The sequence of the second part, **B** (**B′** for the downstream ET-oligonucleotide), can be, for example, a protein tag (such as a *His*-, c-*myc*- or HA-tag), one or more restriction sites, a site-specific-recombinase-target-site or whatever else you may want to introduce into your desired construct. Whatever you choose for **B** or for **B′**, it is co-introduced (in frame, if wanted) into the recombinant construct, at no additional effort. If no operational sequences need to be introduced, parts **B** and **B′** are simply not included in the ET-oligonucleotides. Finally, the third part, **C** (**C′** for the downstream ET-oligonucleotide), primes the PCR amplification of the DNA of interest.

3. When ET recombination is used for DNA cloning and subcloning (*see* **Fig. 2**), the orientation of the ET-oligonucleotides is different as compared to the orientation of ET-oligonucleotides that are used for modification of DNA molecules (*see* **Fig. 1**). Care should be taken to ensure that the ET-oligonucleotides which are used for DNA cloning are in the correct orientation. Also, the linear targeting DNA is a linear vector in this case, and regions **C/C′** will minimally amplify an origin of replication and a selectable marker gene. Several published examples of DNA cloning by ET recombination exist, and it may be a good idea to work through one of these examples *(14)*.

4. The concentrations of the template DNA, oligonucleotides, dNTPs and Taq polymerase, as well as buffer composition, can be chosen as needed. As in any PCR-amplification, finding the optimal conditions may require some trial and error. We use Taq polymerase if the DNA of interest on the linear targeting DNA is a selectable gene (such as an antibiotic selection marker gene). If the DNA of interest present on the linear targeting DNA contains a sequence that cannot be rapidly selected for, a polymerase with proofreading activity (such as the Taq/Pwo mixture supplied in the High Fidelity Amplification kit [Roche]) is usually used.

5. If an antibiotic selection scheme is used to identify correct recombinant products (for review, *see* **ref. *1***), the template used in the PCR amplification can give rise to background colonies and thus needs to be separated from the linear targeting molecule. However, if the PCR-template carries an origin of replication that does not function under normal growth conditions (such as temperature-sensitive ori-

gins, or origins that require the expression of a protein which is not expressed in commonly used *E.coli* strains), or if genomic DNA or linear plasmids (or plasmid fragments) are used as a template, the linear targeting molecule does not need to be separated from the template. If ET recombination is used for DNA cloning (*see* **Fig. 2**), the template usually needs to be separated from the linear vector.

6. *Dpn*I has a 4-basepair recognition site, and only digests methylated DNA. Because a PCR product (such as a linear targeting molecule or linear vector) is not methylated, and template DNA usually is (upon growth in most commonly used bacterial cloning strains), *Dpn*I only digests the template DNA and not the linear targeting molecule.

7. Most commonly used cloning and host strains are not ET proficient. In case ET recombination needs to be carried out directly in the host strain harboring the target molecule (which we recommend if the target molecule is large, such as a BAC, PAC, or P1-vector), ET proficiency needs to be provided via ET proficiency plasmids (*see* **Note 8**). These plasmids need to be transformed into the host strain (*see* **Subheading 3.4.**). Alternatively, the linear targeting molecule and the target can be co-electroporated into a specialized strain, which expresses RecE and RecT (or Redα and Redβ) endogenously, from the chromosome. Examples of such specialized strains include YZ2000 *(14)* and variations thereof. We recommend cotransformation into a specialized strain for modification of small targets (such as plasmids; *12,14*) and for cloning from genomic DNA *(14)*. In case ET recombination is carried out in such a specialized strain, no ET-proficiency plasmid needs to be transformed, and **step 1** of **Subheading 3.4.** is omitted.

8. Several examples of ET-proficiency plasmids have been published: For example, pBAD-ETγ *(2)*, pBAD-αβγ *(3)*, and pR6K-αβγ *(14)*. By transforming an ET-proficiency plasmid into the host strain carrying the target molecule, such strains are made inducibly ET proficient (*see* **Note 10**). The ET-proficiency plasmids allow tightly regulated, inducible expression of at least one of the required recombinases of a given recombinase pair (RecE/RecT or Redα/Redβ, usually by adding L-arabinose (the abovementioned plasmids are based on the pBAD24 backbone *[15]*). Such tightly regulated expression is a big advantage, as it reduces the window in which recombination takes place to the absolute minimum required. Thus, after recombination has taken place, the recombinases are no longer expressed, thereby greatly reducing the risk of unwanted rearrangements (for review, *see* **ref. *1***). ET proficiency plasmids also usually allow expression of the Redγ protein (*see* **Note 9**).

9. Redγ is required to inhibit the RecBCD activity present in most commonly used host and cloning strains (for review, *see* **ref. *1***). The addition of Redγ allows the use of ET recombination in host strains that are RecBCD+.

10. Expression of the required recombinases from the ET-proficiency plasmids mentioned above (*see* **Note 8**) relies on L-arabinose induction. Be sure that L-arabinose is used, as D-arabinose does not work.

11. In case the linear targeting molecule and the target are co-electroporated into a specialized strain like YZ2000, no induction with L-arabinose is required. In that

case, the cells are simply grown until OD_{600} of 0.3 to 0.45 without induction. The preparation of competent cells of these specialized strains, which are mainly used for modification of small plasmids and for cloning from genomic DNA, is otherwise identical to the protocol of **Subheading 3.4.**

12. From here on, all steps have to be carried out quickly, and on ice. Thus, the tubes containing the cells must be kept on ice as much as possible (also during washing). Precooled pipets, a precooled rotor and ice-cold 10% glycerol must be used for washing steps. The Eppendorf tubes to which the cells are finally transferred must be precooled on ice, and the electroporation cuvets and DNA solutions (*see* **Subheading 3.5.**) must be pre-cooled on ice.

13. It is important that during the washing steps, no significant amount of cells is lost. Should this be the case, then the rpm in the centrifugation steps may need to be increased. However, do not increase the centrifugation force too much, as this will result in a very tight bacterial pellet, which is hard to resuspend. The washing steps should be carried out gently, and bubbles should be avoided. When a centrifugation step is finished, the cells should be processed immediately to avoid resuspension while sitting in the rotor.

14. Most important for successful ET recombination is the quality and, in particular, the concentration of the ET proficient cells. From a 35-mL culture of OD_{600} of 0.4, approx 60 μl of highly concentrated competent cells are obtained after all washing steps.

15. The concentration of the linear targeting DNA or linear vector, which can be estimated by gel electrophoresis, *see* **Subheading 3.3., step 9**, should be at least 0.5 μg/μL. If the linear targeting DNA and the target are co-electroporated into a specialized strain (*see* **Note 7**), 1 μL carrying 0.5 μg of a solution of the target DNA must be added as well. The DNA solutions used must be in ddH$_2$O or in buffer EB. If a DNA solution carries too much salt, it may arch during electroporation. Before and after electroporation, the cells may be left waiting on ice briefly, although we recommend to work quickly.

16. Plates are stored at 4°C. Plates containing tetracycline, ampicillin, and hygromycin B should not be used when they are older than 1 mo; plates containing other antibiotics can be kept for at least 3 mo at 4°C. The required antibiotics depend on the cloning task at hand. For example, if the linear targeting DNA molecule carries the kanamycin resistance gene, kanamycin should be added to the LB plates at this stage (at working concentration). However, if the linear targeting DNA molecule for example does not carry a selectable marker, the plates only need to contain the antibiotic to which the target molecule gives resistance (for example for BACs, this is usually chloramphenicol). For a full review on the possibilities of ET recombination based DNA engineering, please refer to **ref. *1***. If no further ET recombination steps are planned, selection for the ET-proficiency plasmid is no longer necessary at this stage, and should be omitted.

17. In some cases, incubation should be extended to at least 24 h before colonies are visible. Incubation overnight after electroporation is usually at 37°C, however, in some cases, temperature-sensitive plasmids are modified by ET recombination:

in that case, incubation after electroporation must be carried out at the appropriate temperature.

18. The efficiency of ET recombination is very high; typically at least 80% of the obtained colonies carry the desired recombinant product. The most significant source of background is usually the template used in the generation of the linear targeting DNA. It is thus important to ensure that the linear targeting DNA is efficiently separated from the template (*see* **Subheading 3.2.**).

19. The strategy for restriction enzyme verification of recombinants is chosen according to the task at hand.

20. In some cases, the ET-proficiency plasmid is still present at this stage, even though no antibiotic selection was maintained for it. Conveniently, plasmids which constitutively express the Redγ protein (for example, pBADETγ, 2; or pBADαβγ, 3) were found to be rapidly lost in the absence of continued selection pressure. It may be necessary to retransform the obtained recombinant product (into whichever strain desired) to completely get rid of the ET-proficiency plasmid. Such retransformation can be done by electroporation **Subheading 3.4.**), or by any other suitable means of transformation (Sambrook and Russell, Molecular Cloning: A Laboratory Manual, 3rd Edition (2001), Cold Spring Harbor Laboratory Press). The obtained recombinant products can be modified further, or can directly be applied for biological studies.

Acknowledgments

The authors would like to thank Inhua Muyrers-Chen and Michael Spiegel for critical readings of the manuscript. The initial development of ET recombination was done in the Francis Stewart laboratory at the European Molecular Biology Laboratories, Heidelberg, Germany, and was sponsored by the Volkswagen Foundation, Program on Conditional Mutagenesis.

References

1. Muyrers, J. P. P., Zhang, Y., and Stewart, A. F. (2001) Recombinogenic engineering: new options for cloning and manipulating DNA. *Trends Bioch. Sci.* **26,** 325–331.

2. Zhang, Y., Buchholz, F., Muyrers, J. P. P., and Stewart, A. F. (1998) A new logic for DNA engineering using recombination in Escherichia coli. *Nature Genet.* **20,** 123–128.

3. Muyrers, J. P. P., Zhang, Y., Testa, G., and Stewart, A. F. (1999) Rapid modification of bacterial artificial chromosomes by ET-recombination. *Nucl. Acids Res.* **27,** 1555–1557.

4. Narayanan, K., Williamson, R., Zhang, Y., Stewart, A. F., and Ioannou, P. A. (1999) Efficient and precise engineering of a 200 kb beta-globin human/bacterial artificial chromosome in E. coli DH10B using an inducible homologous recombination system. *Gene Ther.* **6,** 442–447.

5. Yu, D., Ellis, H. M., Lee, E. C., Jenkins, N. A., Copeland, N. G., and Court, D.L. (2000) An efficient recombination system for chromosome engineering in Escherichia coli. *Proc. Natl. Acad. Sci. USA* **97,** 5978–5983.

6. Lee, E. C., Yu, D., Martinez de Velasco, J., et al. (2001) A highly efficient Escherichia coli-based chromosome engineering system adapted for recombinogenic targeting and subcloning of BAC DNA. *Genomics* **73**(1), 56–65.

7. Angrand, P. O., Daigle, N., van der Hoeven, F., Scholer, H. R., and Stewart, A. F. (1999) Simplified generation of targeting constructs using ET recombination. *Nucl. Acids Res.* **27,** e16.

8. Datsenko, K. A. and Wanner, B. L. (2000) One-step inactivation of chromosomal genes in Escherichia coli K-12 using PCR products. *Proc. Natl. Acad. Sci. USA* **97,** 6640–6645.

9. Hill, F., Benes, V., Thomasova, D., Stewart, A. F., Kafatos, F. C., and Ansorge, W. (2000) BAC trimming: minimizing clone overlaps. *Genomics* **64,** 111–113.

10. Muyrers, J. P. P., Zhang, Y., Buchholz, F., and Stewart, A. F. (2000) RecE/RecT and Redα/Redβ initiate double-stranded break repair by specifically interacting with their respective partners. *Genes Dev.* **14,** 1971–1982.

11. Muyrers, J. P. P., Zhang, Y., Benes, V., Testa, G., Ansorge, W., and Stewart, A. F. (2000) Point mutation of bacterial artificial chromosomes by ET recombination. *EMBO Reports* **1,** 239–243.

12. Muyrers, J. P. P., Zhang, Y., and Stewart, A. F. (2000) ET cloning: Think recombination first. Genetic Engineering, Principles and Methods (Setlow, J. K., ed.), 22, 77-98 Kluwer Academic/Plenum, NY.

13. Nefedov, M., Williamson, R., and Ioannou, P. A. (2000) Insertion of disease-causing mutations in BACs by homologous recombination in Escherichia coli. *Nucl. Acids Res.* **28**(17), E79.

14. Zhang, Y., Muyrers, J. P. P., Testa, G., and Stewart, A. F. (2000) DNA cloning by homologous recombination in Escherichia coli. *Nature Biotech.* **18,** 1314–1317.

15. Guzman, L. M., Belin, D., Carson, M. J., and Beckwith, J. (1995) Tight regulation, modulation, and high-level expression by vectors containing the arabinose PBAD promoter. *J. Bacteriol.* **177,** 4121–4130.

16. Sambrook, J. and Russell, D. W. (2001) *Molecular Cloning: A Laboratory Manual, 3rd Edition,* Cold Spring Harbor Laboratory, Woodbury, NY.

10

BAC Engineering for the Generation of ES Cell-Targeting Constructs and Mouse Transgenes

Giuseppe Testa, Kristina Vintersten, Youming Zhang, Vladmir Benes, Joep P. P. Muyrers, and A. Francis Stewart

1. Introduction

Bacterial artificial chromosomes (BACs) have become a central tool in functional genomics. This is due to their average cloning capacity (around 150 Kb), which can accommodate most eukaryotic genes along with their full set of regulatory elements, and to their greater convenience in handling over other large cloning vectors like P1-based artificial chromosomes (PACs) and yeast artificial chromosomes (YACs). Two key advances for harnessing the full power of BACs have been the development of methods to modify them with precision and ease (see below), and the ability to use them for transgenesis in higher model organisms, like mouse and zebrafish (for review, see (*1*)).

Recently, a homologous recombination-based method (termed Red/ET recombination) was developed that enables the full range of BAC modifications to be quickly and reliably engineered in *Escherichia coli* (*E. coli*) (*2–4*) (for review *see* **ref. 5**). Its general principles and applications are discussed in Chapter 9 by Muyrers et al.

Here, we describe three specific variations of Red/ET recombination to generate mouse targeting constructs or transgenes starting from BACs. The first is the construction of very large targeting constructs (greater than 60 Kb) for use in homologous recombination in mouse ES cells (*6*). Large targeting constructs enable simultaneous mutagenesis of two loci which lie farther apart than the distance covered by conventional targeting constructs. The second is a quick way to assemble targeting constructs of conventional length by Red/ET

From: *Methods in Molecular Biology, vol. 256:*
Bacterial Artificial Chromosomes, Volume 2: Functional Studies
Edited by: S. Zhao and M. Stodolsky © Humana Press Inc., Totowa, NJ

subcloning from BACs. The third describes BAC engineering to make a large transgene for microinjection into oocytes. The chapter by Vintersten et al. describes the microinjection methodology.

1.1. Assembly of BAC-Based Knock-In Constructs for Homologous Recombination in ES Cells

With the rising tide in genomic knowledge, it is increasingly desirable to study gene function in the mouse with more sophisticated approaches than those employed for conventional knock-out or knock-in vectors, such as, introducing two or more modifications in the same locus. These can be either mutations which alter gene function (conferring a loss or gain of function) or "neutral" modifications which enables study of gene expression, such as, the introduction of a reporter gene (most commonly β-galactosidase or GFP) to monitor the expression profile, or a protein tag to allow biochemical chracterization of the target protein at endogenous levels of expression. Introducing two modifications is relatively straightforward with conventional targeting approaches if the two sites are within approx 10 Kb of each other. However, for large genes, this goal has usually been achieved by designing dedicated vectors for each of the mutations to be introduced, which are then transfected into ES cells in separate rounds of targeting. This is a very time-consuming process, and exposes ES cells to loss of totipotency during sequential rounds of transfection and selection. Plus, for every subsequent mutation to be introduced, only 50% of the homologously recombined clones will have integrated the second mutagenic cassette on the same allele already harboring the first. Particularly for genes that display low rates of homologous recombination, the impact of a 50% reduction in the amount of successfully targeted ES clones is undesirable.

We applied Red/ET recombination *(2,4)* to engineer large targeting vectors from BACs so that two or more mutations can be simultaneously targeted to a genomic locus over a wide range of distances (more than 60 kb) *(6)*. The overall strategy is outlined in **Fig. 1**. It consists of sequential rounds of BAC modification in *E. coli* via Red/ET recombination to engineer a complex targeting construct, which is then used in a single round of homologous recombination in ES cells.

The first two steps are termed here "BAC shaving" They refer to the exercise whereby the original BAC is sequentially shortened so that the two loci where the mutagenic cassettes will be placed are flanked, respectively on the 5′ and 3′ side, by flanking regions of only approx 5 Kb. In fact, for the sole purpose of homologous recombination in ES cells, one could use the whole BAC as such, after appropriate placement of the desired cassettes. However, it is possible that one or both cassettes will be "buried" in the middle of the clone at a large distance from the end of the targeting construct. Consequently, it may be impossible to find an appropriate restriction enzyme for a Southern strategy to

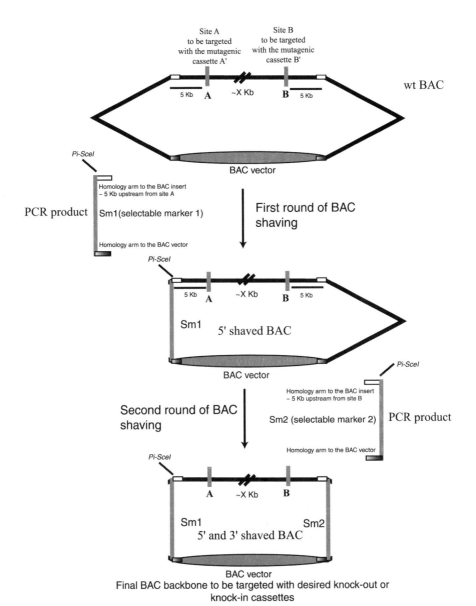

Fig. 1. Sequential "BAC shaving" to yield the backbone for ES targeting constructs. Schematic representation of Red/ET mediated BAC shaving. Two sequential rounds of Red/ET recombination delete from the original BAC the unwanted regions on the 5′ and 3′ side of, respectively, the mutagenic cassettes to be placed at sites A and B. White rectangles indicate regions of homology mediating Red/ET recombination between the BAC insert and the PCR products. Gray gradient rectangles indicate regions of homology mediating Red/ET recombination between the BAC vector and the PCR products. Sm1 and Sm2 indicate the selectable markers used, respectively, in the first and second round of BAC shaving. X Kb indicate any distance between sites A and B which greatly exceedes the length accomodated in conventional targeting constructs.

distinguish homologous recombinants from random integrants. Therefore, the BAC is "shaved" to eliminate the unwanted flanking DNA to leave regions of homology of sufficient length to promote recombination and to enable Southern blot screening. The easiest way to perform BAC shaving is to replace the superfluous portion of genomic sequence with a selectable marker by Red/ET recombination. This is a straightforward Red/ET recombination reaction (described in Chapter 9 by Muyrers et al.), in which a selectable marker is amplified by polymerase chain reaction (PCR) with oligos containing short arms of homology (40–60 nt) to the target region present in the BAC molecule.

After shaving, the desired mutagenic cassettes are placed at the chosen spots in the BAC. This is also a straightforward Red/ET recombination reaction, in which the mutagenic cassette is flanked on both sides by short arms of homology (40–60 nt) to the chosen locus in the BAC molecule.

However, often these cassettes need to be preassembled according to the experimental goals. In many cases, Red/ET recombination can also simplify these preassembly tasks *(7)* (*see* also Chapter 9 by Muyrers et al. for advice).

Following these four assembly steps, the BAC-based targeting construct is electroporated into mouse ES cells according to the standard protocols for conventional targeting constructs.

1.2. Subcloning From a BAC to Create ES Cell-Targeting Constructs

The assembly of ES cell targeting constructs of regular size (10–15 Kb) can also be simply accomplished by subcloning a chosen region from a BAC using Red/ET recombination. Traditionally, such constructs are built starting from smaller genomic clones (for example, λ-phage clones), in which the site to be targeted with a knock-out or knock-in cassette is restricted by the genomic region included in the clone at hand, by laborious cloning steps to assemble the final construct. Another major limitation of this approach is that the genomic clone at hand often limits the choice of restriction enzyme cleavage for the Southern blot strategy to confirm the intended homologous recombination.

Red/ET-mediated BAC subcloning circumvents both these problems *(4)*. As shown in **Fig. 2**, two strategies can be used. Following the order schematized in **Fig. 2A**, the desired genomic region for the targeting construct is first subcloned from a BAC. In so doing, one can freely decide the length of both the 5′ and 3′ homology arms. Plus, one can first choose the most convenient Southern blot strategy, based on the sites at end, and then define the boundaries of the targeting construct so that they will fit with the Southern strategy. Once the chosen fragment has been subcloned into a high or middle copy plasmid, it can then be targeted, in the second step of Red/ET recombination, with the relevant knock-out or knock-in cassette.

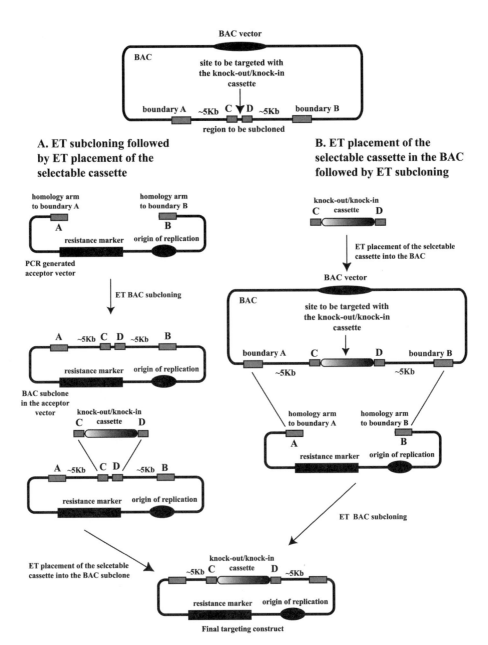

Fig. 2. Schematic representation of Red/ET mediated BAC subcloning to yield the backbone for a knock-out/knock-in targeting construct. The site to be targeted with the knock-out/knock-in cassette is indicated by an arrow. Gray boxes A and B indicate the homology regions for the Red/ET BAC subcloning step. Gray boxes C and D indicate the homology regions used for the Red/ET placement of the knock-out/knock-in cassette. In the strategy depicted in panel A, the desired genomic fragment is first subcloned from the BAC and then targeted with the selectable cassette. Panel B illustrates the converse strategy.

Fig. 2B illustrates the converse strategy, in which the knock-out/knock-in cassette is first targeted to the BAC, and then the targeted genomic region is subcloned from the BAC into the final acceptor vector. The two strategies are equivalent. One advantage of following the order shown in **Fig. 2B** is that in the BAC subcloning step one can positively select recombinant colonies for the presence of the selectable marker already inserted in the BAC. This can eliminate almost completely the background, which is mostly owing to circularization of the PCR amplified acceptor vector.

The subcloning step involves amplifying by PCR a suitable vector, consisting of a bacterial origin of replication and a selectable marker, with oligos containing short arms of homology (50–60 nt) to the sites in the BAC flanking the region to be subcloned. The reaction that follows is a standard Red/ET recombination reaction, and the target region is copied from the BAC into the acceptor vector.

1.3. Red/ET Recombination for the Production of BAC Transgenes

Red/ET recombination can be successfully employed to introduce many kinds of modifications into a BAC (insertion of a mutagenic and/or reporter cassette; insertion of a point mutation; deletion of an undesired region). For the engineering of BACs for transgenes, the same principles described above apply. To illustrate these principles, here we describe a protocol for the elimination of the *lox*P site present in all BAC vectors (because of its role in BAC vectorology) *(9)*. This *lox*P site is a significant problem in genomic engineering since its inclusion in the course of BAC transgenesis places a loose *lox*P site into the genome. This *lox*P site could lead to undesired interchromosomal aberrations upon crossing with mouse lines also harboring *lox*P sites. As the use of Cre-*lox*P conditional approaches to mouse mutagenesis is constantly expanding, loose *lox*P sites compromise such a transgenic line. Red/ET recombination can be easily applied to replace this *lox*P site with a selectable marker. As this is a straightforward Red/ET recombination exercise, analogous to the one described in **Subheading 3.1.2.**, it is here described in detail given its value in routine BAC engineering for oocyte microinjection, and it also illustrates the principles of Red/ET engineering of BAC transgenes.

2. Materials

2.1. Propagation of BACs

1. *E. coli* strain harbouring the BAC of interest (DH10β or HS996).
2. Luria-Bertani (LB) medium for growth of bacteria (Sambrook and Russell, Molecular Cloning: A Laboratory Manual, 3rd Edition (2001), Cold Spring Harbor Laboratory Press): 10 g bacto-tryptone, 5 g bacto-yeast extract, and 10 g NaCl in 1 L water to pH 7.0.

3. LB agar plates are made by adding 15 g agar (Difco) to 1 L LB medium. After boiling, allow to cool to approx 50°C, add the required antibiotics to the working concentration and pour into plate.

4. Antibiotics (chloramphenicol, ampicillin, kanamycin, hygromycin, gentamycin, tetracycline). All antibiotics are purchased from Sigma, and stock solutions are stored at –20°C. For selective LB medium, the required antibiotic(s) is/are dissolved in LB medium to the indicated working concentration.

   a. Chloramphenicol stock solution: a working concentration of 12.5 μg/mL is used for BACs and other large molecules, and a working concentration of 50 μg/ml is used for high-copy plasmids; stored as 34 mg/mL stock in EtOH.

   b. Ampicillin stock solution: a working concentration 50 μg/mL is used for BACs and other large molecule, and a working concentration of 100 μg/mL is used for high-copy plasmids; stored as 1000X stock in 50% EtOH.

   c. Tetracycline stock solution: Working concentration 25 μg/mL, stored as 5 mg/mL stock in 75% EtOH.

   d. Kanamycin stock solution: a working concentration of 15–20 μg/mL is used for BACs and other large molecules, and a working concentration of 50 μg/mL is used for high-copy plasmids; stored as 30 mg/mL stock in ddH_2O.

   e. Gentamycin stock solution: working concentration of 3 μg/mL is used for BACs and other large molecules; stored as 10 mg/mL stock in ddH_2O.

2.2. Preparation of Mini and Maxi Preparations of BAC DNA

1. Resuspension solution: 50 mM glucose, 20 mM Tris-HCl, pH 8.0, 10 mM ethylenediamine tetraacetic acid (EDTA), pH 8.0.
2. Lysis solution: 0.2N NaOH, 1% sodium dodecyl sulfate (SDS).
3. KAcetate solution.
4. Large construct kit from Qiagen (Catalog number 12462).
5. Isopropanol 100%.
6. Ethanol 70%.
7. 7 TE buffer: Tris-HCl, 10 mM pH 8.0, EDTA 1 mM pH 8.0.
8. PBS: 137 mM NaCl, 2.7 mM KCl, 4.3 mM Na_2HPO_4, 1.4 mM KH_2PO_4, pH 7.3.

2.3. PCR Amplification of Selectable Marker Cassettes

1. Plasmid containing a selectable marker (βlactamase, hygromycin phosphotransferase, gentamycin acetyltransferase, neomycin phosphotransferase) serving as a template in the PCR to generate the recombinogenic linear DNA fragment.
2. HPLC purified oligonucleotides consisting of a PCR primer annealing to the selectable marker cassette of choice and an homology arm to the target region in the BAC. Store oligonusleotides at –20°C.
3. Taq polymerase and 10X reaction buffer with $MgCl_2$ (Roche).
4. Deoxynucleotidetriphosphates (dNTPs) (Gibco).
5. PCR purification kit (Qiagen).

2.4. Restriction Enzyme Digests and Purification of Digested DNA

1. Restriction enzyme with suitable 10X reaction buffer (New England Biolabs).
2. 3 *M* NaAcetate.
3. Ethanol 100%.
4. Ethanol 70%.
5. Phenol.
6. Chloroform.

2.5. Preparation of Competent Cells for Red/ET Recombination

1. Plasmid expressing the recombinogenic pair of proteins (RecE/RecT or Redα/ Redβ) under the control of an arabinose inducible promoter (*see* **Note 8**, Chapter 10 by Muyrers et al. for a description of these plasmids).
2. 10% L-arabinose solution. The solution is prepared by dissolving 1 L-arabinose (Sigma) in LB-medium to a total volume of 10 mL. This solution is prepared freshly every time.
3. LB medium.
4. Antibiotics (chloramphenicol, ampicillin, kanamycin, hygromycin, gentamycin, tetracycline). *See* **Subheading 2.1.**
5. Ice-cold 10% glycerol solution. Made by mixing double-distilled water (ddH$_2$O) plus glycerol (Merck) to 10%, and prechilling on ice for at least 4 h (preferably overnight).

2.6. Gel Electrophoresis for Restriction Fragment Fingerprinting

1. Agarose, electrophoresis grade.
2. 1X TAE: 20 m*M* Tris-acetate, pH 8.0, 1 m*M* EDTA.
3. Loading buffer with dye: 10% Ficoll, 0.1% bromophenol blue, 10 m*M* EDTA.
4. DNA size standard (1 Kb DNA ladder from Gibco).

3. Methods

3.1 Assembly of BAC Based Knock-In Constructs for Homologous Recombination in ES Cells

3.1.1. Preliminary Characterization of the BAC

1. Obtain sequence information on the insert/vector boundary. End sequencing with primers reading from the BAC vector into the insert *(10)* will provide sequence information for the two ends of the BAC clone. If this sequence is already present in the database, it will enable to orient the BAC insert with respect to the BAC vector (*see* **Note 1**).
2. Obtain limited amount of **sequence information** on the regions located around 5 Kb, respectively, upstream and downstream, from the two target regions (*see* **Note 2**).

3.1.2. First and Second Round of "BAC Shaving"

1. Choose the boundaries defining the region to be deleted. For the boundaries of the BAC vector (pBeloBAC11), we have chosen the following sequences:
   a. TTCACACAGGAAACAGCTATGACCATGATTACGCCAAGCTATTTAGGT GACACTATAGAATAC and
   b. CTCTGTCGTTTCCTTTCTCTGTTTTTGTCCGTGGAATGAACAATGGAA GTCCGAGCTCATCGCTA
2. For the boundaries lying internally in the BAC insert, choice will depend on the amount of sequence information you have. Avoid repetitive sequences, as they are likely to be found in other spots of the BAC as well.
3. Choose a **selectable marker** to replace the superfluous BAC region. For the two steps of BAC shaving (5′ and 3′ end), it is best to use selectable marker only active in *E. coli*, so as to spare the double selectable markers (active in both prokaryotic and eukaryotic systems) for the mutagenic cassettes. The βlactamase gene, conferring resistance to ampicillin, and the gentamycin acetyltransferase gene, conferring resistance to gentamycin, are both very useful.
4. Design two oligonucleotides composed of a PCR primer portion, annealing to the selectable marker, and an arm containing 50–60 nt of homology to the site in the BAC immediately flanking the region to be deleted. Care should be taken to include in the 5′ oligo, immediately downstream of the homology arm, a restriction site for a rarely occurring enzyme, in order to release the final targeting construct from the modified BAC vector, prior to electroporation into ES cells (*see* **Notes 3** and **4**).
5. Amplify by PCR the selectable marker with the two oligos. PCR conditions should be optimized case by case. Our standard protocol includes: 1 cycle of initial denaturation at 94°C for 5 min, 35 cycles of denaturation at 94°C for 1 min, annealing at 62°C for 1 min, and extension at 72°C for 1 min, followed by a final extension cycle of 10 min at 72°C.
6. Extract the PCR product from the PCR. We use the PCR purification kit from Qiagen and we routinely pool 5 PCR tubes (of 50 µL each) into 1 PCR purification column.
7. Digest the purified PCR product with DpnI (from New Englan Biolabs) 2 h incubation at 37°C are sufficient, but the samples can also be digested overnight. 20 to 40 U/µg of DNA should be used (*see* **Note 5**).
8. Inactivate *Dpn*I by incubation at 80°C for 20 min.
9. Precipitate digested DNA by salt/ethanol precipitation and resuspend in a small volume of bidistilled water (dH₂O). We use a total volume of 10 µL for the DNA pooled from 5 PCR tubes.
10. Check the approximate DNA concentration by gel electrophoresis loading 0.5 µL of the resuspended DNA next to a reference marker. For each electroporation use at least 300 ng DNA.
11. Make competent the cells harbouring your BAC (usually they belong to the strain DH10β or HS996) for either chemical transformation or electroporation. Any protocol you routinely use to prepare competent cells can be followed.

12. Transform your BAC-containing competent cells with the appropriate plasmid expressing the recombinogenic proteins RecE/RecT or Redα/Redβ (*see* **Note 6**) and plate on double selection plates containing chloramphenicol (12.5 μg/m) and the drug to which the "Red/ET plasmid" confers resistance.

13. Pick 6 to 10 colonies, grow miniprep cultures, and check the actual presence of the Red/ET plasmid by restriction digest with an appropriate enzyme.

14. Prepare competent cells for Red/ET recombination from 4 to 6 independent colonies. Follow the protocol described in detail in Chapter 10 by Muyrers et al., **Subheading 3.4.**).

15. Electroporate at least 300 ng of the *Dpn*I digested PCR product generated in **step 10** (up to 2 to 2.5 μL) into each vial of competent cells harbouring the BAC and the plasmid expressing the recombinogenic proteins RecE and RecT, or Redα and Redβ. The conditions for electroporation are as follows: V 2.3, capacitance 10 μF; resistance 600 Ohm. You should obtain a time constant higher than 4.5 ms.

16. Add 1 mL LB medium (without selection drugs) to the cuvet in which the electroporation was performed, pipet once or twice to take up most of the cells and transfer to an Eppendorf tube.

17. After 1 to 1.30′ h shaking incubation at 37°C, plate cells on L-agar plates containing the appropriate selection drugs (*see* **Note 7**). Ten to twenty colonies should be picked and grown overnight in 5 mL L-broth supplemented with the appropriate antibiotics, at the same concentrations as those used for solid culture.

18. Minipreparations of DNA should be made according to the protocol described (*see* **Note 8**). One fourth of the total DNA obtained (5 μL) is normally enough for restriction digest fingerprinting analysis by agarose gel electrophoresis. The aim is to check the restriction pattern of all colonies with **three** different frequently occurring restriction enzymes. EcoRI, BamHI, and HindII can be employed. We recommend from 3 h to overnight incubation at 37°C. (*see* **Notes 9–10**).

19. Run the digests on a big 0.4% agarose gel (0.8 × 20 × 20 cm) at low speed [1.2–1.5 V/cm (24-30 V)] in 1X TAE. The running time is usually 12–18 h. Occasionally, 48–72 h runs can be very informative to distinguish the larger bands.

20. Maxi DNA preparations should be made from two independent colonies for sequence analysis *(10)*. We use routinely the large construct kit from Qiagen. Sequence should be obtained for both sites of integration of the selectable marker into the BAC. Furthermore, the integrity of the rare cutter target site (Pi-SceI or I-*Sce*-I) should also be confirmed, as both oligo depurination and PCR amplification are a potential source of mutations.

21. Choose 4 to 6 independent colonies for the second round of "BAC shaving," all displaying the same restriction pattern, including the two colonies that were sequenced.

22. To perform the second round of "BAC shaving," on the other side of the BAC molecule, repeat **steps 12–18** for each of the 4–6 colonies harboring the modified "first round shaved" BAC (*see* **Note 11**).

3.1.3. First and Second Round of "BAC Targeting"

The end product of the previous two rounds of BAC shaving is the modified BAC backbone ready to be targeted with the desired mutagenic cassettes. Although targeting the mutagenic cassettes does not usually involve deletion of large portions of the BAC, the principle and the methods employed are exactly the same described above for the BAC shaving. Simply repeat **steps 12–19**. In **step 14**, instead of the DpnI digested PCR product, you should electroporate the appropriate mutagenic cassette, which you have previously assembled *(7)* Additional suggestions for the "BAC targeting" procedure are given in **Notes 12–14**.

3.1.4. Preparation of the BAC DNA for ES Cell Electroporation

1. At the end of all targeting steps, choose for ES cell electroporation those colonies that have lost the recombinogenic plasmid. This is usually lost spontaneously upon withdrawal of selection pressure. The easiest way to check it is to digest DNA from the relevant colonies with suitable restriction enzymes, which would produce a known pattern upon cleaving the recombinogenic plasmid.
2. Inoculate at least 2 L of bacterial culture with the relevant colony, in the presence of all drugs for which the BAC contains resistance markers and grow for 16 h.
3. Prepare DNA using commercially available kits (*see* **Note 15**).
4. Quantitize the maxipreparation of DNA by both ultraviolet (UV) spectrophotometry and gel electrophoresis. The yield can vary and is between 10 and 50 µg/500 mL bacterial culture. We usually resuspend the DNA in 100 µL TE, achieving a final concentration of 300–500 ng/µL.
5. Digest DNA with the relevant restriction enzyme to release the targeting construct (insert) from the BAC vector. In the case of homing endonucleases, which are provided at low concentrations (typically 1 U/ µg), we use 2–4 U of enzyme/µg of DNA in a first overnight digestion. On the following day, the same amount of enzyme is again added to the tube and the contents are properly mixed. Digestion is allowed to proceed for an additional day.
6. Check the digestion of the construct by agarose gel electrophoresis. Given the average length of these constructs (> 60 Kb) we suggest running agarose gels as described in **step 17, Subheading 3.1.2.** (*see* **Note 16**).
7. Extract the DNA from the digestion reaction with phenol and chloroform and resuspend in PBS under sterile conditions. You should aim towards a final amount of 80 µg DNA resuspended in 80 µL phosphate-buffered saline (PBS). All extraction steps can be performed on the bench, however, it is best to resuspend DNA under the hood to assure sterility. Store the ready targeting construct at 4°C.
8. Grow ES cells and perform electroporation and selection according to standard protocols *(7)*.

3.2. Subcloning From a BAC the Target Region for Conventional ES Cell-Targeting Constructs

1. Define in the BAC the region that you would like to use as the backbone of your ES cell targeting construct.
2. Define the boundaries of your targeting construct so that your mutagenic cassette will be flanked on both sides by at least 5 Kb of genomic sequence (*see* **Note 17**).
3. Design a convenient Southern blot strategy based on sequence and/or mapping information (*see* **Note 18**).
4. Choose the vector (hence, called acceptor vector) in which you would like to subclone your target region (*see* **Note 19**).
5. Design oligonucleotides containing a PCR primer annealing to the acceptor vector and a homology arm to the regions in the BAC defining the fragment to be subcloned. You should include, between the PCR primer and the homology arm, a restriction site for an enzyme which does not cleave within your subcloned fragment, so as to be able to release the targeting construct from the vector before ES cell electroporation (*see* **Note 20**).
6. Follow **steps 6–15, Subheading 3.1.2.**
7. After 1–1.30′ h shaking incubation at 37°C, plate cells on L-agar plates containing only the drug to which the acceptor vector confers resistance (you do not want to keep selection for the BAC).
8. Ten to twenty colonies should be picked and grown overnight in 5 mL L-broth supplemented with the appropriate antibiotic, at the same concentrations as those used for solid culture.
9. Check DNA minipreparations with a battery of restriction enzymes to confirm the correct identity of your subcloned fragment (*see* **Note 21**).
10. Digest your subclone with the enzyme for which you inserted the target site in the oligos, and check that complete cleavage occurs (*see* **Note 22**).
11. You can now insert in this subclone your desired knock-out or kncok-in cassette by standard Red/ET recombination as described in Chapter 9 by Muyrers et al.

3.3. Red/ET Recombination for the Production of BAC Transgenic Mouse Lines

1. Choose a **selectable marker** to replace the *lox*P site. We use the gentamycin resistance gene (gentamycin acetyltransferase). Any selectable marker active in *E. coli* can be used (*see* also **step 2, Subheading 3.1.2.**).
2. Design two oligonucleotides composed of a PCR primer portion, annealing to the selectable marker, and an arm containing 50–60 nt of homology to the sites in the BAC vector immediately flanking the loxP site (*see* **Notes 23** and **24**).
3. Follow **steps 6–18** described in **Subheading 3.1.2.** (*see* **Note 25**).
4. Choose for oocyte microinjection those colonies which have lost the recombinogenic plasmid (*see* **step 1** in **Subheading 3.1.4.**).
5. Inoculate 500 mL of bacterial culture with the relevant colony, in the presence of all drugs for which the BAC contains resistance markers and grow for 16 h.

6. Prepare DNA using commercially available kits (*see* **Note 14**).
7. Quantitize the maxipreparation of DNA by both UV spectrophotometry and gel electrophoresis. The yield can vary and is between 10 and 50 µg/500 mL of bacterial culture. We usually resuspend the DNA in 100 µL TE, achieving a final concentration of 300–500 ng/µL.

Quantitation for oocyte microinjection is crucial, with the final concentration for injection varying between 0.5 and 2 ng/µL (*see* also Chapter 11 by Vintersten et al). Therefore, UV spectrophotometry should be used only as a starting point for setting an exhaustive series of dilutions for agarose gel quantitation. To this end, run a thin agarose gel (0.6–0.8% w/v) with one or more known standards of DNA amount. We find it best to use 5 ng as the reference standard. As for the BAC DNA, we suggest using 1/10, 1/20, 1/50, 1/100 dilutions, although this will obviously depend on the concentration of the maxipreparation. In any case, for better visualization and determination of the optimal concentration, the dilutions should be loaded next to one another on the same gel and should be enough to produce a full gradient.

4. Notes

1. The starting point is a BAC containing all regions which need to be targeted. This is best achieved by screening a BAC library (we have used the CITB mouse BAC library available from Research Genetics, Inc.) with a probe specific for one of the target regions in the gene of interest. This library utilises the BAC vector pBeloBAC11; more information on the vector and the construction of the library is available at the following web site: http://www.tree.caltech.edu. Commonly, more than one clones will be isolated, which need to be screened with an additional probe for the presence of the second target region. Given the average length of BACs (130 Kb), it should be straightforward, for most genes, to isolate BACs which harbor all target regions of interest.
2. This can be the most time-consuming step of the whole procedure. In the luckiest cases, the particular BAC has been already sequenced or the whole genomic region is already present in the database. If not, the simplest thing to do is to look for exons, which are likely to be located at the desired distance from the target site. Such information should be available from the literature for most genes, even in the absence of a detailed genomic map.
3. Unless one is confident about the absence of conventional eight-cutters target sites (*Not*I, *Pme*I, *Asc*I, *Pac*I) within the BAC insert, we recommend using target sites for the newly available homing endonucleases, like *Pi-Sce*I, I-*Sce*I, and *Ceu*I, that have very large, asymmetric recognition sites. They are available from New England Biolabs.
4. When using particularly long oligos, it is advisable to have them high-performance liquid chromatography (HPLC) purified. Moreover, a relatively

common problem with long oligos is the occurrence of depurination of one or more residues during synthesis. When these altered residues undergo scrutiny by the *E. coli* replication/repair machinery, they can result in deletions of entire segments (up to 20 nucleotides). Therefore, it is worth alerting the oligo company to this problem and demanding that uppermost care be taken to minimize it.

5. Digestion with *Dpn*I is important to eliminate the template plasmid used in the PCR, which could constitute a source of background (*see* also **Note 6** in Chapter 9 by Muyrers et al.).

6. For the plasmids that can be used to express the Red/ET recombination proteins, please see the chapter by Muyrers et al. (*see* **Note 8**) and references therein. As for the resistance marker of the Red/ET plasmid, we use Tetracycline resistance, in order to keep more commonly used markers, like ampicillin, kanamycin, hygromycin, and gentamycin for the actual BAC targeting steps.

7. Chloramphenicol, at a concentration of 12.5 µg/mL, should always be included to select for the presence of the BAC. As BACs are present at one copy/cell, upon integration of new resistance marker genes in the BAC, selective pressure should be adjusted accordingly. The following concentrations should be used: ampicillin (50 µg/mL); gentamycin (3 µg/mL); kanamycin (20 µg/mL); hygromycin (50 µg/mL). In individual cases, it might be necessary to increase or lower these concentration, for example if there is cross-resistance between two markers. Selection for the plasmid expressing the recombinogenic proteins should not be exerted at this stage.

8. You can refer to the protocol described in **Subheadings 3.3.** and **3.6.** of Chapter 9.

9. A fast alternative to screen putative recombinants is colony PCR. For this, two PCR primers should be designed, the first which anneals in the selectable marker cassette introduced by Red/ET recombination, and the second which anneals on the original BAC molecule outside of the site chosen for homologous recombination, so that a product can only be generated if integration has occurred correctly. Design your primers so as to have relatively small products (up to 1 Kb). PCR conditions must be determined case by case. This strategy can be used either directly on colonies, or on minicultures or on miniprep DNAs. When trying colony PCR, after aliquoting the PCR mix, we found it best to pick the colony and deliver it first into the PCR vial, and then into the tube containing LB medium, which will be used for overnight culture.

   Although this is a fast and reliable method to identify correct recombinants, positive colonies should still be screened by restriction digest fingerprinting, since the occurrence of correct integration does not exclude the possibility of undesired intramolecular rearrangements in other regions of the BAC molecule.

10. It is very important to run, alongside the DNA from the candidate colonies, a digest of the wild-type BAC DNA as a reference standard. This will make it easy to visualize which bands have "disappeared" during the deletion reaction. In most cases, only one restriction pattern will be visible for all colonies with all enzymes tested, demonstrating that no undesired intramolecular rearrangements occurred

during the recombination reaction. If more than one pattern emerges, there will be a predominant one, most likely representing the correct product, which can be further analyzed with an additional set of enzymes. Patterns occurring in only one or few colonies can be safely discarded as the results of undesired rearrangements.

11. Throughout the procedure, we keep selection for all selectable markers currently present on the BAC. However, during the preparation of competent cells for Red/ET recombination, it is **very important** to remove selection with any drug inhibiting protein synthesis, in order not to interfere with the production of the recombinogenic proteins. For example, if the **gentamycin** resistance gene has already been targeted to the BAC, in this step, **gentamycin** selection can and should be removed, and be applied back later, after picking the colonies from the subsequent round of Red/ET recombination.

12. As soon as a new recombination step is achieved, cells should be grown in the presence of all drugs for which they now contain a resistance gene. We were able to grow cells with the appropriate concentrations of all five drugs mentioned above (chloramphenicol, ampicillin, gentamycin, hygromycin, and kanamycin).

13. When performing restriction digest footprinting for every new step of homologous recombination, it is advisable to run alongside both a digest of the wild type BAC and the BAC originating from the previous targeting step as a reference standard.

14. Upon electroporation into ES cells, the possibility exists that such a large construct breaks and the two mutagenic cassettes are integrated independently of each other. Therefore, detection of the homologously recombined fragments at both the 5′ and 3′ ends of the targeting construct does not indicate whether the two cassettes have been integrated on the same allele. A strightforward way to address the problem is to include, in each of the two mutagenic cassettes, a restriction site which is absent from the wild type allele of the targeted gene. Southern blotting can then be performed on a pulsed-field electrophoresis (PFGE) gel, and the expected band will be observed only if the two cassettes are on the same allele. Thus, care should be taken to include the appropriate restriction site (either an eight-cutter or a homing endonuclease) while assemblying the mutagenic cassettes.

15. For BAC DNA preparation, we use the large-construct kit from Qiagen and every 500 mL of culture are processed separately. The protocol supplied should be strictly followed. While many other purification methods are likely to be equally effective, the large-construct kit from Qiagen includes an additional step, in which exonuclease I is used to completely eliminate fragments of genomic DNA and nicked BAC molecules. The resulting preparation is therefore highly enriched in supercoiled BAC DNA.

16. The hallmark of efficient *Pi-SceI* digestion is the appearance of two bands of the BAC: the very large insert (targeting construct) and the BAC vector. It should be noted that the BAC vector will no longer be 7.2 Kb long, but around 10 Kb for the presence of the two selectable markers which have been added during the two steps of "BAC shaving."

17. You need limited amount of sequence information around the sites which will become the ends of your targeting construct. If your BAC has not been sequenced, and you rely on exon sequence present in the literature (*see* also **Note 2**), we suggest that you confirm this sequence in your BAC before designing oligos, since even single base mismatches (a frequent occurrence in most databases) can substantially lower the efficiency of Red/ET recombination.

18. When possible, it is best to choose relatively common restriction enzyme for Southern strategies because they reliably digest the often dirty preparations of DNA from ES cells and mouse tails (especially in large-scale experiments). After restriction enzyme digestion, the homologously recombined fragment should be readily distinguished from the wild-type one. It is advisable that the correctly recombined band be smaller than the wild-type one, so as to avoid partial digestions (which produce larger fragments) as a potential source of false-positive clones. It is also best to avoid enzymes which produce very large fragments (15 Kb or longer), as the blotting efficiency is reduced. We found it particularly convenient to develop a Southern strategy in which the same enzyme could be used to screen both the 5′ and the 3′ side. In this way, the same blot can be sequentially hybridized, not a minor advantage given the limiting amounts of DNA in ES targeting experiments.

19. We suggest using a middle-copy vector, like for example pACYC 177 or pACYC 184. However, unless your fragment is particularly large, a high copy plasmid could serve the purpose just as well.

20. Usually 8-cutters are best suited for this (*Not*I, *Pac*I, *Pme*I, *Asc*I).

21. The most significant source of background in this experiment results from circularization of the PCR product (which then behaves as an autonomously replicating plasmid). To keep it at the lowest levels, the two oligos should not have any shared sequences near their ends, which could promote recombination and circularization (**4**).

    Anyway, it is always straightforward to distinguish by electrophoresis the circularized vector from the correct subclone.

22. If long oligos have been used, the restriction site might have been mutated as a result of the depurination problem discussed in **Note 4**.

23. If using the genatmycin resistance gene, these two oligos can be used:
    GentF: 5′**GCCCCGACACCCGCCAACACCCGCTGACGCGAACCCCTT
    GCGGCCGCATCGAAT***TGAAGGCACGAACCCAGTTGACATAAGCC* 3′
    Residues 1–53 (in bold) constitute the 5′ arm of homology to the BAC backbone, immediately upstream of the loxP site. The portion of the oligo which anneals to the gentamycin resistance gene includes residues 54–83 *(italics)*.
    GentR:
    **5′CTCTGTCGTTTCCTTTCTCTGTTTTTGTCCGTGGAATGAACAATG-
    GAAGTCCGAGCTCATCGCTA***TCGGCTTGAACGAATTGTTAGGTGGC* 3′
    Residues 1–65 constitute the homology arm to the BAC backbone downstream of the *lox*P site (in **bold**). The portion of this oligo annealing to the gentamycin resistance gene comprises residues 66–90 *(italics)*.

24. At this stage, you should consider to include in the 5′ oligo, immediately downstream of the homology arm, a restriction site for a rarely occurring enzyme, in order to linearise the modified BAC prior to oocyte microinjection, in case you may wish to do so. This is better than relying on the *Not*I sites already present in the BAC vector, which are most commonly used for this purpose, because some BACs are likely to have an additional *Not*I site in the insert. The safest strategy is to include a target site for the newly available intein-coded exonuclease, like *Pi-Sce*I, *I-Sce*I, and *Ceu*I.
25. When using gentamycin, we have noticed that on average only about half of the colonies picked do actually grow in liquid culture. This is not a major limit, as most of the colonies which grow are correctly recombined.

References

1. Giraldo, P. and Montoliu, L. (2001) Size matters: use of YACs, BACs and PACs in transgenic animals. *Transg. Res.* **10,** 83–103.
2. Zhang, Y., et al. (1998) A new logic for DNA engineering using recombination in Escherichia coli. *Nat. Genet.* **20,** 123–128.
3. Muyrers, J. P., et al. (1999) Rapid modification of bacterial artificial chromosomes by ET-recombination. *Nucl. Acids Res.* **27,** 1555–1557.
4. Zhang, Y., et al. (2000) DNA cloning by homologous recombination in Escherichia coli. *Nat. Biotechnol.* **18,** 1314–1317.
5. Muyrers, J. P., Zhang, Y., and Stewart, A. F. (2001) Techniques: Recombinogenic engineering—new options for cloning and manipulating DNA. *Trends Biochem. Sci.* **26,** 325–331.
6. Testa, G., Zhang, Y., Vintersten, K., et al. (2003) Engineering the mouse genome with bacterial artificial chromosomes to create multipurpose alleles. *Nat. Biotechnol.* **21,** 443–447.
7. Angrand, P. O., et al. (1999) Simplified generation of targeting constructs using ET recombination. *Nucl. Acids Res.* **27,** e16.
8. Joyner, A. L. ed. *Gene Targeting: A Pratical Approach.* second edition, 1999, The Practical Approach Series, Oxford Univ. Press.
9. Shizuya, H., et al. (1992) Cloning and stable maintenace of 300-kilobase-pair fragments of human DNA in Escherichia coli using an F-factor-based vector. *Proc. Natl. Acad. Sci. USA* **89,** 8794–8797.
10. Benes, V., Kilger, Ch., Paabo, S., and Ansorge, W. (1997) Direct primer walking on P1 plasmid DNA. *Biotech.* **23,** 98–100.

11

Microinjection of BAC DNA into the Pronuclei of Fertilized Mouse Oocytes

Kristina Vintersten, Giuseppe Testa, and A. Francis Stewart

1. Introduction

Microinjection of DNA into the pronuclei of fertilized oocytes is one of the two most commonly used methods for gene transfer into the mouse genome *(1,2)*. The first successful attempt to perform this technique was carried out by Lin in 1966 *(3)*, who could show that the early fertilized embryo could survive the mechanical damage of inserting a glass needle into the pronucleus. However, it was not until 1981 that small DNA fragments were integrated into the genome *(4)*. This technique is well described, and has now become a standard procedure *(5,6)*. More recently, the use of larger DNA fragments has been established. Yeast artificial chromosome (YAC) *(7,8)*, P1 artificial chromosome (PAC) *(9,10)*, and bacterial artificial chromosome (BAC) DNA *(11,12)* can all be used for the generation of transgenic mice. A comprehensive review of BAC and YAC transgenesis is given by Giraldo and Montoliu *(13)* extensively comparing these applications. Although the basic technique for microinjection of large constructs are similar to those used for shorter DNA segments, there are some special requirements *(14)*. In this chapter, we describe relevant steps for microinjection using large DNA constructs.

It is important to keep in mind that, as with so far in any kind of pronuclear DNA injection, the integration takes place in a random manner, hence, the integration site and even the copy number of the integrated transgene is not possible to predict. All the resulting founder animals will all carry a unique integration pattern. In most cases, the integration will take place at only one site; however, occasionally separate integration sites on two different chromo-

From: *Methods in Molecular Biology, vol. 256:*
Bacterial Artificial Chromosomes, Volume 2: Functional Studies
Edited by: S. Zhao and M. Stodolsky © Humana Press Inc., Totowa, NJ

somes can occur *(15)*. When these founder animals are mated for transmission of the construct through the germline, segregating integration patterns may be observed among the offspring. The integration will generally take place shortly after microinjection, before the DNA replication in the one cell stage embryo. In some cases though, the integration is delayed, resulting in a mosaic transgenic embryo/animal *(16)*. Occasionally mosaicism could interfere with the germline transmission of the transgene.

The BAC construct can be injected in either one of three forms: supercoiled, linearized and as a purified insert released from the BAC vector *(13)*. We have used supercoiled BAC DNA. Beyond being the fastest method, this presents the advantage that the DNA, being more compacted in its supercoiled form, may be less prone to shearing and fragmentation during the injection procedure. Furthermore, suitable restriction sites for cleavage within the BAC vector are not required. Finally, solutions of linearized BAC DNA tend to be more viscous, and, hence, more difficult to inject, than when the BAC is in its supercoiled form.

The main drawback of using intact (supercoiled) BACs is that integration of the BAC transgene will occur upon random linearisation, thus possibly interrupting the transgene in a deleterious manner. On the contrary, injection of a linearized transgene is more predictable for reliable integration of the intact transgene. A variation of this approach has been the use of linearized BAC inserts released from the vector and purified by pulsed-field gel electrophoresis (PFGE). This is admittedly the cleanest experimental avenue; however, it is fairly laborious and apparently unnecessary, because, as it has been shown with YACs, the presence of vector sequences within the integrated transgene does not seem to affect expression *(17)*.

If one wishes to linearise the BAC in the absence of suitable restriction sites, ET recombination *(18)* represents a fast and convenient way to precisely introduce a unique site at any desired place in the BAC. This technique, which can also serve to delete the *lox*P site present in the BAC vector, is described in Chapter 9 by Testa et al.

The handling and general conditions surrounding both embryo donor and recipient mice are of major importance for the successful outcome of microinjection experiments. For further reading about optimal husbandry conditions, the reader is referred to Foster et al. *(19)*.

Handling and culture of embryos, the microinjection process, and the final transfer of the embryos back in vivo, all require experience, and optimal culture conditions *(6,20)*. The microinjection process requires expensive, specialized equipment. Therefore, these experiments should not be undertaken without careful consideration of required recourses (*see* **Note 1**).

1.1. Production of Embryos for Microinjection

The number of animals needed per experiment will depend on several factors: age and quality of the embryo donors and stud males, physical parameters in the animal facility and the microinjection lab, embryo culture conditions, DNA purity and concentration, and the manual skills of the operator.

Generally, 25–30% of the injected oocytes will immediately lyse owing to the mechanical damage. On average, about 30% of the surviving embryos will develop to term, and only 15–20% of those will be transgenics. Therefore, if all conditions are met, it is usually sufficient to inject 150 fertilized oocytes in order to produce 3–6 founder animals.

The choice of genetic background of the host embryo should be considered carefully. In most cases, a cross between two F1 animals, such as (C57BL/6J × CBA) F1, (C57BL/6J × SJL) F1, or (C57BL/6J × DBA/2) F1 are used, because these combinations have proven to provide large numbers of good quality embryos. In these cases, however, the genetic background of the founder will be mixed (*see* **Note 2**). It has been shown that transgene expression can be modulated or suppressed by the genetic background of the donor embryo *(21,22)*.

Embryo donor females should be superovulated in order to increase the number of embryos obtained per mouse (*see* **Subheadings 3.1.** and **3.2.**). The exact timing of the developmental stage of the oocytes is of major importance, and can only be achieved by careful adjustment of the hormone treatment timing, light cycle in the animal facility, and time of embryo collection (*see* **Note 3**).

1.2. Recovery, Handling, and In Vitro Culture of Preimplantation Stage Embryos

Great care should be taken to assure the best possible in vitro culture conditions for the embryos. A humidified 5% CO_2 incubator, 37°C is used for the embryo culture. This incubator should be reserved for embryos only, and not shared with other tissue culture. The most commonly used culture medium in 5% CO_2 is M16, and the HEPES buffered equivalent for use at the bench is M2 *(23)*. These media can be purchased from commercial suppliers (Sigma, Specialty Media), or prepared according to published protocols *(5)* (*see* **Note 4**).

The embryos are easiest handled (collected, moved, washed, sorted, and manipulated) by careful mouthpipeting, using a special devise *(24)* (*see* **Subheading 3.3.**) (*see* **Note 5** and **Fig. 1**).

Mice tend to ovulate and mate at the midpoint of the dark period (night). The oocytes will soon thereafter reach the swollen ampulla region of the oviduct,

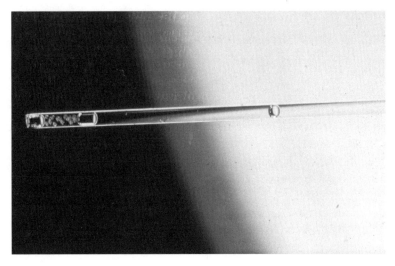

Fig. 1. Glass capillary for embryo handling, loaded with air bubbles and embyos.

where fertilization takes place. In the following morning (12 h after fertilization, at embryonic stage E0.5), the ampulla is greatly enlarged, and can easily be located among the oviduct coils, which makes the recovery of the embryos fairly easy (*see* **Subheading 3.4.** and **Note 7**).

The embryos are tightly packed together and surrounded by a large amount of cumulus cells at the time of recovery. Cumulus cell masses must be removed from the embryos before injection can take place (**Subheadings 3.5.** and **3.6.**) (*see* **Note 8**).

1.3. Microinjection Equipment

An inverted microscope with either Differential Interference Contrast (DIC) or Hoffman optics are necessary in order to locate the pronuclei. A low magnification lens (×2.5 or ×5) is used to get an overview of the injection, and a high magnification lens (×32 or ×40) is used for the actual injection process. Two micromanipulators are needed for the movement of the holding and injection capillaries. Commercially available micromanipulators such as Leitz, Eppendorf, or Narishige are all suitable, and the choice of brand is a matter of personal preference. However, we prefer the Eppendorf TransferMan NK because it is an electronic system with which set positions can be preprogrammed. This feature greatly enhances the efficiency and speed of injection. The control of the holding capillary (capture and release of the embryos) can be achieved by using a micrometer screw controlled, oil-filled glass syringe, or a commercially available control unit (Eppendorf CellTram air), or by simple mouthpipeting. The flow of DNA in the injection needle is best controlled by an injector (Eppendorf FemtoJet), but can also be performed (by experienced experimentators) with the use of a glass syringe. In all the above cases, an absolutely airtight connection has to be established between the control device and capillary. Air-filled thin polythene tubing should be used between electronic injectors and injection needle. If injection is performed by hand, a thick hard silicon tubing (Tygon R3603) is the best alternative. For the connection of the holding capillary, hard silicon (Tygon R3603) or thin polythene tubing can be used, either oil- or air-filled. Oil-filled tubings provide a finer (slower) control, but have to be free of any airbubbles to work well. Air-filled tubings provide faster movements, which can be compensated by filling oil only in the holding capillary itself.

The injection takes place in an injection chamber: a small drop of M2 medium covered by embryo tested light paraffin oil. We recommend to use a simple aluminium frame attached to a clean glass slide with high vacuum grease (Dow Corning) to hold the media drop and oil in place. It is also possible to use a glass slide with compression well, or even plastic tissue culture dishes. Microinjection needles can be obtained from commercial suppliers (Eppendorf, FemtoTipII), or prepared on a needle puller such as the model P97 from Sutter Instruments (*see* **Note 10**). A comprehensive reference about glass capillaries and needle pullers has been published by Flaming and Brown *(25)*.

1.4. Dilution of BAC DNA Solution

The DNA should be diluted in BAC injection buffer (*see* **Subheading 3.8.**), which aids to stabilize BAC DNA *(26)*. It is very important to prepare an ultrapure buffer, completely free of any particles, solvents, detergents, and so on. Even the smallest debris will inevitably clog the microinjection needle and any other substances then DNA or buffer may be toxic to the embryos. The final concentration should be kept as close as possible to 1.5 ng/µL. It has been shown that integration rate increases, but embryo viability decreases when the DNA concentration is raised *(6)*. It is to some extent possible to increase the rate of single-copy integrations by decreasing the DNA concentration *(5)*. However, overall efficiency will be reduced.

1.5. Microinjection Process

Groups of embryos are moved to the microinjection chamber, not more then can easily be injected within 20 min. The embryos are picked up one by one with the holding needle, and orientated so that the two pronuclei are readily visible.

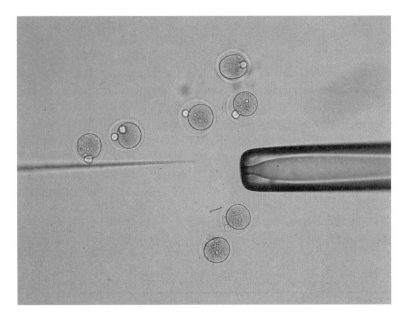

Fig. 2. Microinjection chamber. Positioning of holding capillary to the left, embryos in the middle, and injection needle to the right in the microinjection chamber.

The injection needle is inserted through the cell membrane and cytoplasm into the target pronucleus. DNA is injected until a clear swelling of the pronucleus can be observed (*see* **Notes 11–16**) (*see* **Figs. 2** and **3**).

1.6. Transfer of Injected Embryos to Pseudopregnant Female Mice

Most female mice have an ovulation cycle of 3–5 d, but in order to get into the hormonal status of pregnancy, there is a need for physical mating to take place. Female mice, which are to be used as recipients for transferred embryos, are therefore mated to sterile (vasectomized) male mice. By detection of a copulation plug in the following morning, pseudopregnant females can be selected for transfer surgery. The vasectomy of male mice *(5)* is a simple surgical procedure, where an incision is made in the scrotal sac or abdominal wall, and an approx 10-mm long piece of the vas deferens is removed by cauterisation. Males can be used for mating two weeks after surgery.

Injected embryos are placed in the reproductive tract of pseudopregnant recipient females *(27)*. Anesthesia is induced by an intraperitoneal (ip) injection of a suitable anaesthetic solution (*see* **Subheading 3.10.**). The oviduct is exposed by a surgical procedure (*see* **Subheading 3.11.**) and the embryos placed in the ampulla region, through the infundibulum(*see* **Notes 17** and **19**).

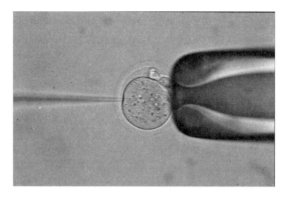

Fig. 3. Injection into the pronucleus. Holding capillary can be seen to the right. The embryo is captured in the middle, with the injection needle inserted into one of the pronuclei.

2. Materials

2.1. Preparation of Hormones for Superovulation

1. PMSG, Pregnant Mares Serum Gonadotropin (Intervet, Intergonan).
2. HCG, Human Corionic Gonadotropin (Sigma CG-10).
3. Sterile distilled H_2O, chilled to 4°C.

2.2. Superovulation of Young Female Mice

1. PMSG, ready made solution 0.05 IU per µL.
2. HCG, ready made solution 0.05 IU per µL.
3. Disposable 1-mL syringe and 27- or 30-G needle.

2.3. Preparation of Transfer Capillaries

1. 50-µL glass capillaries (Brand).
2. Bunsen burner.
3. Capilette (Selzer, Germany, Capilette pipetierhilfe).
4. Flat mouthpieces (HPI Hospital products, Altamore Springs, USA 1501).
5. Flexible silicon tubing with an inner diameter of 3 mm.

2.4. Dissection of Oviducts

1. Anatomical forceps.
2. Fine straight scissors.
3. Fine straight forceps.
4. 3-cm tissue-culture dish.
5. M2 culture medium.

2.5. Preparation of Hyaluronidase Solution

1. Hyaluronidase type IV from bovine testis (Sigma H-4272).
2. M2 embryo culture medium.

2.6. Recovery and In Vitro Culture of Fertilized Cocytes

1. 3.5-cm Falcon tissue culture dishes (35-3001).
2. M2 culture medium.
3. M16 culture medium.
4. Embryo tested light paraffin oil (Merck, Sigma).
5. Hyaluronidase solution (*see* **Subheading 2.5.**).
6. Watchmakers forceps.
7. Disposable 1-mL syringe with 30-G needle.

2.7. Preparation of Holding Capillary and Injection Needle

1. Borocilicate glass capillary (Leica 520119) for holding capillary.
2. Bunsen burner.
3. Diamond point pen.
4. Microforge (Bachofer, Alcatel, Narishige), alternatively Bunsen burner.
5. Borocilicate glass capillary with inner filament (Clark Electronic Instruments, GC100TF) for injection needle.
6. Horizontal glass capillary puller (Sutter Instruments, P97).
7. Alternatively, purchased needles/capillaries can be used (Eppendorf VacuTips and FemtotipsII).

2.8. Preparation of Injection Buffer and Dilution of BAC DNA

1. Spermine (Sigma, tetrahydrochloride S-1141).
2. Spermidine (Sigma, trihydrochloride S-2501).
3. 1 M Tris-HCl, pH 7.5, autoclaved.
4. 0.5 M EDTA, pH 8.0, autoclaved.
5. 5 M NaCl, autoclaved.
6. Sterile ddH$_2$O.

2.9. Microinjection

1. Inverted microscope as described.
2. Two micromanipulators.
3. Holding capillary (*see* **Subheading 2.7.**).
4. Injection needle (*see* **Subheading 2.7.**).
5. Microloader (Eppendorf Microloader)
6. M2 culture medium.
7. Fertilized oocytes.
8. DNA solution, 1–2 ng per µL.
9. Injector (Eppendorf FemtoJet), alternatively, 10-mL glass syringe (Becton Dickinson, 10CC, 2590,6458).

2.10. Preparation of Ketamine/Xylazine Anesthetics

1. Ketamine hydroclorid (Ketanest, Park Davis).
2. Xylazine 2% (Rompune, Roche).
3. Sterile ddH$_2$O.

2.11. Embryo Transfer to the Oviduct of Pseudopregnant Female Mice

1. Ketamine/Xylazine (*see* **Subheading 2.10.**).
2. Sterile PBS.
3. Transfer capillary.
4. M2 medium.
5. Embryos.
6. 70% EtOH.
7. Paper tissues.
8. Sterile blunt forceps.
9. Two sterile watchmakers forceps.
10. Sterile fine scissors.
11. Sterile seraffin clip.
12. Wound clip with applicator (Clay Adams) or suture needles with ligature.
13. Pseudopregnant female mouse 8–12 wk old, E0.5.
14. Stereo microscope with cold light lamp.

3. Methods

3.1. Preparation of Hormones for Superovulation

1. Thaw the hormones by adding prechilled sterile H$_2$O directly to the frozen vial (*see* **Note 6**).
2. Mix carefully and dilute the hormones to a final concentration of 50 IU per mL.
3. Aliquot and freeze immediately at –20°C until use.

3.2. Superovulation of Young Female Mice

3.2.1. Day 1, 11 AM (see **Note 6**)

1. Thaw an appropriate amount of PMSG immediately prior to use—do not keep the solution at room temperature for longer then maximum 30 min.
2. Inject each female mouse intra peritonealy with 100 µL (5IU) PMSG solution.

3.2.2. Day 3, 10 AM (see **Note 6**)

1. Thaw an appropriate amount of HCG as described above.
2. Inject each female mouse intra-peritonealy with 100 µL (5IU) HCG solution.
3. Mate the superovulated females with stud males, and check for copulation plugs the following morning.

3.3. Preparation of Transfer Capillaries

1. Hold a glass capillary in the flame of a Bunsen burner until the glass starts to melt.
2. Withdraw the capillary from the flame, and immediately apply a sharp pull to the ends. The faster the pull the thinner will the diameter of the capillary be.
3. Break the thin part of the capillary approx 5 cm from the thicker shaft.
4. Using a diamond pen, make a mark on the capillary straight across the thin end, 2–3 cm from the thickening shaft.
5. Break the thin end at the mark.
6. Hold the end of the capillary in the flame of the Bunsen burner until it is no longer sharp (*see* **Note 5**).
7. Capillaries that will be used for embryo transfer should be sterilized by baking in an oven at 180°C for 3 h.

3.4. Dissection of Oviducts

1. Sacrifice the plug positive female mice by cervical dislocation or CO_2 gas (*see* **Note 7**).
2. Wet the abdominal wall with 70% EtOH.
3. Wipe off excess EtOH with a paper tissue.
4. Make a cut through the skin across the midline at the lower part of the abdomen with scissors.
5. Tear the skin off the abdomen by grasping the two skin flaps above and below the cut, and simply pulling them apart.
6. Using the fine scissors and forceps cut open the abdominal wall until all the organs are visible.
7. Move away the intestines by scooping them aside with the forceps.
8. Grasp the top part of the uterus with the forceps.
9. Carefully tear away the connective tissue on the ovary and oviduct with the scissors.
10. Make a cut between the ovary and the oviduct coils, and a second cut across the top part of the uterus horn (*see* **Note 7**).
11. Place the oviducts in a tissue culture dish with M2 medium.

3.5. Preparation of Hyaluronidase Solution

1. Dissolve the hyaluronidase in M2 culture medium obtaining a stock solution of 10 mg/mL.
2. Filter sterilize, aliquot, and store at −20°C (the stock solution is stable for several months).

3.6. Recovery of Fertilized Oocytes From the Oviducts

1. Place the oviducts in a tissue culture dish with 2 mL of fresh M2 medium.
2. Grasp an oviduct at a time with watchmakers forceps, and locate the swollen ampulla (*see* **Note 8**).

3. Rip up the ampulla with the tip of the needle, and make sure that the oocytes surrounded by the cumulus masses swim out in the medium.
4. Thaw the hyaluronidase solution, and add to the embryo containing M2 medium to a final concentration of 300 μg per mL.
5. Incubate at room temperature for 2–3 min.
6. Collect the fertilized embryos using a transferpipet (*see* **Note 8**).
7. Move the embryos through 3 drops of M2 medium.
8. Repeat **step 7**, but using M16 medium, and place the embryos in the CO_2 incubator.

3.7. Preparation of Holding Capillary

1. Follow **steps 1–4** in **Subheading 3.3.**, but use borosilicate glass capillaries. Pull the glass relatively thin, and absolutely straight.
2. Measure the dimensions of the capillary at the cut end: the outer diameter should be 100 μm.
3. Melt the tip of the capillary on the microforge (or alternatively in the flame of the Bunsen burner) until only a small 15-μm wide opening remains.
4. Pull injection-needles on a microcapillary puller. The shaft should be long and very thin, with an inner diameter (id) of 0.5 μm at the tip (*see* **Note 10**).
5. Alternatively, use purchased needles and capillaries (Eppendorf VacuTips and FemtotipsII).

3.8. Preparation of Injection Buffer and Dilution of BAC DNA

1. 100 mM polyamine mix (1000X stock solution) (*see* **Note 9**): Dissolve the Spermine and Spermidine together in sterile ddH_2O so that the end concentration is 30 mM Spermine and 70 mM Spermidine. Filter sterilize (0.22 μm) aliquot and store at –20°C.
2. Basic injection buffer: Mix the following in a plastic disposable 50-mL Falcon tube:
   a. Add 0.5 mL of 1 M Tris-HCl, pH 7.5.
   b. 10 μL of 0.5 M EDTA, pH 8.0.
   c. 1 mL of 5 M NaCl.
   d. Add sterile H_2O up to 50 mL.
   e. Aliquot, filter sterilize (0.22 μm) and store at 4°C.
3. Ready to use injection buffer: 50 mL basic injection buffer, add 50 μL polyamine mix. Use directly, do not store.

3.9. Microinjection

1. Fill approx 5 μL of the BAC DNA solution into the injection capillary using a microloader.
2. Attach the injection needle to the right hand side micromanipulator handle, and position it correctly in the media drop in the microinjection chamber (*see* **Fig. 2**).
3. Attach the holding capillary to the other micromanipulator handle, and position it in the media drop.

4. Move a group of 10–15 embryos from the incubator into the media drop.
5. Open up the tip of the injection needle by carefully tapping it on the holding needle (*see* **Note 11**). It is important to break up the tip to a size where the DNA readily flows without that too high pressure has to be applied. However, the size should not be so large that it damages the embryos.
6. Clear the needle tip by flushing a small amount of DNA at high pressure. Set a constant flow of DNA to a low level (*see* **Note 12**).
7. Pick up one embryo with the holding capillary, and position it so that both pronuclei can be seen at the midplane of the embryo.
8. Set the focus on the nearest pronucleus, making sure that the border can be seen clearly (*see* **Note 13**).
9. Move the injection needle to the same y-axis position as the targeted pronucleus (either 6 o'clock or 12 o'clock of the embryo) and adjust the height of the needle so that the tip of the needle appears completely sharp (without changing the focus!). This procedure will allow for the needle to exactly target the pronucleus.
10. Move the injection needle to a 3 o'clock position.
11. Insert the injection needle straight into the embryo, pushing through the zona pellucida, cell wall, cytoplasm, and into the targeted pronucleus (*see* **Notes 14** and **15** and *see* **Fig. 3**).
12. Apply a higher pressure to the DNA flow, and keep injecting until a clear increase in the size of the pronucleus can be seen.
13. Withdraw the injection needle, and release the embryo.
14. Repeat with the remaining embryos, and move them back into the incubator, placed in M16 media (*see* **Note 16**).

3.10. Preparation of Ketamine/Xylazine Anesthetics

1. Dissolve 100 mg Ketamin hydrochloride and 0.8 mL Xylazine 2% in 11 mL sterile ddH$_2$O.
2. Mix and store at 4°C for up to 2 wk.

3.11. Embryo Transfer to the Oviduct of Pseudopregnant Female Mice

1. Anesthetize the mouse by injecting 10 µL Ketamine/Xylazine solution per gram bodyweight (*see* **Note 18**).
2. Load the transfer capillary with M2 medium, adding two large air bubbles for better control of the movement. Add one small air bubble, pick up the embryos, and add one more small air bubble. The embryos should now be enclosed in a small amount of medium, close to the tip of the capillary, and surrounded by two small air bubbles (*see* **Note 19**).
3. Wait until the mouse has reached surgical anaesthesia. This can be checked by pinching the tail or a hind leg. The mouse should not react to these stimuli.
4. Moisten the eyes of the mouse with a drop of PBS (*see* **Note 20**).
5. Place the mouse on its right side, with the legs toward the operator.
6. Remove the fur around the area just behind the last rib.

7. Wipe with 70% EtOH.
8. Make a 12-mm incision in the skin right under the muscles surrounding the vertebra, and immediately behind the last rib. A blood vessel and a nerve should be seen running in the abdominal wall, vertically across this area.
9. Make a 10-mm incision in the abdominal wall. Locate the fat pad between the ovary and the kidney.
10. Lift out the fat pad, and secure it with a seraffin clip outside the body wall. Take care not to touch the ovary or oviduct coils!
11. Place the mouse under a stereo microscope, and focus on the oviduct area.
12. Tear the bursa surrounding the ovary and oviduct coils with the two pairs of watchmaker forceps.
13. Locate the infundibulum. If bleeding is obscuring the view, place a very small piece of paper tissue in the cleft between ovary and oviduct coils.
14. Insert the transfer capillary into the infundibulum, and expel the embryos. Check that the two small airbubbles have entered into and behind the first turn of the coils (the air bubbles can easily be seen through the wall of the oviduct).
15. Remove the seraffine clip, and place the organs back into the abdominal cavity.
16. Seal the skin incision with a wound clip.
17. Repeat the procedure on the other side.
18. Place the mouse in a quiet, dark, and warm area until it recovers (*see* **Notes 21** and **22**).

4. Notes

1. The reader is strongly recommended to gather further information and experience with mouse embryo micromanipulation before attempting to perform the actual microinjection. Useful protocols, advice, and consideration are available at the following Websites: Thom Saunders, http://www.med.umich.edu/tamc/BACDNA.html, and Dr. Lluis Montoliu, http://www.cnb.uam.es/~montoliu/prot.html and in **ref. 5**.
2. We have used embryos obtained by crossing CD1 outbred females with C57 inbred males, and this combination has worked very efficiently in our hands. The number of embryos that can be obtained per donor female in these crosses is usually high, the embryo viability is good, the embryos are easy to inject and survive the microinjection process well. If an inbred background is desirable for the transgene, it is possible to use for example C57BL/6J embryos, although the microinjection process will be more difficult, and the viability of the embryos is lower.
3. The highest numbers are usually obtained from very young female mice, just at weaning age. However, the embryo quality is often variable and low, and very small females may suffer significantly during the stressful experience of being mated by a large and often less careful male. For this reason, we recommend using 5- to 6-wk-old females. Both the time-point at which each of the hormone injections are given, and the light cycle in the room where the animals are kept will have influence on the ovulation. The optimal time during which microinjection can take place is as short as 2–4 h. If the embryo development is not advanced enough, the pronuclei will be very small and difficult to target. In this case, the

hormones should be given earlier in the morning, and/or the light cycle should be set back. On the other hand, if the embryos are at too late a stage of development, the pronuclei will already be fused, and the first division take place. The solution here is to give the hormones later, and/or set the light cycle forward. We use a light cycle of 14 h light/10 h dark, with the light switching on at 5:00 AM.

4. Each batch of oil, and all ingredients in homemade embryo culture media should be tested for supporting optimal embryo development prior to use. M16 should be preequilibrated in 5% CO_2 incubator for minimum 2 h before use.

5. Care should be taken to use glass capillaries with a smooth end, as sharp edges would harm the zona pellucida. It is also important to minimize the time, which the embryos spend outside the incubator as much as possible. One-cell stage embryos should not spend more then 30 min in room air and temperature

6. It is essential to prepare the hormones as soon as possible after thawing, to work fast, and to freeze the solution as soon as it is prepared. Repeated freeze-thawing should be avoided. Hormones are stable in –20°C for at least 3 mo.

7. Care should be taken not to damage the ampulla during dissection, as this would inevitably result in loosing the embryos. The choice of method to sacrifice the donor mice (cervical dislocation or CO_2 gas) does not have any influence on embryo viability, and is, therefore, a matter of personal preference and local legislation.

8. The cumulus masses can easily be seen through the wall of the swollen ampulla. By grasping the oviduct with watchmaker's forceps, and tearing the wall of the ampulla with a sharp 27-G needle, the embryos are allowed to swim out in the M2 media filled tissue-culture dish. It is essential to remove the embryos from the hyaluronidase solution as soon as the cumulus masses are dissolved.

9. Polyamine mix is stabile at –20°C for several months. The microinjection buffer without polyamines added is stabile at +4 for several months. The ready-to-use buffer (polyamines added) should be prepared fresh for each experiment, and not stored.

10. Settings for P97:
    a. Borosilicate glass capillaries with inner filament, outer diameter 1.0 mm, inner diameter 0.78 mm (Clark Electronical Instruments GC100TF-15).
    b. Filament = B032TF.
    c. Pressure = 100.
    d. Heat = 305 (approx 10 values below ramp test result).
    e. Pull = 100.
    f. Velocity = 150.
    g. Time = 100.

11. Before attempting to inject the embryos, the tip of the injection needle should be broken up to a larger size. Because BAC DNA is more viscous then solutions with smaller DNA fragments, it is more likely to clog the needle during injection. There is also a risk for shredding the DNA in case it is pushed through a small opening with a high pressure. The final size of the needle tip should be around 3–4 μm.

12. If the Eppendorf FemtoJet is used, the Pc should be set to 5–10, and the Pi to 40–50. The injection should take place in "Manual" mode, where the injection time is adjusted to each pronucleus, and kept until a clear swelling is achieved.

13. It is easiest to target the larger of the two pronuclei, and/or the one, which is closest to the injection needle. If the embryos are well timed, the pronuclei should both appear in the center of the embryo, both be large and clearly visible. At the optimal time for injection, it is often difficult to tell which pronucleus is the male, and which is the female. There is no advantage in choosing one or the other. The only criteria for the choice should be the ease of injection: the pronucleus which is largest and/or nearest to the injection needle should be targeted. If the pronuclei are small, the embryo has not yet reached the optimal time-point for injection. It may help to bring the embryos back to the incubator for an hour, and start the injection process later. In case the embryos are left for too long at 37°C before the injection is started, the two pronuclei will have fused, and injection is impossible. The timing of the developmental stage can be influenced by the superovulation protocol (*see* **Subheading 3.2.**).

14. Care should be taken to avoid touching any of the nucleoli within the pronucleus. These structures are extremely sticky, and will immediately attach to the needle. If this would happen, the injected embryo should be discarded, and the injection needle changed.

15. The cell membrane is sometimes very difficult to penetrate. Because it appears very elastic, it will simply follow the tip of the injection needle into the embryo, and sometimes all the way into the pronucleus. If this happens, a small "bubble" will form at the tip of the needle, and the pronucleus will not expand. The needle can in these cases be pushed further on, through the pronucleus, and then slowly withdrawn into the pronucleus again, which usually solves the problem.

16. Embryos of minor quality should be discarded. Under the high magnification during injection, it is easy to detect those embryos with only one or more then two pronuclei, embryos with sperm under the zona, and so on. It is important to already at the injection sort away these embryos, and those which could not be successfully injected. Later, it will be impossible to distinguish between well-injected healthy embryos and those of poor quality.

17. Embryos can be transferred immediately after injection. It is also possible to culture the embryos overnight in vitro, and transfer only the well developing 2-cell stage embryos the next day. Great care should be taken to make sure all embryos are well in the oviduct, and to minimize the damage during the surgery. The anesthesia should be carefully optimized to give a short, but deep sleep, and the mouse should be kept warm until recovery.

18. We recommend the use of a mixture of Ketamine and Xylazine, as this combination has proven to be nontoxic, have a high safety marginal in terms of overdoses, and give a deep but short surgical anaesthesia. These measures will all contribute to increasing the possibility of maintaining a pregnancy.

19. Generally not more then 25–30% of the transferred 1-cell stage embryos (slightly more if 2-cell stage embryos are transferred) will develop to term. In order to

keep the litter size at an optimal level (6–8 pups); the number of embryos to trans-fer should be kept in the range of 25–30. We highly recommend transferring half of the embryos to each oviduct, although it is possible to put all into one side.

20. During anesthesia, the eyes of the mouse will stay wide open. The cornea will quickly dry out, leading to severe pain when the animal recovers. It is, therefore, highly recommendable to moisten the eyes with a drop of PBS as soon as the mouse has fallen asleep.

21. The normal gestation time in most mouse strains is 19–20 d. Outbred CD1—which is one of the most commonly used mice as recipients for embryo transfer—tend to deliver very reliably on E19. A commonly encountered problem following embryo transfer is low implantation rates, and dying/reabsorbing featuses. If the number of featuses which develop to term is too low, the size of each conceptus will be larger then normal, which often results in difficulties to give birth. Litter sizes of less then 4 pups often causes delivery difficulties in recipient females. These pups generally do not survive, and the mother is caused much pain and suffering unless caesarean section is performed. The procedure is fairly simple, and the survival rate of the sectioned pups very high (5). During the first half of the pregnancy, the weight gain of the mother is moderate, but weight increases dramatically during the last week. By weighing the recipients every other day, it is possible to predict in which cases the litter size could be expected to be small, since the curve of weight gain will reliably indicate litter size and possible problems during preg-nancy. However, a female that has not given birth on E20, should promptly be opened for caesarean section in any case.

22. It is essential to get the pups dry and warm immediately after sectioning. The breath-ing can be stimulated by wiping them roughly with a soft paper tissue. A common mistake is to be too soft with the pups. Because they have not gone through the physical stress of birth, they usually do not start breathing regularly unless they are turned around, pinched, and gently massaged with cotton tips or paper tissue pieces. Place the pups with a new fostermother when they are completely freed from all blood, regularly breathing, and have gained a normal body temperature.

Acknowledgments

We gratefully acknowledge Dr. Lluis Montoliu (Centro Nacional de Biotec-nologia, Madrod, Spain) and Dr.Thom Saunders (University of Michigan, USA) for advice and protocols for BAC DNA purification and microinjection. We would also like to thank the EMBL Laboratory Animal Resources for great support with all aspects of animal care and maintenance.

References

1. Gordon, J. W., et al. (1980) Genetic transformation of mouse embryos by microin-jection of purified DNA. *Proc. Natl. Acad. Sci. USA* **77,** 7380–7384.

2. Gordon, J. W. and Ruddle, F. H. (1981) Integration and stable germ line transmis-sion of genes injected into mouse pronuclei. *Science* **214,** 1244–1246.

3. Lin, T. P. (1966) Microinjection of mouse eggs. *Science* **151,** 333–337.
4. Brinster, R. L., et al. (1981) Somatic expression of herpes thymidine kinase in mice following injection of a fusion gene into eggs. *Cell* **27,** 223–231.
5. Nagga, A., et al. (2002) *Manipulating the Mouse Embryos, a Laboratory Manual.* 2nd ed. Cold Spring Harbor Laboratory Press, New York.
6. Pinkert, C. A. and Polites, H. G. (1994) "Transgenic Animal Production Focusing on the Mouse Model", in *Transgenic Animal Technology* (Pinkert, C. A., ed.), Academic: San Diego, CA.
7. Peterson, K. R., et al. (1993) Transgenic mice containing a 248-kb yeast artificial chromosome carrying the human beta-globin locus display proper developmental control of human globin genes. *Proc. Natl. Acad. Sci. USA* **90,** 7593–7597.
8. Schedl, A., et al. (1993) A method for the generation of YAC transgenic mice by pronuclear microinjection. *Nucl. Acids Res.* **21,** 4783–4787.
9. Sternberg, N. L. (1992) Cloning high molecular weight DNA fragments by the bacteriophage P1 system. *Trends Genet.* **8,** 11–16.
10. Smith, D. J., et al. (1995) Construction of a panel of transgenic mice containing a contiguous 2-Mb set of YAC/P1 clones from human chromosome 21q22.2. *Genomics* **27,** 425–434.
11. Nielsen, L. B., et al. (1997) Human apolipoprotein B transgenic mice generated with 207- and 145- kilobase pair bacterial artificial chromosomes. Evidence that a distant 5′-element confers appropriate transgene expression in the intestine. *J. Biol. Chem.* **272,** 29,752–29,758.
12. Yang, X. W., Model, P., and Heintz, N. (1997) Homologous recombination based modification in Escherichia coli and germline transmission in transgenic mice of a bacterial artificial chromosome. *Nat. Biotechnol.* **15,** 859–865.
13. Giraldo, P. and Montoliu, L. (2001) Size matters: use of YACs, BACs and PACs in transgenic animals. *Transgen. Res.* **10,** 83–103.
14. Camper, S. A. and Saunders, T. L. (2000) Transgenic rescue of mutant phenotypes using large DNA fragments., in *Genetic Manipulation of Receptor Expression and Function* (Acilli, D., ed.), Wiley: New York. p. 1–22.
15. Lacy, E., et al. (1983) A foreign beta-globin gene in transgenic mice: integration at abnormal chromosomal positions and expression in inappropriate tissues. *Cell* **34,** 343–358.
16. Wilkie, T. M., Brinster, R. L., and Palmiter, R. D. (1986) Germline and somatic mosaicism in transgenic mice. *Dev. Biol.* **118,** 9–18.
17. Kaufman, R. M., Pham, C. T., and Ley, T. J. (1999) Transgenic analysis of a 100-kb human beta-globin cluster-containing DNA fragment propagated as a bacterial artificial chromosome. *Blood* **94,** 3178–3184.
18. Zhang, Y., et al. (2000) DNA cloning by homologous recombination in Escherichia coli. *Nat. Biotechnol.* **18,** 1314–1317.
19. Foster, H. L., Small, J. D., and Fox, J. G. (1983) *The Mouse in Biomedical Research.* Academic, New York.
20. Brinster, R. L., et al. (1985) Factors affecting the efficiency of introducing foreign DNA into mice by microinjecting eggs. *Proc. Natl. Acad. Sci. USA* **82,** 4438–4442.

21. Harris, A. W., et al. (1988) The E mu-myc transgenic mouse. A model for high-incidence spontaneous lymphoma and leukemia of early B cells. *J. Exp. Med.* **167,** 353–371.
22. Chisari, F. V., et al. (1989) Molecular pathogenesis of hepatocellular carcinoma in hepatitis B virus transgenic mice. *Cell* **59,** 1145–1156.
23. Whittingham, D. G. (1971) Culture of mouse ova. *J. Reprod. Fertil. Suppl.* **14,** 7–21.
24. Refferty, K. A. (1970) *Methods in Experimental Embyology of the Mouse.* Johns Hopkins Press, Baltimore, MD.
25. Flaming, D. G. and Brown, K. T. (1982) Micropipette puller design: form of the heating filament and effects of filament width on tip length and diameter. *J. Neurosci. Methods* **6,** 91–102.
26. Montoliu, L., et al. (1995) Visualization of large DNA molecules by electron microscopy with polyamines: application to the analysis of yeast endogenous and artificial chromosomes. *J. Mol. Biol.* **246,** 486–492.
27. Bronson, R. A. and McLaren, A. (1970) Transfer to the mouse oviduct of eggs with and without the zona pellucida. *J. Reprod. Fertil* **22,** 129–137.

12

Generation of BAC Transgenic Mice

Vikki M. Marshall, Janette Allison, Tanya Templeton, and Simon J. Foote

1. Introduction

Transgenic mice provide an important experimental system for studying the effects of gene overexpression and for analyzing the expression of well-characterized "single" genes that contribute to a particular trait. However, single-gene expression in transgenic (*Tg*) mice is not appropriate for the functional analysis of the many genes that may reside in a typical genetically mapped locus of the order of 1 cM. Bacterial artificial chromosome (BAC) vectors, which can accommodate large genomic intervals of up to 300 kb are appropriate for such studies (*1*). BACs of this size may contain several contiguous genes along with their *cis*-acting regulatory elements. As such, BACs are valuable tools in complementation or "rescue" experiments for identifying genes contributing to a genetically mapped quantitative trait of interest.

In addition, it is known that the important regulatory elements of many single mammalian genes may be located tens of kilobase pairs up- or downstream of the gene coding sequence. Hence, large genomic BAC inserts are also invaluable for the analysis of expression of some single genes, where timing of expression from endogenous promoters and enhancers is required. An example is the human apolipoprotein *B* gene, where a distant element, located between 33 and 70 kb upstream of the *apoB* gene, is required for expression of *apoB* in intestinal tissue (*2*).

BACs offer several advantages over yeast artificial chromosomes (YACs) as integrants, including the relative ease with which they can be manipulated in vitro, and the fact that they can be purified from their host (*Escherichia coli*) genome. As transgenes, BACs generally offer stable integration, and show a

From: *Methods in Molecular Biology, vol. 256:*
Bacterial Artificial Chromosomes, Volume 2: Functional Studies
Edited by: S. Zhao and M. Stodolsky © Humana Press Inc., Totowa, NJ

linear relationship between copy number and the level of expression of integrated genes. In addition, BACs can be modified by homologous recombination in *E. coli* prior to microinjection, further enhancing their utility as transgenes *(3–5)*.

BAC transgenic mice are generated by pronuclear injection of fertilized mouse oocytes with a purified BAC fragment, followed by transfer of the injected oocytes to pseudopregnant foster mothers. Pups that arise from a transfer can be tested by various methods for the presence of the BAC transgene. These include BAC vector-specific polymerase chain reaction (PCR), fluorescent PCR using microsatellite markers that are polymorphic between the mouse host strain and BAC strain, and Southern blot analysis. Each method must be capable of distinguishing between DNA derived from the integrated BAC transgene and that of homologous DNA in the host genome.

Previous studies that utilize BAC transgenesis can be broadly divided into two types: 1) those that aim to analyse the consequences of BAC overexpression in vivo, e.g., the trisomy of Down syndrome, overexpression of precursor amyloid protein *(6)*, overexpression of human apolipoprotein *B (2)* and 2) those that attempt to elucidate gene function by functional complementation or so-called "BAC rescue" experiments, e.g., the *Clock* circadian rhythm genes *(7,8)*, correction of deafness in *shaker*-2 mice *(9)*, and, rescue of the "tattered" phenotype *(10)*.

Although the generation of BAC transgenic mice is necessarily complex and lengthy (*see* **Notes 1** and **2**), it provides an invaluable system for analyzing gene function and regulation in vivo. A broad outline of procedures is shown in **Fig. 1**. This chapter will provide detailed techniques for the purification of BAC DNA for microinjection, a discussion of microinjection techniques, the rationale behind breeding schemes for transgenic lines, and the screening of BAC transgenic progeny.

2. Materials
2.1. Buffers

All solutions used for purification of the linearized BAC, including 1X Agarase buffer and 1X Microinjection buffer (solutions 12–16), must be made up in embryo grade pure water, and using tissue culture-grade plasticware for all procedures. Glassware and metal weighing instruments may be contaminated with detergents and other products that are toxic to mouse oocytes. All other solutions should be prepared in high-quality, molecular biology grade water (e.g., MilliQ H_2O).

Fig. 1. *(see facing page)* Outline of procedures involved in the generation of BAC transgenic mice for functional analysis.

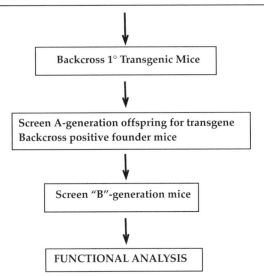

LARGE-INSERT BAC CLONE

1. Verify BAC clone by PCR
2. test BAC-end PCR screening oligos on BAC clone

Purify circular BAC DNA (alkaline lysis)

Linearise:-
1. NotI digestion
2. PI-SceI digestion (pBACe3.6)
3. λ-terminase (pBeloBAC11)
4. Cre-lox linearisation (pBeloBAC11)

Purify BAC insert

PFGE, band excision, β-Agarase digestion

Microinject fertilised mouse oocytes with purfied BAC DNA

Screen primary transgenic offspring (tail biopsy DNA)

1. PCR BACvector sequence or vector-insert junction
2. Southern blot analysis (or dot blot)
3. Fluorescent microsatellite markers polymorphic between host and BAC strain

Backcross 1° Transgenic Mice

Screen A-generation offspring for transgene
Backcross positive founder mice

Screen "B"-generation mice

FUNCTIONAL ANALYSIS

1. TE (10:1): 10 m*M* Tris-HCl pH 8.0, 1 m*M* ethylene diaminetetraacetic acid (EDTA), pH 8.0. Dissolve 1.21 g Tris-HCl base and 0.372 g EDTA Na$_2$.2H$_2$O in 1 L dH$_2$O, adjust pH to 8.0 with concentrated HCl. Dispense into 2 × 500 mL bottles, autoclave, and store at room temperature (RT). Alternatively, dilute from 1 *M* Tris-HCl, pH 8.0, and 0.5 *M* EDTA pH 8.0 stock solutions, autoclave, and store at RT.

2. Buffer P1 (Resuspension Buffer, supplied in Qiagen Midi Kit) : 50 m*M* Tris-HCl, pH 8.0, 10 m*M* EDTA, 100 µg/mL RNaseA. Dissolve 6.06 g Tris-HCl base, 3.72 g Na$_2$EDTA.2H$_2$O in 800 mL dH$_2$O. Adjust to pH 8.0 with concentrated HCl. This solution can be autoclaved. Store at +4°C after the addition of RNaseA.

3. Buffer P2 (Lysis Buffer, supplied in Qiagen Midi Kit): 0.2 *M* NaOH, 1% sodium dodecyl sulfate (SDS). Dissolve 8.0 g NaOH pellets in 950 mL dH$_2$O, add 50 mL 20% SDS to a final volume of 1 L. Store at RT.

4. Buffer P3 (Neutralization Buffer, supplied in Qiagen Midi Kit): 3.0 *M* potassium acetate, pH 5.5. Dissolve 294.5 g potassium acetate in 500 mL dH$_2$O. Adjust pH to 5.5 with glacial acetic acid (requires approx 110 mL). Adjust the volume to 1 L with dH$_2$O. Store at RT or +4°C for convenience (Solution P3 is used chilled).

5. Buffer QBT (Equilibration Buffer, supplied in Qiagen Midi Kit): 750 m*M* NaCl, 50 m*M* MOPS, pH 7.0, 15% isopropanol, 0.15% Triton X-100. Dissolve 43.83 g NaCl, 10.46 g MOPS (free acid) in 800 mL dH$_2$O. Adjust pH to 7.0 with NaOH. Add 150 mL propan-2-ol and 15 mL 10% Triton X-100 solution. Adjust volume to 1 L with dH$_2$O. Store at RT.

6. Buffer QC (Wash Buffer, supplied in Qiagen Midi Kit): 1.0 *M* NaCl, 50 m*M* MOPS, pH 7.0, 15% isopropanol. Dissolve 58.44 g NaCl, 10.46 g MOPS (free acid) in 800 mL dH$_2$O . Adjust pH to 7.0 with NaOH. Add 150 mL propan-2-ol. Adjust volume to 1 L with dH$_2$O. Store at RT.

7. Buffer QF (Elution Buffer, supplied in Qiagen Midi Kit): 1.25 *M* NaCl, 50 m*M* Tris-HCl pH 8.5, 15% isopropanol. Dissolve 73.05 g NaCl, 6.06 g Tris-HCl base in 800 mL dH$_2$O and adjust the pH to 8.5 with concentrated HCl. Add 150 mL propan-2-ol. Adjust volume to 1 L with dH$_2$O. Store at RT.

8. 3 *M* sodium acetate (anhydrous NaOAc) pH 5.3: Dissolve 24.61 g sodium acetate powder in 80 mL dH$_2$O. Adjust pH to 5.3 using glacial acetic acid. Adjust volume to 100 mL. Autoclave and store at RT.

9. 10X PI-*Sce*1 Buffer (supplied with enzyme): 100 m*M* Tris-HCl pH 8.6, 1 *M* KCl, 100 m*M* MgCl$_2$, 100 m*M* dithiothreitol (DTT).

10. 10X *Not*1 restriction enzyme buffer (supplied with enzyme): 500 m*M* Tris-HCl pH 7.5, 100 m*M* MgCl$_2$, 1 *M* NaCl, 0.2% Triton X-100, 1 mg/mL BSA.

11. 5X TBE Stock Buffer (for PFGE): Dissolve 54 g Tris-HCl base, 27.5 g boric acid and 20 mL 0.5 *M* EDTA (pH 8.0) in 1 L dH$_2$O. Can be autoclaved. Store at RT. Dilute 1/10 to 0.5X TBE in dH$_2$O for PFG electrophoresis.

12. 1X Agarase Buffer: 10 m*M* *bis* Tris-HCl pH 6.5 (at RT), 0.2 m*M* EDTA, 100 m*M* NaCl. Weigh the following salts into a sterile 500-mL tissue culture flask: 1.045 g *bis*-Tris-HCl, 37.2 mg EDTA(Na)$_2$, 2.92 g NaCl. Make up to 500-mL with Sigma embryo pure water. Adjust pH with 100–200 µL concentrated HCl to

pH 6.5 (using pH 5–10 paper sticks initially, then a pH meter). Filter-sterilize through a 0.2-μm Nalgene filter (or equivalent), discarding the first few milliliters of buffer. Store in 40-mL aliquots in 50-mL Falcon tubes (or equivalent) at –20°C. If facilities are not available for weighing milligram amounts of EDTA, prepare 0.2 *M* EDTA(Na)$_2$ by dissolving 3.72 g EDTA (Na)$_2$ in 40 mL embryo pure water, add two pellets of NaOH and use a pH meter to adjust to pH 8.0 using 0.1 *M* NaOH. Add 0.5 mL of 0.2 *M* EDTA(Na)$_2$ to the tissue culture flask containing preweighed *bis*-Tris-HCl and NaCl.

13. 5 *M* NaCl in embryo pure H$_2$O: Dissolve 14.61 g NaCl in 50 mL embryo pure water in a 50-mL Falcon tube. Store at –20°C.

14. 0.5 *M* EDTA (disodium ethylenediaminetetra-acetate.2H$_2$O) pH 8.0 in embryo pure H$_2$O: Dissolve 9.30 g Na$_2$EDTA.2H$_2$O in 40 mL embryo pure water in a tissue culture flask. Adjust pH to 8.0 using NaOH pellets (approx 1 g). Adjust volume to 50 mL with embryo pure water. Store at –20°C.

15. 1.0 *M* Tris-HCl pH 7.4 in embryo pure H$_2$O: Dissolve 6.05 g Tris-HCl base in 40 mL embryo pure water. Adjust pH to 7.4 with concentrated HCl. Add embryo pure water to 50 mL. Store at –20°C.

16. YAC/BAC Microinjection Buffer: 100 m*M* NaCl, 10 m*M* Tris-HCl pH 7.4, 0.2 m*M* EDTA: Add 10 mL 5 *M* NaCl, 200 μL 0.5 *M* EDTA pH 8.0, and 5 mL 1.0 *M* Tris-HCl pH 7.4 to 484.8 mL embryo pure water in a 500-mL tissue culture flask. All stock solutions must be prepared in embryo pure water. Filter-sterilize the solution after rinsing a 0.2 μm Nalgene filter (or equivalent), 2× with a small volume of embryo pure water. Discard liquid. This removes detergents from the filtration membrane. Store in 30-mL aliquots in 50-mL Falcon tubes at –20°C.

17. 10X Dye Loading Buffer for Analytical PFGE: 50% glycerol, 10 m*M* EDTA pH 8.0, 0.25% bromophenol blue in sterile dH$_2$O. Store in 10-mL aliquots at –20°C. Working aliquots can be stored at +4°C or RT.

18. 10X PCR Gold Buffer (supplied with AmpliTaq Gold enzyme): 150 m*M* Tris-HCl pH 8.0, 500 m*M* KCl.

19. 25 m*M* MgCl$_2$ for PCR (supplied with AmpliTaq Gold enzyme).

20. 10X λ-terminase Reaction Buffer: (10X TA Buffer; supplied with enzyme): 330 m*M* Tris-HCl acetate pH 7.8, 660 m*M* potassium acetate, 100 m*M* magnesium acetate, 5 m*M* DTT.

21. 1% agarose/0.5X TBE for sealing wells: Dissolve 0.5 g DNA grade agarose in 50 mL 0.5X TBE. Heat in a microwave for approx 30 s until agarose is molten. Store at RT. Remelt in a microwave and precool to approx 45°C prior to each use.

2.2. Other Solutions

1. Chloramphenicol Stock Solution: Dissolve 125 mg chloramphenicol in 10 mL absolute ethanol. Store in 1-mL aliquots at –20°C.

2. Absolute (99%) ethanol.

3. 70% ethanol.

4. 100X BSA for PI-*Sce*1 digestion (100 mg/mL BSA, supplied with enzyme).

5. 10 m*M* ATP (for λ-terminase digest): supplied with enzyme.

2.3. Preparations

1. Luria Broth (LB): Dissolve 10 g Tryptone, 5 g yeast extract, 10 g NaCl in 1 L (final volume) of high purity water. Distribute into 100-mL volumes and autoclave at 15 lbs pressure for 30 min. Store at RT.

2.4. Commercially Available Products

1. Qiagen Plasmid Midi Purification Kit (Cat. No. 12143). Store as per manufacturer's instructions.
2. Nylon Cell strainers, 100 μm (Falcon 2360; Becton Dickinson, Franklin Lakes, NJ).
3. 15- or 30-mL Glass Corex tubes (or other suitable tube capable of centrifugation at 12,000g).
4. 200G "genomic" filter tips (ART cat. no. 3532, 10 pkts sterile, racked tips; Molecular BioProducts Inc., San Diego, CA).
5. 1000G "genomic" filter tips (ART cat. no. 3592, 10 pkts sterile, racked tips; Molecular BioProducts Inc.).
6. Nescofilm sealing film (or equivalent, such as Parafilm); Bando Chemical Ind. Ltd., Kobe, Japan.
7. PI-*Sce*1 endonuclease: (New England BioLabs, Beverly, MA) cat. no. 696L (500 U). Store at –20°C.
8. *Not*1 restriction endonuclease; (MBI Fermentas; Vilnius, Lithuania) cat. no. ER0592 (1500 U). Store at –20°C.
9. λ-terminase: Epicentre Technologies (Madison, WI) cat. no. LT4450 (50 U), LT44200 (200 U).
10. Agarose for analytical PFGE : DNA Grade agarose, (Progen Industries Ltd., Queensland, Australia) cat. no. 200-0010, or equivalent, e.g., Molecular Biology Certified Agarose, 100 g, (Bio-Rad, Hercules, CA) cat. no. 162-0133.
11. Agarose for preparative PFGE; SeaPlaque low melting temperature agarose, (FMC Bioproducts, Rockland, ME) cat. no. 50101 (25 g).
12. Low Range PFG Marker: (New England BioLabs) cat. no. N0350S.
13. Mid Range I PFG Marker: (New England BioLabs), cat. no. N3551S.
14. Millipore filters for spot dialysis: 100/pk, 0.05 μm, white VMWP, 25 mm (Millipore, Bedford, MA) cat. no. VMWP02500. Safety caution: flammable solid.
15. Ultrapure water for embryo transfer (embryo tested): Sterile filtered, endotoxin tested (500 mL). Cat. no. W-1503 (Sigma-Aldrich Co. Ltd., Irvine KA12 8NB, UK). Store at RT.
16. *bis*-Tris-HCl: Sigma-Aldrich Co. Ltd., Irvine KA12 8NB, UK.
17. Disposable 50-mL tubes: Falcon cat. no. 352070 (Becton Dickinson, Franklin Lakes, NJ).
18. AmpliTaq Gold (250U): (Applied Biosystems, Foster City, CA, cat. no. 4311806). Store at –20°C in a frost-free freezer.
19. β-Agarase: cat. no. 392S or 392L, 1 U/μL (New England BioLabs).
20. Ethidium Bromide stock solution, 10 mg/mL (Progen Industries Limited, cat. no. 200-0418).

21. Agarose-1000 (Gibco BRL, Life Technologies Rockville, MD) (100 g) cat.no. 10975-035.
22. Nalgene 500-mL disposable Filter Units (Nalgene, Rochester, NY) cat. no. 166-0020.
23. Ultrapure dNTPs for PCR: 100 m*M* dNTP set, 4X100 µmol (Amersham Pharmacia Biotech, Piscataway, NJ) cat. no. 27-2035-02.
24. Pregnant mare serum (PMS), Serum Gonadotrophin (5X1000 IU, brand name Folligon, Intervet Australia Pty. Ltd, Castle Hill, NSW, Australia). Resuspend vial in 5 mL solvent (supplied). Store at +4°C. A freshly prepared vial of Folligon can be used for up to 8 d.
25. Human chorionic gonadotrophin (hCG) 5000 IU, brand name APL (Wyeth-Ayerst Laboratories, PA). Resuspend vial in solvent (supplied). Store at +4°C. Solution can be used for up to 3 mo.
26. Two Mouse BAC Libraries are commercially available from BACPAC resources (http://www.chori.org/bacpac/) or Research Genetics (ResGen, Invitrogen) (http://mp.invitrogen.com/newbacpacorder.php3)
    a. Female Mouse BAC library (129S6/SvEvTac), cat. no. RPCI-22, average insert size 154 kb, cloned into the *Eco*R1 site of pBeloBACe3.6.
    b. Female Mouse BAC library (C57BL/6J), cat. no. RPCI-23, average insert size 197 kb. Human male and female BAC libraries are also available, (cat. no. RPCI-11 and RPCI-13, respectively) with average insert sizes of 178 and 166 kb. Another C57BL/6 mouse library is available from Genome Systems (www.genomesystems.com), cat. no. BAC-9541, average insert size 120 kb.

3. Methods

3.1. Purification of Circular BAC DNA (see Note 3)

1. Pick a single BAC colony and inoculate 100-mL of LB medium containing 12.5 µg/mL Chloramphenicol (add 100 µL of Chloramphenicol stock to 100 mL of LB). Grow at 37°C for 14 h (must be less than 16 h) in a 250-mL conical flask with vigorous shaking at approx 250 rpm (*see* **Notes 4–6**).
2. Harvest the cells by centrifugation at 3000 rpm (2060*g*) for 30 min at 4°C (using Beckman GS6K-R floor model centrifuge or equivalent) in a 50-mL disposable tube. Dispose of supernatant appropriately, and refill the same tube with the remaining 50 ml overnight culture and spin at 3000 rpm (2060*g*) for 30 min.
3. Remove the second supernatant and drain the excess LB by inverting the tube(s) on to tissues (*see* **Note 7**).
4. Resuspend cells thoroughly in 10 mL of Buffer P1 containing 100 µg/mL RNase A (provided with kit). There should be no clumps of *E.coli* remaining in the tube.
5. Add 10 mL Buffer P2 to the tube, mix gently but thoroughly by inverting 4–6 times, and incubate at RT for exactly 3 min (*see* **Note 8**).
6. Add 10 mL chilled Buffer P3 to each tube, mix immediately but gently by inverting 4–6 times, and incubate on ice for 15 min.
7. Centrifuge at 3000 rpm (2060*g*) for 30 min–1 h at +4°C. Remove supernatant promptly.

8. Filter the supernatant through a nylon 100-μm cell strainer (Falcon). Collect the liquid in a 50 mL tube (*see* **Note 9**).

9. Equilibrate a QIAGEN-tip 100 by applying 4 mL Buffer QBT and allowing the column to empty by gravity flow.

10. Apply the BAC lysate to the QIAGEN 100-column and allow it to pass through the resin by gravity flow.

11. Wash the QIAGEN-column with 2 × 10 mL Buffer QC.

12. Support the column over a 15- or 30-mL Corex tube and elute DNA with five aliquots of 1 mL Buffer QF, prewarmed to 65°C (*see* **Note 10**).

13. Precipitate BAC DNA by adding 3.5 mL propan-2-ol (at RT) to the eluted DNA and mix gently by inversion after sealing the tubes securely with Parafilm. Centrifuge in a Sorvall at 10,000 rpm (12,000*g*) for 30 min at 4°C. Prior to centrifugation, mark the side of each tube with ink and load the rotor with the mark at the position where the pellet will form (pellet may not be visible) (*see* **Note 11**).

14. Carefully decant the supernatant, and drain the tubes by inverting on tissues for 5–10 min (reposition them 1–2 times on the tissue for thorough drainage). Then invert tubes (right way up) in a rack and allow to air-dry for a further 10 min. The pellet may be visible as a light smear on the inside of the tube.

15. Dissolve DNA by squirting 500 μL of TE (10 : 1) onto the pellet. DO NOT pipet the DNA at all. Then swirl the 500 μL of TE over the pellet gently, seal tubes with Parafilm and leave at 4°C overnight or longer to dissolve.

16. Transfer the DNA sample (500 μL TE) to a sterile microfuge tube using a 200-gage filter tip (*see* **Note 12**). Precipitate the BAC DNA by adding 1/10 the volume (50 μL) of 3 *M* NaOAc pH 5.3 and 2X the volume (1 mL) of 99% ethanol and invert the tube gently several times. BAC DNA should be visible as "strands" at this stage. Do not mix the BAC DNA vigorously at any stage.

17. Centrifuge BAC DNA at 13,200 rpm (16,110*g*) in a microfuge for 2 min at RT. Remove supernatant using a 1-mL filter tip.

18. Wash the pellet by adding 500 μL of 70% ethanol, invert tube gently several times and spin 2–5 min at 13,200 rpm (16,110*g*) in a microfuge at RT.

19. Remove supernatant, drain, and air-dry the pellet.

20. Resuspend DNA in 50 μL of TE (10 : 1) by squirting 50 μL TE on to the pellet, flick gently 2–3 times, and dissolve DNA at +4°C overnight.

21. Determine the concentration of your DNA by optical density on a spectrophotometer at 260 nm. Typical yields range from 15 to 35 μg from a 100-mL O/N culture, depending on the clone.

3.2. Enzyme Digestion and Pulsed Field Gel Electrophoresis (PFGE) Analysis of BAC DNA on an Analytical Scale (see Note 13)

Following purification of circular BAC DNA, small-scale digests should be prepared for analysis of the clone prior to preparative DNA digests, in order to determine the presence of internal *Not*1 sites (if not already known) and to check the integrity of the DNA. For clones in the pBeloBAC11 vector, perform a *Not*1 and λ-terminase digest; for clones in pBACe3.6, perform a *Not*1 and PI-*Sce*1 digest.

3.2.1. Not1 Digestion of BAC DNA

1. Set up reaction as follows:
   a. 250 ng BAC DNA.
   b. 1.5 µL 10X *Not*1 buffer.
   c. 0.5 µL *Not*1 enzyme (10 U/µL).
   d. dH₂O to 15 µL.
2. Incubate at 37°C for 1–2 h.
3. Stop the reaction by adding 2 µL 10X Gel Loading Dye to the reactions. Store at +4°C prior to PFGE.

3.2.2. λ-Terminase Digestion of BAC DNA Cloned into pBeloBAC11

The recognition sequence for lambda terminase cleavage is 156 bp at the cos region of the pBeloBAC11 vector.

1. Set up reaction as follows:
   a. 250 ng BAC DNA.
   b. 1.5 µL 10X TA buffer.
   c. 1.5 µL 10 m*M* ATP.
   d. 1.5 µL λ-terminase (3U).
   e. dH₂O to 15 µL.
2. Incubate at 30°C for 3 h.
3. Stop the reaction by adding 2 µL 10X Gel Loading Dye to the reactions. Store at +4°C until PFGE.

3.2.3. PI-Sce1 Digestion of BAC DNA Cloned into pBACe3.6

1. Set up reaction as follows:
   a. 250 ng BAC DNA.
   b. 3 µL 10X PI-*Sce*1 buffer.
   c. 0.3 µL 100X BSA solution (10 mg/mL).
   d. dH₂O to 30 µL.
2. Incubate at 37°C for at least 2 h.
3. Stop the reaction by adding 2 µL 10X Gel Loading Dye to the reactions. Store at +4°C until PFGE (*see* **Note 14**).

3.2.4. PFGE of Digested BAC DNA

1. Melt 1 g of DNA certified agarose in 100 mL 0.5X TBE Electrophoresis Buffer. DNA-certified agarose can be used for analytical gels, but SeaPlaque low melting point agarose must be used for the preparative gels below.

2. Cool to 55°C and pour molten agarose in gel casting module with 15-well comb in place (supplied with pulsed field gel apparatus) and allow to set at RT.
3. Precool at least 2 L of 0.5X TBE in PFGE tank e.g., Bio-Rad CHEF Mapper. Circulate buffer using pump.
4. Cut a slice of Mid RangeI Pulsed Field Gel Marker (approx 1 mm thick) and equilibrate in 1 mL 0.5X TBE.
5. Heat preprepared 1% agarose/0.5X TBE preparation in a microwave until molten.
6. Remove comb from cast gel and apply MidRange I Pulsed Field Gel Marker to one well using flamed, cooled forceps.
7. Set the marker plug in the well by applying sufficient molten agarose (precooled to approx 40°C) to just fill the well (approx 75 μL). Allow to set at RT.
8. Place gel in PFGE tank in holding frame as per manufacturer's instructions.
9. Load all of BAC analytical digests using standard tips.
10. Allow PFG to run under the following conditions:
    a. Run time: 20 h.
    b. Included angle: 120°.
    c. Pulse time: 5–15 s.
    d. Ramp: linear.
    e. Voltage gradient: 6 V/cm.
    f. Buffer temperature: 15°C.
11. When the run is complete, remove gel and stain in approx 800 mL 0.5X TBE running buffer containing approx 1 μg/mL ethidium bromide (EtBr).
12. Agitate gently on a rocking platform at RT for 30 min–1 h.
13. Pour off staining solution (discard EtBr appropriately; it is a powerful mutagen), and replace with more 0.5X TBE running buffer or dH_2O.
14. Agitate gently on a rocking platform at RT for 30 min–1 h.
15. Visualize PFG by ultraviolet transillumination and photograph. Inspect gel for any BAC DNA degradation and for the presence of internal *Not*1 sites. Estimate the size of the BAC insert by comparing to the size standard(s).

3.3. Preparative Not1 Digestion and Purification of Linearized BAC DNA for Pronuclear Injection

3.3.1. Preparative Not1 Digestion

The preparative digest given below is for *Not*1 endonuclease digestion of purified BAC DNA. If λ-terminase or PI-*Sce*1 digestion is required, as a preparative digest, the volumes given for analytical digests (above) can be up-scaled.

1. Prepare digest as follows:
    a. BAC DNA: 9–12 μg.
    b. 10X *Not*1 buffer: 10 μL.
    c. *Not*1 enzyme: 5 μL.
    d. q.s. dH_2O: 100 μL.

2. Incubate at 37°C for at least 3 h, up to 12 h.

3. Prepare a 1% low melting point (LMP) agarose gel using SeaPlaque agarose (FMC Bioproducts cat. no. 50101) in 0.5X TBE buffer in a casting mold for the CHEF PFGE apparatus. Use 150 mL of molten agarose for the 140 × 127 mm Bio-Rad mold.

4. Heat the digested BAC DNA (from **step 2** above) to 50°C.

5. Add 1/2 volume (50 µL) of preheated (70°C) 2% LMP SeqPlaque agarose.

6. Mix up and down one time using a warmed ART 200G wide-bore tip and load into several (4–6) individual wells until full. Two wells are used for EtBr staining and the remaining wells are used for band excision.

7. Load LowRange I size standards in the two outside lanes.

8. Allow molten agarose in wells to set at room temperature. Make careful note of the exact loading of DNA in each well.

9. Rinse the PFG apparatus electrophoresis box with several liters of 0.5X TBE diluted in high-quality, e.g., Milli-Q H$_2$O.

10. Transfer gel to PFGE tank containing pre-cooled (15°C), circulating 0.5X TBE buffer. Precool the gel before turning on the electrophoresis power.

11. Run gel under the same conditions outlined in **Subheading 3.2.4., step 10**.

12. Excise the central strips of gel (longitudinally) either side of the wells containing BAC DNA, but include at least half a well of BAC DNA from the outer BAC-containing lanes. Mark the corners of all three strips of agarose such that the gel can be accurately reconstructed several hours later. Preserve the central strip of gel by wrapping in Saran wrap and storing at +4°C.

13. Stain the two flanking agarose strips by gentle agitation in 0.5X TBE running buffer containing 1 µg/mL ethidium bromide at RT for 30 min–1 h.

14. Destain the strips by removing EtBr solution (dispose of appropriately) and replace with fresh 0.5X TBE solution and gently agitate at RT for a further 30–60 min.

15. Photograph the gel slices under UV transillumination and visualise the size standards and digested BAC DNA. Mark the position of the BAC DNA by cutting triangular nicks accurately around the DNA band.

16. Reassemble the entire BAC pulsed field gel on plexi-sheet and locate by alignment the position of the digested BAC DNA in the central slice.

17. Using a clean scalpel blade, excise the digested BAC DNA in a slice of agarose approx 3 mm wide (*see* **Notes 15** and **16**). Transfer to a 50-mL tube containing at least 20 mL 1X agarase buffer.

18. Stain the remainder of the gel from the central strip to ensure that the correct DNA band has been excised.

3.3.2. Isolation of Linearized BAC DNA From Gel Slice

1. Equilibrate the gel slice in 1X agarase buffer at +4°C for approx 6 h with occasional agitation. Change the 1X agarase buffer at least one more time and continue to equilibrate the slice overnight.

2. Cut a small piece from the gel slice and put aside to test for DNA integrity. Cut the remaining gel into small pieces and place a maximum of 400 mg of agarose gel slice per 1.5-mL microfuge tube. (Check by preweighing an empty microfuge tube first.) Be careful to remove all of the agarase buffer.

3. Spin the agarose pieces to the bottom of the tube by centrifugation in a microfuge for 6 s at approx 6000g.

4. Rapidly place the tube at 68°C in a water bath for 10 min, and visually inspect to ensure the agarose is completely melted.

5. Immediately transfer the tube to a 40°C water bath for 5 min (the gel near the outside of the tube will cool very quickly and form a solid skin if the tube is left at RT for more than a few moments).

6. After the gel has equilibrated to 40°C, add 1 µL (1.0 U) agarase enzyme per 100 mg of agarose gel. Mix gently by pipeting up and down once slowly with a sterile, warmed 200-gage wide-bore tip.

7. Incubate at 40°C in a water bath for 2 h.

8. Remove a small aliquot and put aside to test DNA integrity. Centrifuge tubes at 13,000 rpm at RT for 20 min in a microfuge, ensuring that the centrifuge is not warm from a previous run. Excessive temperatures could shear the DNA.

9. Using a cut-off tip, slowly remove the top two-thirds of the solution, avoiding the lower third. Do not be tempted to take more solution, as the lower third will contain debris and solids that could block the micro-injection needle. Set the lower solution aside to test for DNA integrity.

3.3.3. Dialysis of BAC DNA for Microinjection

1. Preequilibrate several Millipore filters (VMWP 02500, type VM, 0.05 µm) by floating them (shiny side up) on the surface of 30 mL of 1X microinjection buffer in a 30-mL Petri dish at RT for 6 h to overnight.

2. Transfer preequilibrated filters to a fresh dish containing a fresh 30-mL 1X microinjection buffer.

3. Carefully pipet 200 µL of agarase-digested BAC DNA using a 200-gage wide-bore tip onto the preequilibrated filters, leaving 20 µL as a control to run on a PFG. Dialyze for at least 6 h.

4. Recover the final BAC DNA preparation from the filters using 200-gage wide-bore tips and pool into a sterile 1.5-mL screw-cap microfuge tube. Take an aliquot to test for DNA integrity. Store at +4°C. (Do not drop or shake the tube, as long molecules will be damaged. Never freeze the DNA.) Although the BAC DNA is stable at 4 °C for several months, it tends to block the microinjection needle after about 3–4 wk storage. It is therefore best to inject the DNA as soon as possible after isolation.

5. Take all the aliquots that were put aside and run them on a pulse field gel to check for DNA integrity during the isolation procedure (*see* **Fig. 2**).

3.3.4. Estimation of Purified BAC DNA Yield by PFGE

DNA used for microinjection should be quantitated by PFGE against known quantity standards, e.g., purified BAC inserts from previous successful microinjection experiments. Quantitation of purified BAC insert is vitally important to the success of the microinjection process and effects integration efficiency (*see* **Note 17**). The optimum concentration of DNA for microinjection is approx 1 ng/µL.

3.3.5. Microinjection of Fertilized Oocytes With Purified BAC DNA (see **Note 18**)

1. Microinjection needles should be filled by placing them into a microfuge tube containing BAC DNA, and allowing the needles to fill by capillarity.
2. When pipeting the purified BAC insert, it is essential to use cut-off tips at all times, and to pipet the DNA extremely slowly to avoid shearing the molecules. Do not shake, flick, or drop tubes containing the BAC DNA. If microinjection is to be performed by a facility or technician not familiar with BAC DNA, ensure they understand how to handle the DNA properly by observing the above procedures. Do not centrifuge the DNA if there are problems with needle blockage. A larger hole can be put in the needle tip if there are problems with blockage.
3. As it is difficult to estimate the exact concentration of BAC DNA for microinjection, try using neat, 2X and 3X dilutions of estimated DNA at different microinjection sessions rather than persisting with the same DNA concentration for every session.
4. Normally, oocytes for microinjection are removed early in the morning. With the NOD strain, however, we have found that removing oocytes at 12:00 improves the number of usable fertile oocytes. SJL/JaxARC and NODLt/Jax strains provide good numbers of oocytes after superovulation, but up to 40% of the oocytes may be unfertile. To obtain sufficient fertilized oocytes for microinjection, mate at least 20–24 female mice with stud males.

3.3.6. Utility of Different Inbred Mouse Strains for BAC Transgenesis

Egg donor strains that are suitable for transgenesis with small DNA fragments are equally amenable to transgenesis with BAC DNA. We have used the following mouse strains as egg donors for transgenesis with BAC DNA preparations: SJL/JaxARC, NODLt/Jax, C57BL/6/Jax, and C57BL/6 × SJL (F2). In addition, we have found that BALB/CAnBradleyWehi mice provide reasonable numbers of fertilised eggs for microinjection and tolerate BAC DNA.

Not all inbred mouse strains are suitable for superovulation, and a considerable amount of "trial and error" is involved in establishing procedures. The following procedures can be attempted as a starting point.

1 2 3 4 5 6 7 8

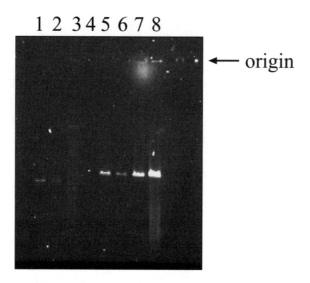

← origin

Fig. 2. Pulsed field gel electrophoresis of purified, *Not1* digested BAC133 fragment at various stages of purification along with control fragments. Lane 1, BAC82 (20 μL of 1/2 dilution of original stock) for quantitation; lane 2, BAC82 (20 μL of 1/4 dilution of original stock) for quantitation; lane 3, 20 μL neat BAC407 (prepared several months prior to running gel); lane 4, 20 μL neat BAC458 (prepared several months prior to running gel); lane 5, 20 μL neat BAC133 purified fragment; lane 6, 20 μL (1/2 dilution) BAC133 purified fragment; lane 7, BAC133 (20 μL of preparation before spinning at agarase step); lane 8, 20 μL BAC133-*Not*1 preparation prior to purification. Note the absence of smaller, degraded DNA in the lanes containing purified BAC133 DNA (lanes 5 and 6), as well as the degradation of older BAC insert preparations (stored at +4°C for approx 6 mo), lanes 3 and 4.

1. For mice maintained on a 14/10 h light/dark cycle:

    To test the egg donor potential of inbred strains, females at 4, 5, 7, and 8 wk are injected with 5 U of PMS intraperitoneally on day 1 at 14:30, then on day 3 at 14:30 with 5 U of hCG. Mice are mated at this time. The following morning at 9:00, the fertilized eggs are recovered in M2 medium then changed to M16 medium as described (*11*). As well as testing age dependence, different hormone concentrations should also be tested.

2. We have found that the following protocols were most suitable for the egg strains we have used.

    a. Strain SJL, age of female 31–35 d, PMS 10 U, hCG 10 U.

    b. Strain NOD, age of female 8–9 wk, PMS 2.5–5 U, hCG 5 U.

    c. Strain C57BL/6, age of female 31–35 d, PMS 5 U, hCG 5–10 U.

d. Strain B6 × SJL (F2), age of female 28–31 d, PMS 5 U, hCG 5 U.

e. Strain BALB/C, age of female 4.5 or 8–9 wk, PMS 5 U, hCG 5 U.

3. Determine the efficiency of mating, egg yield, number of fertile eggs, and ability of eggs to survive in culture overnight.

4. Test viability of fertilized eggs after injection with a small DNA insert. Test survival of injected eggs in culture overnight.

5. We have found that eggs of certain strains (SJL, NOD) do not culture well overnight, especially after being injected with BAC DNA. Therefore, oocytes from these strains are always transferred on the day of injection.

6. We use CBA × C57BL/6 (F1) pseudopregnant mothers as recipients of injected eggs, although any suitable foster strain can be used. BAC-injected eggs are less viable than eggs injected with smaller DNA fragments and, therefore, at least 20–25 eggs are transferred per foster female. This ensures a litter size of four or more pups. Pseudopregnant mothers often kill litters consisting of only one or two pups.

3.4. Screening BAC Tg Mice

3.4.1. PCR Screening Offspring of Mice Microinjected With BAC Inserts (see **Note 19**)

Offspring can be tested for integration of the BAC transgene by PCR amplification of tail snip DNA from pups at weaning (approx 21 d of age). Genomic DNA is purified according to established procedures *(12)*. All PCR screening must be performed in a PCR "clean" room, using filter tips and dedicated pipetors. False positives owing to PCR contamination are extremely wasteful of valuable animal resources. pBeloBAC11 vector PCR primer sequences for screening transgenic offspring are shown below, and their position on the vector map is shown in **Fig. 3**.

3.4.1.1. pBeloBAC11 Vector Primer Sequences

1. BACVEC_T_F, 5′ GTGCGGGCCTCTTCGCTATT 3′, forward, T7.
2. BACVEC_T_R′, 5′ CAGGTCGACTCTAGAGGATC 3′, reverse, T7.
3. BACVEC_S_F, 5′ TGTCACCTAAATAGCTTGGC 3′, forward, SP6.
4. BACVEC_S_R, 5′ ACGCAATTAATGTGAGTTAGC 3′, reverse, SP6.
   BACVEC_T and BACVEC_S primer pairs give PCR products of 180 and 150 bp, respectively. As suggested by Chrast et al., this product size differential allows one to carry out multiplex screening of mouse tail biopsy DNA, using both vector-directed primer pairs in the same PCR *(13)*. If multiplexing is not required, an alternative set of primers to the BACVEC_S_F/BACVEC_S_R pair can be used, which result in PCR products of higher specificity in our hands (*see* **Note 20**). An alternative strategy for PCR screening of pBeloBAC11-integrants is to amplify vector-BAC insert junctions using either the SP6 or T7 primer, paired with a primer internal to the appropriate BAC end. For such vector-insert screens of pBeloBAC11 integrants we use SP6 and T7 primer sequences shown in the notes (*see* **Note 21**).

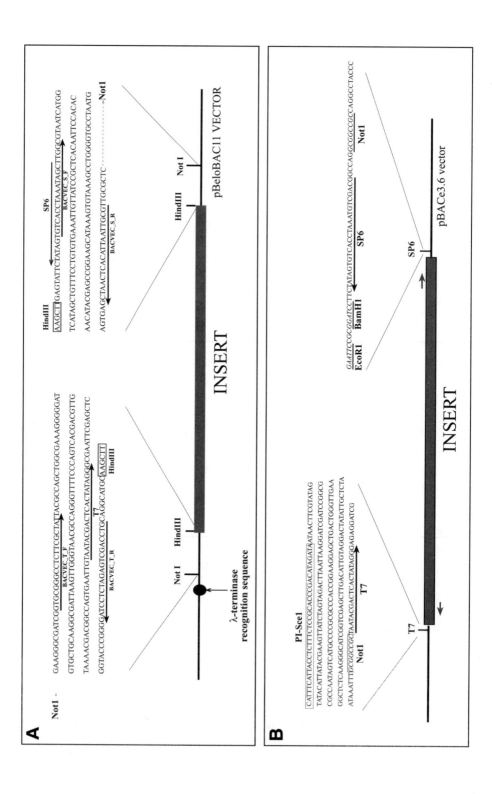

A

Not1

GAAGGGCGATCGGTGCGGGCCTCTTCGCTA`TAC`CAGCTGGCGAAAGGGGGAT
GTGCTGCAAGGCGATTAAGTTGGGTAACGCCAGGGTTTTCCCAGTCACGACGTTG
`BACVEC_T_F`
TAAAACGACGGCCAGTGAATTGTAATACGACTCACTATAGG`GCGAATTCGAGCTC`
GGTACCCGGGGATCCTCTAGAGTCGACCTGC`AGGCATGC`A`AGCTT`
`BACVEC_T_R`   T7   **HindIII**

HindIII
`AAGCTT`GAGTATTCTATAGTGTCACCTAAATAGTTG`GCGTAAATCATGG`
`BACVEC_S_F`   SP6
TCATAGCTGTTTCCTGTGTGAAATTGTTATCCGCTCACAATTCCACAC
AACATACGAGCCGGAAGCATAAAGTGTAAAGCCTGGGGTGCCTAATG
AGTGAG`CTAACTCACATTAATTGCGTTGCGCTC`- - - - - - - - - - - - ->**Not1**
`BACVEC_S_R`

Not I   HindIII   HindIII   Not I

INSERT

λ-terminase
recognition sequence

pBeloBAC11 VECTOR

B

PI-Sce1

`CATTTCATTACCTCTTTCTCCGCACCCGACATAGATA`ATAAACTTCGTATAG
TATACATTATACGAAGTTATCTAGTAGACTTAATTAAGGATCGATCCGGCG
CGCCAATAGTCATGCCCCGGCCCACCGGAAGGAGCTGACTGGGTTGAA
GGCTCTCAAGGGCATCGGTCGAGCTTGACATTGTAGGACTATATGTCTCTA
ATAAAATTT`GCGGCCGC`GTAATACGACTCACTATAGGGAGAGGATCG
Not1   T7

T7   SP6

INSERT

`GAATTCCGCGGATCC`TT`TATAGTGTCACCTAAATAG`TCGACGGCCAGGCGGCCGCCGCCAGGCCTACCC
EcoR1   **BamH1**   SP6   Not1

pBACe3.6 vector

3.4.1.2. pBACe3.6 Vector Primer Sequences

Note that the SP6 primer sequences for pBACe3.6 and pBeloBACII are not interchangeable, although the T7 primer sequence is identical for the two vectors.

1. PBACe3.6_SP6 , 5′ CGTCGACATTTAGGTGACACTATAG 3′, forward, SP6 side.
2. INTERNAL BAC Primer (1), Varies depending on the BAC used as integrant, reverse, SP6 side.
3. PBACe3.6_T7, 5′ TAATACGACTCACTATAGGG 3′, forward, T7 side.
4. INTERNAL BAC Primer (2), Varies depending on the BAC used as integrant, reverse, T7 side.

3.4.1.3. Protocols

1. Purify genomic DNA from tail biopsies of each pup according to Laird et al.; dissolve in 100–200 µL TE (10:1) and store at +4°C.
2. Prepare PCRs as follows (volumes shown for one DNA sample only). Because many transgenic mouse DNAs will be screened with several primer sets over the course of the experiment, DNA should be arranged in panels in 96-well microtiter plates, and the PCRs carried out in 96- or 384-well PCR plates using mastermixes of the reagents shown below.
   a. Purified tail Biopsy DNA: 30 ng.
   b. 10X AmpliTaq Gold Buffer: 2 µL.
   c. 25 mM MgCl$_2$: 2 µL.
   d. 2 mM dNTPs: 2 µL.
   e. AmpliTaq Gold: 0.072 µL.
   f. Primer 1: 3.3 pmol.
   g. Primer 2: 3.3 pmol.
   h. dH$_2$O: q.s. to 20 µL.
   Note that other Taq polymerases can be used (*see* **Note 22**).
   Include three important controls along with each BAC transgenic screen. These are
   a. Genomic DNA from the same background strain used as the transgene recipient.
   b. A dH$_2$O control to check for contamination of oligonucleotides, PCR buffers, or nucleotides.
   c. Approx 1 ng of BAC DNA used for BAC transgenesis. A convenient source of such DNA is a colony of the BAC cloned in *E. coli*. Take a small amount of colony on a sterile tip and swirl into a tube containing 10 mL dH$_2$O, mix by inverting the tube several times. Store indefinitely at +4°C. Use 10 µL of this

---

Fig. 3. *(see opposite page)* Schematic diagram showing vector sequences that flank insert cloning sites in pBeloBAC11 and pBACe3.6 BAC vectors. The position of oligonucleotide primers used for screening BAC transgenic pups is shown.

BAC DNA in a 20 μL PCR. Extreme care must be taken, however, to avoid contaminating genomic DNA from *Tg* mice with the BAC clone.

3. Cycle under the following conditions:
   a. 93°C for 10 min 1×.
   b. 94°C-15 s/65°C-30 s/72°C-30 s 1×.
   c. 94°C-15 s/62°C-30 s/72°C-30 s 1×.
   d. 94°C-15 s/59°C-30 s/72°C-30 s 1×.
   e. 94°C-15 s/56°C-30 s/72°C-30 s 1×.
   f. 94°C-15 s/53°C-30 s/72°C-30 s 1×.
   g. 94°C-15 s/50°C-30 s/72°C-30 s 40× cycles.
   h. 72°C-2 mins 1×.
   i. Finish cycle at 25°C (*see* **Note 23**).
4. Load 10 μL of each PCR on to 3% NuSieve or Agarose1000 analytical agarose gels (containing 0.5 μg/mL EtBr), prepared in 0.5X TBE. Twelve-channel pipets can be used to load agarose gels set with combs that have 9-mm tooth spacing. Photograph gel and record results.
5. Because transgenic breeding schemes are lengthy, and a large degree of effort is required to generate transgenic mice, it is essential to keep good records of transgenic mice and breeding schemes. A computerised data recording system for tracking the results of *Tg* DNA screens, as well as breeding is appropriate. Microsoft Excel spreadsheets may be used, but such databases are inadequate for tracking more than a few transgenic lines. Several different databases have been designed specifically for tracking transgenic (and congenic) lines, including MICE, which was developed on FileMaker Pro *(14)* and LAMS, developed on Microsoft Access *(15)*.
6. Choose primary BAC transgenic mice that are positive for both BAC ends. Primary positive animals are backcrossed to mice of an appropriate genetic background. Transgenic positive offspring (A-generation) from the primary animal become founders for the transgenic line, and can be bred for several generations to accumulate a cohort of animals for functional analysis.

4. Notes

1. The establishment of BAC transgenic lines can take several months to years, an important consideration in the planning of such experiments. Following microinjection of BAC DNA and the transfer of fertilised oocytes to foster mothers, any primary offspring are screened for the presence of the BAC transgene by various methods. Typically, primary *Tg*-positive mice are backcrossed at 7–8 wk of age to mice of an appropriate genetic background to allow transmission of the transgene to their progeny, which are denoted the "A"-generation. These transgenic-positive offspring become the founders of the transgenic line. The progeny of A-generation backcross, transgene positive mice are denoted "B"-generation; their progeny are denoted "C"-generation, and so on.

   Although statistically 50% of the offspring of a primary transgenic mouse may be expected to be transgene positive, mosaicism in the primary transgenic positive animal may necessitate the screening of numerous offspring (in our hands, up to 80) before transmission of the transgene is evident. This becomes difficult if the primary transgenic positive animal is female, since only 8–12 pups can be generated every 21 d. As a further complexity, integration-site effects make it necessary to establish several (usually 2–3) transgenic lines from different founders for each BAC insert to be tested.

2. The production of BAC *Tg* mice should only be attempted when a researcher has in place an appropriate strategy for screening *Tg* offspring, both for the presence of the BAC transgene and for detecting a change in phenotype of BAC transgenic positive vs transgenic negative littermates. Strategies for determining BAC transgene copy number are also required, as are numerous expensive resources, including animal facilities that are adequate to produce and house the various transgenic lines. Others factors to consider prior to commencement:
   a. Transgenic progeny may need to be a certain age before the testing of a BAC-derived phenotype can commence, and the end-point of the experiment may be 12 mo or more beyond this.
   b. Often, a phenotype expressed by BAC transgenic mice may be subtle, necessitating the production of large numbers of transgenic positive offspring for demonstrating a phenotype with any statistical significance.

3. This protocol for the purification of BAC DNA is based on the Qiagen Plasmid Midi Kit, with some modifications. It is essential to verify BAC clones by at least one of a number of methods prior to DNA purification. Such methods include PCR, BAC-end sequencing (discussed in other chapters), or Southern blot analysis. Many previously published protocols for the preparation of circular BAC DNA by alkaline lysis demand larger culture volumes as well as larger volumes of resuspension, lysis, and neutralization buffer than are indicated here. However, we have purified >100 BAC clones with consistently high yield and quality using the method outlined here. At least 10 of these BAC DNAs have been used successfully to generate a large number of BAC transgenic lines, and the yield from one 100 mL BAC-containing *E. coli* culture is sufficient for several microinjection sessions. The modifications to standard protocols given (**steps 5**, **8**, and **15**) are critical to the success of this method.

4. Before preparing BAC DNA on this scale, you should have assayed several single colonies by PCR, e.g., with microsatellite markers that reside on this BAC to confirm its identity. BAC colonies are stored in close proximity in 384-well trays, hence it is possible for adjacent colonies to contaminate colonies of interest in the source library.

5. To ensure adequate aeration, ensure that your culture flask is at least 250 mL in size, preferably 500 mL, i.e., 5× the culture volume.

6. BAC vectors pBeloBAC11 and pBACe3.6 both contain chloramphenicol resistance genes. Check other BAC vectors for the presence of different antibiotic resistance genes.

7. Cells can be stored at this stage at –70°C in 50-mL tubes and the DNA prepared at a later stage. Ensure the pellets are thoroughly drained prior to freezing.

8. Check Buffer P2 before use for any signs of SDS precipitation, which can occur at low storage temperatures. If necessary, dissolve the SDS by warming to 37°C. It is crucial to time this step precisely; maximum total exposure time to NaOH in buffer P2 is 5 min. Longer incubation times increase the chance of nicking BAC DNA.

9. Other methods for separating cellular debris from BAC lysate include filtering through cheesecloth or sterile gauze. We have found that such methods result in considerable loss of lysate volume and, therefore, final yield. Nylon cell strainers are sterile, disposable, and result in minimal volume loss. If the lysate starts to overflow in the cell strainer, it is because the perimeter of the cell strainer has been "sealed" by lysate, reducing liquid flow. If this occurs, simply lift the cell strainer gently and slightly to release the vacuum.

10. Prewarming the elution buffer (QF) may help to increase yield by improving the efficiency of elution from the Qiagen-100 column. Eluting in five aliquots of 1 mL instead of one aliquot of 5 mL helps to reduce temperature loss from the elution buffer.

11. Unsiliconized tubes should be used, so that the BAC DNA adheres to the inside wall of the glass Corex tube; a compact pellet, as would be formed if tubes were siliconised, may be lost when the supernatant is decanted.

12. To avoid shearing of BAC circular DNA, a 200-gage "genomic" filter tip (ART cat. no. 2069G) should be used. This tip is sterile, and has a very wide bore; alternatively, one can cut off existing filter tips with flamed scissors.

13. In principle, intact circular BACs could be used as a source of DNA for microinjection. We have found, however, that such DNA contains "broken molecules", and it is preferable to purify intact, linear molecules by PFGE. Our approach is essentially based on the YAC isolation protocol of Clare Huxley with some modifications *(16)*.

    *Not*1 recognition sequences flank the cloning site in the two commonly used BAC vectors, pBeloBACII and pBACe3.6. Hence, restriction enzyme digestion with *Not*1 can be used to release a linear genomic fragment for use in microinjection experiments. However, some BAC clones will contain internal *Not*1 sites. In such cases, λ-terminase or the intron-encoded endonuclease PI-*Sce*1, can be used to linearize the pBeloBAC11 or pBACe3.6 cloning vectors, respectively, by cutting at unique sites. Alternatively, Cre recombinase can be used to linearize pBeloBAC11 at its unique *lox*P site *(17)*. BAC vector sequence (11.5 kb in the case of pBACe3.6, 7.5 kb in the case of pBeloBAC11) is contiguous and, therefore, comicroinjected with the large genomic insert in such cases. We have not systematically studied the effect of integrating a bacterial cloning vector into the mouse genome or its possible effect on the integration or expression of an attached transgene. However, studies by Jakobovits et al. *(18)* showed that integration of whole YAC genomes did not affect the expression of a human HPRT gene locus, and studies by Kaufman et al. *(19)* showed that expression of the human β-globin cluster was probably not affected by retained YAC vector sequences. Several studies have shown that DNA fragment size, rather than attached vector sequence, affects the propensity of an insert to rearrange upon integration, and this situation applies equally to YACs and BACs *(19,20)*. Larger BAC inserts of 190–200 kb are certainly more prone to mechanical breakage in vitro than smaller inserts of 100–120 kb, and should be handled extremely gently, to ensure that intact molecules integrate.

14. Add SDS to a final concentration of 1% to PI-*Sce*1 digested DNA prior to electrophoresis.

15. Under no circumstances should the central slice of agarose be exposed to UV trans-illumination or ethidium bromide, as this will nick and fragment the large DNA, possibly resulting in the integration of truncated DNA into the mouse genome.

16. The volume of DNA excised should be kept to an absolute minimum as it determines the final concentration of DNA that is to be microinjected. However, it is also important to ensure that most of the band is excised. It can be easily missed if the gel alignment is fractionally off.

17. The optimum concentration of DNA for microinjection (0.5–1 ng/µL), has been determined empirically by a number of researchers (19). Higher concentrations up to 6 ng/µL may be toxic to mouse oocytes (3). Low DNA concentration is especially important for oocytes of certain mouse strains (SJL, NOD).

18. Pronuclear injection can be carried out using procedures described elsewhere *(11)*. An overview of the microinjection procedure, as well as technical tips based on our experience with generating several BAC transgenic lines, will be presented in this section. Microinjection of mouse oocytes with BAC DNA is essentially identical to that of microinjection with small DNA fragments, the main difference being the concentration and handling of the DNA to be injected. Increasing the DNA concentration leads to a reduction in the number of offspring produced, while decreasing the concentration results in a greater number of offspring but fewer transgenic animals.

    It is possible to comicroinject two BAC inserts in the same microinjection session. This strategy halves the number of animals required in BAC rescue experiments. Once a phenotype is rescued, the two BACs can be microinjected separately at a later date to determine which of the two BACs is responsible for complementation. We have obtained efficiencies of 5–20% primary transgenics for SJL oocytes injected with two 200-kb BACs.

19. Because the host mouse genome contains endogenous alleles of the gene of interest, PCR screening of transgenic mice requires a screening protocol that is BAC-specific. If the BAC integrant is derived from the pBeloBAC11 vector, the vector-derived sequence that flanks a *Not*1-digested fragment is of sufficient length for the design of two sets of vector-specific PCR primer pairs, one on each side of the insert (*see* **Fig. 3**). If the BAC integrant is derived from the pBACe3.6 vector, however, the vector-derived flanking sequence is of insufficient length to design opposing primer pairs. Hence, it is necessary to screen the vector-insert junction using a BAC vector-specific forward primer and a reverse primer that is internal to the BAC insert. This internal primer should be screened, e.g., using the RepeatMasker database to rule out homology with interspersed repetitive elements such as LINE or SINE repeats. An advantage of screening for BAC vector ends is

that one can assume integration of the intact BAC insert. However, one can not assume from this that the DNA is not rearranged.

20. Alternative primer sequences to BACVEC_S_F and BACVEC_S_R are 5'-AGCTTGGCGTAATCATGGTC-3' and 5'-ATTAATGCAGCTGGCACGAC-3', respectively (not shown in **Fig. 3**). This primer pair gives a PCR product of 192bp.

21. pBeloBAC11_SP6 primer: 5' AGCTATTTAGGTGACACTATAG and 3' pBeloBAC11_T7 primer: 5' TAATACGACTCACTATAGGG 3'. A vector-insert specific screening strategy is essential if two BACs have been comicroinjected into the same fertilised oocyte. A PCR screen of vector sequence alone will not distinguish between integration of one, the other or both BACs. If possible, the internal BAC end primer should be designed such that its melting temperature is approximately the same as that of the SP6 and T7 primers (49 and 56°C), respectively.

22. *Taq* polymerase, e.g., AmpliTaq DNA polymerase may equally be used for BAC Tg screening of tail-derived DNA. However, we have achieved very "clean" results (minimal nonspecific amplification products) with AmpliTaq Gold, which also has the advantage that much smaller amounts of target DNA are required than with *Taq* polymerase. It is advisable to perform a titration of the amount of input DNA in the first instance. Using too much target DNA can result in non-specific amplification products.

23. Note that these conditions may need to be optimised for different PCR machines and probably for different primer sets. These "touchdown" amplification conditions *(21)* are useful for when several different internal primers will be used with the universal primers SP6 or T7, as is the case when screening several different BAC transgenic lines.

Acknowledgment

We are deeply indebted to Clare Huxley (Division of Biomedical Sciences, Imperial College School of Medicine London) for teaching us the technique of YAC DNA isolation for microinjection.

References

1. Shizuya, H., Birren, B., Kim, U. J., et al. (1992) Cloning and stable maintenance of 300-kilobase-pair fragments of human DNA in Escherichia coli using an F-factor-based vector. *Proc. Natl. Acad. Sci. USA* **89,** 8794–8797.

2. Nielsen, L. B., McCormick, S. P., Pierotti, V., et al. (1997) Human apolipoprotein B transgenic mice generated with 207- and 145-kilobase pair bacterial artificial chromosomes. Evidence that a distant 5'-element confers appropriate transgene expression in the intestine. *J. Biol. Chem.* **272,** 29,752–29,758.

3. Yang, X. W., Model, P., and Heintz, N. (1997) Homologous recombination based modification in Escherichia coli and germline transmission in transgenic mice of a bacterial artificial chromosome. *Nat. Biotechnol.* **15,** 859–865.

4. Nefedov, M., Williamson, R., and Ioannou, P. A. (2000) Insertion of disease-causing mutations in BACs by homologous recombination in Escherichia coli. *Nucl. Acids Res.* (Online). **28,** E79.

5. Lalioti, M., and Heath, J. (2001) A new method for generating point mutations in bacterial artificial chromosomes by homologous recombination in Escherichia coli. *Nucl. Acids Res.* (Online). **29,** E14.

6. Lamb, B. T., Sisodia, S. S., Lawler, A. M., et al. (1993) Introduction and expression of the 400 kilobase amyloid precursor protein gene in transgenic mice [corrected]. *Nat. Genet.* **5,** 22–30.

7. Antoch, M. P., Song, E. J., Chang, A. M., et al. (1997) Functional identification of the mouse circadian Clock gene by transgenic BAC rescue. *Cell* **89,** 655–667.

8. King, D. P., Zhao, Y., Sangoram, A. M., et al. (1997) Positional cloning of the mouse circadian clock gene. *Cell* **89,** 641–653.

9. Probst, F. J., Fridell, R. A., Raphael, Y., et al. (1998) Correction of deafness in shaker-2 mice by an unconventional myosin in a BAC transgene. *Science* **280,** 1444–1447.

10. Means, G. D., Boyd, Y., Willis, C. R., and Derry, J. M. (2001) Transgenic rescue of the tattered phenotype by using a BAC encoding Ebp. *Mamm. Genome.* **12,** 323–325.

11. Hogan, B., Beddington, R., Costantini, F., and Lacy, E. (1994) *Manipulating the Mouse Embryo*, 2nd ed. Cold Spring Harbor Laboratory Press, Cold Spring Harbor, NY.

12. Laird, P. W., Zijderveld, A., Linders, K., Rudnicki, M. A., Jaenisch, R., and Berns, A. (1991) Simplified mammalian DNA isolation procedure. *Nucl. Acids Res.* **19,** 4293.

13. Chrast, R., Scott, H. S., and Antonarakis, S. E. (1999) Linearization and purification of BAC DNA for the development of transgenic mice. *Transgen. Res.* **8,** 147–150.

14. Boulukos, K. E. and Pognonec, P. (2001) MICE, a program to track and monitor animals in animal facilities. *BMC Genet.* **2,** 4.

15. McKie, M. A. and Webb, S. (1999) LAMS-A laboratory animal management system. *Mamm. Genome.* **10,** 349–351.

16. Huxley, C. (1998) Exploring gene function: use of yeast artificial chromosome transgenesis. *Methods* **14,** 199–210.

17. Mullins, L. J., Kotelevtseva, N., Boyd, A. C., and Mullins, J. J. (1997) Efficient Cre-lox linearisation of BACs: applications to physical mapping and generation of transgenic animals. *Nucl. Acids Res.* **25,** 2539–2540.

18. Jakobovits, A., Moore, A. L., Green, L. L., et al. (1993) Germ-line transmission and expression of a human-derived yeast artificial chromosome. *Nature* **362,** 255–258.

19. Kaufman, R. M., Pham, C. T., and Ley, T. J. (1999) Transgenic analysis of a 100-kb human beta-globin cluster-containing DNA fragment propagated as a bacterial artificial chromosome. *Blood* **94,** 3178–3184.

20. Peterson, K. R., Navas, P. A., Li, Q., and Stamatoyannopoulos, G. (1998) LCR-dependent gene expression in beta-globin YAC transgenics: detailed structural studies validate functional analysis even in the presence of fragmented YACs. *Human Molec. Genet.* **7,** 2079–2088.

21. Don, R. H., Cox, P. T., Wainwright, B. J., Baker, K., and Mattick, J. S. (1991) 'Touchdown' PCR to circumvent spurious priming during gene amplification. *Nucl. Acids Res.* **19,** 4008.

13

BAC Rescue

A Tool for Functional Analysis of the Mouse Genome

Deborah A. Swing and Shyam K. Sharan

1. Introduction

Transgenic mice are widely used for determination of gene function and to generate animal models of human disease *(1)*. Such mice are generated by introducing a gene of interest into the genome and ectopically expressing it using a heterologous promoter. The choice of the heterologous promoter depends upon the tissue or the time at which a particular gene has to be expressed. Although these transgenic mice have been very informative, quite commonly the transgenes do not accurately reflect the expression pattern of the endogenous promoter owing to the lack of certain regulatory sequences that may not be present in the sequence used to drive the expression of the transgene. This disparity may be attributed to the fact that the *cis*-regulatory sequences may be located several kilobases (kb) away from the coding region or may even be embedded in the neighboring gene *(2,3)*. In addition, the presence of the *cis*-acting elements cannot be determined merely by the DNA sequence information. Bacterial artificial chromosomes (BACs) contain large segments of genomic DNA (> 300 kb) and thus, most likely, contain all the *cis*-regulatory elements of a gene *(4)*. This large insert size provides an opportunity to express any gene under the control of the regulatory elements that is likely to mimic the endogenous expression pattern. BACs are, therefore, ideal for generating transgenic mouse lines.

1.1. Functional Complementation Using BACs: Applications and Advantages

The presence of the entire coding sequence of a gene within a BAC can be determined by molecular analysis (sequencing, by polymerase chain reaction

From: *Methods in Molecular Biology, vol. 256:*
Bacterial Artificial Chromosomes, Volume 2: Functional Studies
Edited by: S. Zhao and M. Stodolsky © Humana Press Inc., Totowa, NJ

[PCR] or Southern blotting). However, the presence of the entire regulatory region can be only achieved by testing its ability to function like the endogenous gene. This may be accomplished by comparing the expression pattern of the transgene with that of the endogenous gene or by testing the ability of the BAC to fully complement a loss-of-function mutation in the endogenous gene. If the BAC is able to revert the mutant phenotype back to wild-type, then it would suggest that the entire gene is present in the BAC.

BACs containing all the coding and regulatory sequences of a gene can be used for a wide range of applications. They can be used for expression studies by inserting a reporter gene like *LacZ* or the green fluorescent protein into the coding region of the gene of interest *(5,6)*. In addition, they can provide tools for the characterization of the promoter region. For example, by generating deletions or subtle mutations, the *cis*-regulatory elements can be identified. BACs that contain all the regulatory elements of a gene can also be used to generate transgenic lines to express the site specific recombinase genes *Cre* or *Flp (7,8)*. These recombinases are being widely used to generate tissue specific knockout mice. A transgenic mouse expressing the *Cre* gene under the control of the neuron-specific Enolase (*Eno2*) gene promoter has been recently generated by inserting the *Cre* gene near the 3′ end of the *Eno2* gene in the BAC. The *Cre* transgene has been demonstrated to be in all the mature neurons *(7)*. The pattern of expression is similar to the endogenous gene with no evidence of position effect *(9)*. In contrast, transgenic lines that were generated by using small promoter sequences to express a transgene showed variation between multiple lines and rarely recapitulate the endogenous gene expression completely *(7)*.

BAC rescue in mice also provides an opportunity to test the functional conservation of orthologs of any gene. In spite of the divergence of the nucleotide or amino acid sequences, such orthologs may retain the basic function of the gene. For example, the human breast cancer susceptibility protein, BRCA1 and its murine ortholog, show 59% amino acid sequence identity, which led to the speculation whether the two genes have diverged enough during the course of evolution that their functions have also changed *(10)*. Therefore, we have used the BAC complementation technique to test whether the human *BRCA1* gene can function in mice *(11)*. A BAC clone with a 200-kb insert, containing the 80-kb human *BRCA1* gene, 30-kb downstream, and 90-kb of sequence from the upstream region, was used to generate transgenic lines. Surprisingly, the BAC transgene successfully rescued the embryonic lethal phenotype associated with a targeted mutation in the murine *Brca1* gene. In addition, the expression of the human *BRCA1* gene mirrored the murine *Brca1* gene expression pattern suggesting functional conservation of regulatory elements between the two species *(11)*. This system is now being used to express versions of *BRCA1* mutations found in human cancer patients that will facilitate our understanding of their role in the process of tumorigenicity.

In positional cloning experiments, BAC rescue of a mutant phenotype provides a simple, but definitive method to determine which region of the critical interval contains the mutated gene. This approach is particularly useful when the critical region is several hundred kilobases and spans multiple BAC clones. Functional complementation using BACs helps to narrow down the region to a single BAC clone. BAC complementation has been used to identify the gene involved in various mouse mutations including the *clock, shaker-2, leaden, ashen,* and *tattered* genes *(12–17)*. In this chapter, the methods involved in performing a BAC rescue experiment in mice is described.

2. Materials

2.1. Selecting BAC Clone(s) for the Rescue Experiment

1. BAC library filters or BAC DNA pools and individual clones arrayed in 384/96-well plates can be obtained from BACPAC Resources, CHORI (www.chori.org) or Research Genetics, Inc. (www.resgen.com).
2. DNA probes or PCR primers from the 5′ and 3′ ends of the gene or region of interest.
3. BAC end sequencing primers: Primers to sequence the ends of insert cloned into pBeloBAC11 vector *(18)*: T7 primer (forward): 5′-TAATACGACTCACTATA GGGCGAATTCGAGCTCGG-3′; SP6 primer (reverse): 5′-GATTACGCCAAGC TATTTAGGTGACACTATAGAATAC-3′. Primers to sequence the ends of insert cloned into pBACe3.6 vector *(19)*: T7 primer (forward): 5′-TTTGCGGCCGC TAATACGACTCACTATAGGGAGAG-3′; SP6 primer (reverse): 5′-CGCCTG GCCGTCGACATTTAGGTGACACTATAG-3′.
4. BigDye Terminator ready reaction Kit (cat. no. 4302149, Applied Bio-systems, Inc.).
5. Centri-Sep Columns (cat. no. CS-901, Princeton Separations).
6. SpeedVac Plus (SC110A, Savant).
7. Sequencing machine (e.g., ABI Prism 377 sequencer, Applied Biosystems Inc.).
8. Thermocycler (e.g., PTC, MJ Research, Inc.).
9. *Taq* DNA polymerase and deoxynucleotide 5′-triphosphates (dNTPs).

2.2. BAC DNA Preparation

1. Alkaline lysis solutions I: 25 m*M* Tris-HCl, pH 8.0, 50 m*M* glucose, 10 m*M* ethylenediamine tetraacetic acid (EDTA), pH 8.0, 2.5 mg/mL of Lysozyme (cat. no. 107 255, Roche Molecular Biochemicals) and RNase A (100 µg/mL).
2. Alkaline lysis solution II: 0.2*N* NaOH and 1% Sarcosyl (freshly prepared).
3. Alkaline lysis solution III: 5 *M* potassium acetate, pH 4.8.
4. 1X TE: 10 m*M* Tris-HCl, pH 8.0, 1.0 m*M* EDTA.
5. Ultracentrifuge (L8-M, Beckman).

6. Ultracentrifuge rotor (VTi65.2, Beckman).
7. Quick-seal tubes (cat. no. 344075, Beckman).
8. Cesium Chloride (Omnipure, cat. no. 3030, EM Sciences).
9. Slide-A-Lyzer, dialysis cassettes 10 000 MWCO (Catalog No. 66425, Pierce).
10. 6-mL Falcon tubes (cat. no. 2063, Falcon).

2.3. Generating Transgenic Mice

2.3.1. Mice

Factors that play an important role in generating transgenic mice include the genetic background (strain) of the mice, age, diet, and housing conditions. Mouse rooms should be isolated to avoid loud noises and vibrations. Any kind of stress severely reduces the number of eggs obtained from donor females. Ideal conditions for maintaining mouse colonies include an internal barrier facility with a 12-h day and night cycle, 70–72°F room temperature and a relative humidity of 40–60%. The amount of fat in the diet plays an important role in the egg production. A 9% fat containing diet (Purina 5021, autoclavable mouse chow) is recommended for the strains described below.

1. **Donor females.** We have tested multiple mouse strains and found that the genetic background plays an important role in the number of eggs that can be recovered after superovulation of the donor females. In general, F_1 hybrids produce larger number of eggs, have better sperm, show higher rate of fertilization, and survive the manipulation and rigors of microinjection and consequently, produce greater number of founder animals. The combination of using 4–6 wk old C57BL/6NCr × C3H/HeNMTV$^-$ F_1 hybrid females (B6C3F1/NCr from Charles River at NCI-Frederick) as primary donor and C57BL/6J × C3H/HeJ F_1 stud males (B6C3F1/J from The Jackson Laboratories, Bar Harbor) gives the best results. Such matings typically yield on an average 30 eggs per plugged female.
2. **Recipient females.** C57Bl/6NCr × DBA/2NCr F_1 hybrid females (B6D2F1/NCr from NCI-Frederick animal facility) are used as recipient mothers. Sterile males used to mate with these females are C57Bl/6NCr × DBA/2NCr F_1 hybrids (B6D2F1/NCr from NCI-Frederick animal facility). Males are vasectomized at the age of 6 wk and by 9 wk are ready to mate with recipient females (*see* **Note 1**).

2.3.2. Equipment for Microinjection

1. Stereomicroscope (Wild M5A, Leica Inc.) for egg collection.
2. Surgical Microscope (OPMI 1, Zeis) to perform surgery and oviduct collection.
3. Inverted Microscope (Nikon Diaphot TMD with Diascopic Differential.
4. Interference Contrast (DIC) Nomarski attachment (*see* **Note 2**) for microinjection.
5. Micromanipulator (Leitz).
6. Vertical Pipet puller (Kopf Model 720).
7. Microforge (DeFonbrune, Model MF80).

2.3.3. Microinjection and Embryo Culture Media

Whitten's 640 media is used for embryo culture (*see* **Note 3**). Because the media has no preservative, it must be kept gassed when not used for egg manipulation. Store at 4°C and use within 3 wk.

Whitten's 640 media is prepared as follows.

1. NaCl (Sigma cat. no. S5886), 640.0 mg /100 mL.
2. KCl, (Sigma cat. no. P5405), 36.0 mg/100 mL.
3. KH_2PO_4 (Sigma cat. no. P5655), 16.0 mg/100 mL.
4. $MgSO_4$ (Sigma cat. no. M2643), 29.4 mg/100 mL.
5. $NaHCO_3$ (Sigma cat. no. S5761), 190.0 mg/100 mL.
6. a-D(+)-Glucose (Sigma cat. no. G6138), 100.0 mg/100 mL.
7. L(+)-Lactic Acid (Sigma cat. no. L2000), 53.0 mg/100 mL.
8. Sodium Pyruvate (Sigma cat. no. P2256), 2.5 mg/100 mL.
9. Penicillin G (Sigma cat. no. P7794), 7.5 mg/100 mL.
10. Streptomycin Sulphate (Sigma cat. no. S6501), 5.0 mg/100 mL.

Mix and dissolve the above reagents in MilliQ purified water. Add 0.2 mL 0.5% Phenol Red solution (diluted in PBS) and 0.1 mL Na_2EDTA solution (prepared by mixing 363.4 mg/100 mL water, pH 7.5). Gas the solution for 10 min with 5% O_2, 5% CO_2, and 90% N_2. Add 300 mg BSA Fraction V (Sigma cat. no. A3311) and disslove into the solution. Filter-sterilize (0.2 μm filter) and aliquot into desired volume.

2.3.4. Hormones

1. Pregnant mare serum (PMS): Prepare working solution of PMS (cat. no. G4877, Sigma), 50 IU/mL saline solution (0.9% NaCl). Inject 5 IU(0.1 cc) per female.
2. Human chorionic gonadotropin (hCG): Prepare working solution of hCG (cat. no. CG-5, Sigma) 50 IU/mL Saline Solution (0.9% NaCl). Inject 5 IU (0.1 cc) per female, 46 h after the PMS injection.

2.3.5. Avertin

Prepare stock solution by adding 15.5 mL ot tert-amyl alcohol (Aldrich cat. no. 24,048-6) to a 25 g bottle of 2,2,tribromoethanol (Aldrich, cat. no. T4,840-2). Wrap the bottle with aluminum foil. Mix the bottle well and place at 37°C for 15–20 min. Store in a dark bottle at 4°C. To prepare the working solution, dilute 0.5 mL of the stock in 40.5 mL of 0.9% saline solution in a brown glass bottle (Avertin is light-sensitive). Shake the bottle vigorously and warm at 37°C for 30 min. Store in the dark at room temperature for 6–8 mo. Dosage: inject 0.03 mL per gram of body weight.

2.3.6. Hyaluronidase

Type III from Sheep testis (cat. no. H2251, Sigma).

3. Methods

3.1. Screening BAC Genomic Library

The first step in performing a BAC rescue experiment using transgenic mice is to identify the BAC(s) that spans the critical region. In the case of a known gene, the BAC should contain at least the 5′ and 3′ ends of the gene (*see* **Note 4**). In the case of an unknown gene, the molecular markers that define the proximal and distal limits of the region containing the gene should be present within the BAC clone(s).

Screen a BAC genomic library using appropriate probes. The details of a BAC library screening method are described in Chapter 6 of Volume 1. Depending upon the library and type of DNA probes available, identify the positive clone by hybridizing the library filters with radioactive-labeled probe or screen by a PCR-based method. Based upon the complexity of the library, i.e., the number of haploid genomes represented in the library, multiple positive clones will be obtained. Identify the BAC clones that contain the 5′ as well 3′ ends of the gene or the two markers that define the limits of the region of interest. To identify BACs that span a large interval, screen the BAC library with multiple probes that map within this region.

When several BAC clones are obtained, it is important to select a BAC that contains the most upstream region of the gene. As most of the *cis*-regulatory elements are present in this region, selecting such a BAC is desired to increase the probability of including all regulatory elements within the transgene. Identifying the BAC with most upstream sequence can be accomplished by mapping the BAC ends relative to each other. A PCR-based method to map the BAC ends is described in **Subheading 3.4.** BACs that contain only one end of the gene are not suitable in complementation studies. However, they are useful for orienting the BAC ends with respect to the gene.

3.2. Rapid Extraction of BAC DNA

The mini prep method described *(20)* below works well for rapid extraction of DNA and produces DNA that is suitable for restriction digest, PCR analysis and sequencing.

1. Inoculate a single colony into 3 mL LB containing chloramphenicol (20 µg/mL) and grow overnight at 37°C in a shaking incubator at 250 rpm.
2. The next day, pellet 1.5 mL culture by centrifugation in a 1.5-mL tube at 10,000*g* for 1 min.

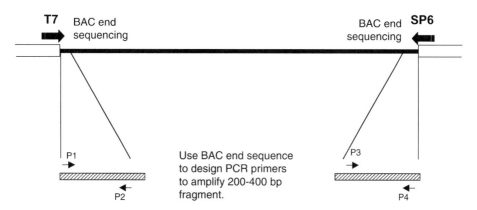

Fig. 1. Generating BAC end-specific primers. The bold line represents the insert of a BAC clone with each of the two open-end boxes representing the T7 and SP6 ends. To generate PCR primers specific for each end of the BAC, sequence the BAC end using primers specific for each end (shown as bold arrow). The sequence generated from each end (shown as hatched box) is used to generate end specific primer sets, P1-P2 and P3-P4.

3. Resuspend in 100 μL chilled alkaline lysis solution I by mixing the cells up and down using the pipet tip. Place on ice for 5 min.
4. Add 200 μL alkaline lysis solution II. Mix by gently inverting the tubes until the lysate is clear. Place on ice for 5 min.
5. Add 150 μL alkaline lysis solution III. Mix gently until a white precipitate appears. Place on ice for 5 min. Centrifuge the tubes at 10 000g for 5 min and then transfer the supernatant to a clean 1.5-mL tube.
6. Add 2 vol 95% ethanol to the supernatant. Mix and place on ice for 30 min. Centrifuge at 10 000g for 10 min. Discard supernatant and wash the pellet with 70% ethanol. Air-dry the pellet and resuspend in 30–50 μL 1X TE.

3.3. Generate PCR Primers From the T7 and SP6 Ends of Each BAC

A general scheme to generate PCR primer specific to the two ends of a BAC insert is shown in **Fig. 1**. Use primers specific for the T7 and SP6 ends to determine the sequence of the ends of the BAC insert. The sequencing of two ends of a BAC is described below.

1. Mix about 2–5 μg of BAC DNA (*see* **Note 5**), 1 μL 5 μ*M* sequencing primer and 8 μL of BigDye Terminator Reagent mix (PE Applied Biosystems) per 20 μL reaction volume.
2. Reaction conditions include: initial denaturation at 94°C for 1 min followed by 30 cycles of 94°C–30″; 50°C–15″; 60°C–4′.

3. Remove unincorporated nucleotides by passing sequencing reaction through Centri-Sep (Princeton Separations) columns and dry under vacuum in SpeedVac at room temperature.
4. Generate the sequence data using a sequencing machine and appropriate software (e.g., ABI Prism 377).

3.4. Mapping BAC Ends Using PCR

1. Based on the BAC end sequence generated above, design 18–20-mer PCR primer pairs from each end of the BAC insert to amplify about 200–400 bp fragment (*see* **Note 6**).
2. Using the BAC DNA as template, determine which BAC DNA contains sequences specific for each primer pair. For example, if there are three BACs then a total of 18 PCR reactions will have to be set up, three reactions (one with each BAC as template) for each primer pair. For three BACs there will be six primer pairs. As a rule, a negative control (with no template) should also be included for each primer set. The BAC used to generate the primer pair serves as a positive control.
3. To set up the PCR, use 10–50 ng of BAC DNA and 200–300 ng of each primer, 0.2 mM dNTPs, 1X PCR buffer and 1.25 U of *Taq* DNA polymerase per 50 µL reaction volume. PCR conditions include: initial denaturation at 94°C for 2 min followed by 35 cycles of 94°C–1'; 55°C–1'; 72°C–2'.
4. Run 25 µL of the PCR on a 1% agarorse gel.

The PCR primer set that amplifies the expected product only from the BAC that was used to design it, represents either the most proximal or the distal end. To determine if a particular end represents the upstream (5') or downstream (3') region of the gene, map these ends using the BAC clones that contain only the 5' end or the 3' end of the gene. The BAC that contains the most upstream region and the entire coding region should be selected for BAC rescue experiment (*see* **Fig. 2A**).

In the case of a large region where multiple BACs span the region of interest and the gene is unknown, BACs with maximum overlapping sequence should be used to rescue a mutant phenotype. This increases the likelihood of having the full-length gene in at least one of the BAC clones (*see* **Fig. 2B**).

3.5. Performing BAC Rescue in Mice

To generate transgenic mice using the BAC DNA, freshly prepared super-coiled DNA is used for microinjection (*see* **Note 7**). DNA purified by Cesium chloride ultracentrifugation should be used.

3.5.1. Preparing BAC DNA for Microinjection

1. Start an overnight culture of the BAC clone in 500 mL LB containing chloramphenicol (20 µg/mL). Allow the culture to grow for 16–20 h. Pellet the bacterial culture by spinning at 6000g for 15 min at 4°C. Discard the supernatant.

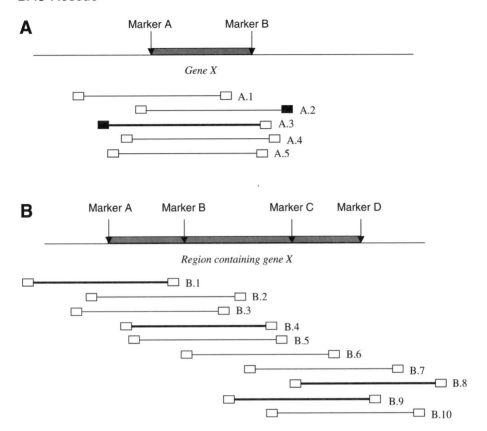

Fig. 2. Identification of BACs to be used in rescue experiments. (**A**) To identify the BAC clone that may rescue the mutation in a known gene *X*, screen the BAC library with the 5′ and 3′ end probes (A and B, respectively). Of the 5 clones obtained (A.1, A.2, A.3, A.4, and A.5), all except A.1 contain both markers. Map the ends (shown as open boxes) with respect to each other to determine the most distal ends (e.g., the ends of A.2 and A.3 shown as solid box). To orient these two BAC ends with respect to the gene, identify which of these ends is present or absent in BAC A.1, which lacks the 3′end of *X*. The distal end of A.3 is present in A.1 thus it represents the 5′ end of the gene where as the distal end of A.2 is not present in A.1 thus it represents the 3′ end of *X*. Therefore, BAC A.3 (insert shown with bold line) has most upstream region of gene *X*. (**B**) To identify BACs that may rescue a mutation in *gene X* that maps between markers A and D, screen a BAC library using each of the fours markers, A, B, C, and D that span the region containing *X*. Map the ends of the BACs (shown as open boxes) with respect to each other as well as the four markers to determine their relative order. Identify the minimum number of BACs that span the entire region, but have sufficient overlap. Based on the map shown here, B.1, B.4, B.8, and B.9 should be used to rescue the mutation in *X*.

2. Resuspend the cell pellet in 35 mL chilled alkaline lysis solution I. Add 70 mL freshly prepared alkaline lysis solution II. Gently mix the solution by inverting the tube 5–6 times. Stand on ice for 5 min.

3. Add 52.5 mL alkaline lysis solution III. Gently invert to mix 5–6 times. Stand on ice for 15–20 min. Centrifuge at 15 000g for 30 min at 4°C.

4. Filter the supernatant through a single layer of cheesecloth. Add 0.6 vol of chilled Isopropanol. Gently mix by inverting the tube 5–6 times and stand on ice for 30 min.

5. Centrifuge at 15 000g for 30 min at 4°C. Discard supernatant and air dry the pellet. Resuspend in 4 mL TE, pH 8.0 (incubate at 65°C for 10 min with intermittent mixing). Resuspended sample may appear cloudy. Transfer supernatant to a 15-mL tube. Add RNase A to a final concentration of 20 µg/mL. Incubate at 37°C for 30 min.

6. Add 4.64 g of cesium chloride (CsCl) per 4 mL TE. Invert to mix until CsCl is completely dissolved. Add 400 µL of Ethidium-bromide (10 mg/mL). Gently mix and centrifuge at 1600g for 10 min.

7. Transfer the supernatant into Quick-seal tubes. Fill to the top with mineral oil (if required). Balance and seal the tubes. Place the tubes into the ultracentrifuge rotor (VTi65.2, Beckman). Run samples at 20–25°C for 16–20 h at 58 000 rpm in a L8-M Beckman ultracentrifuge with zero deceleration.

8. Following centrifugation, gently remove the tubes from the rotor. View tube under long wavelength ultraviolet (UV) source (354 nm). Vent the sealed ultra tube at the top using a 25G needle. Using a 1-cc syringe attached to an 18 gage needle, pierce into the tube just below the lowest band, which contains the supercoiled DNA, and carefully draw out the band. Transfer the sample to a 6-mL tube (cat. no. 2063, Falcon).

9. Extract with water-saturated 1-Butanol to remove Ethidium bromide. Add equal volume of 1-Butanol to the DNA sample, mix and let the tube stand until two layers are formed. 1-Butanol forms the top pink layer while the DNA remains in the lower aqueous layer. Gently discard the top layer and repeat extraction until the aqueous layer is clear.

10. To remove the Cesium chloride, dialyze at 4°C in 1 L sterile 1X TE using Slide-A-lyzer. Dialyze for 24–36 h changing the TE buffer 2–3 times. Check concentration of the DNA. Prepare 100–200 µL multiple dilutions of the DNA (4 ng/µL, 2 ng/µL, 1 ng/µL, and 0.5 ng/µL) in filter sterilized TE (10 mM Tris-HCl, pH 7.4; 0.1 mM EDTA) for microinjection.

3.5.2. Generating Transgenic Mice

A detailed description of the microinjection procedure is beyond the scope of this chapter. Instead some of the key steps involved in egg collection and treatment of embryos after microinjection will be described (for a detailed description of microinjection procedure *see* **ref. *21***). To test the ability of a BAC to rescue a mutant phenotype in the founder transgenic animal, it is ideal to generate transgenic mice in the mutant background. If the homozygous mutants are viable and fertile, then homozygous donor females should be crossed to homozygous mutant males. If the mutation results in lethality or sterility in homozygous state, then fertilized eggs should be collected from heterozygous females mated with heterozygous males. The functional complementation of the mutation can be examined in the homozygous mutant transgenic animals. If the mutation is present in a strain of mice that either yields very low number of eggs after superovulation or the males have low sperm count, one solution is to use wild-type females and males from a strain that provides good results. The transgenic founder animals can then be crossed to the heterozygous/homozygous mutant animals to obtain heterozygous transgenic F_1 animals. The ability of the BAC to rescue the mutant phenotype can be examined in the homozygous mutant F_2 offspring by crossing heterozygous transgenic mice to heterozygous/homozygous mutant animals.

3.5.3. Egg Collection and Microinjection

1. Three days prior to collecting the fertilized eggs, inject 0.1 cc of PMS (5 IU) into the intraperitoneal cavity of 20 four- to six-week-old donor females around 4 PM.
2. After 46 h inject 0.1 cc of hCG (5 IU) and setup mating of each female with a stud male (preferably 2–10 mo old).
3. The following morning, check for copulatory plugs in the morning by 7 AM. Females that have successfully mated as evidenced by the copulatory plug are separated for use.
4. Euthanize five to six females at a time and dissect out the oviducts using no. 5 dumont tweezers and a pair of dissecting scissors under a surgical microscope (OPMI, Zeiss). Place the oviducts into a 35 mm petri dish containing Whitten's 640 medium. Transfer one oviduct at a time into another 35 mm petri dish containing 3 mL of warm (37°C) Whitten's 640 medium and 700 U/mg/mL of hyaluronidase. Using a Wild5A dissecting microscope, free the eggs from the oviduct using tweezers to hold the oviduct and a 28-G needle to prick open the oviduct. Allow the eggs to empty into the media and then discard the emptied oviduct. Repeat this procedure with rest of the oviducts.
5. After all the oviducts have been emptied (approx 10–15 min) start collecting the eggs from the first oviduct emptied. By now, the eggs are free of all the adhering cumulus cells and supernumeracy sperm. Avoid incubating the eggs for more that 30 min in the hyaluronidase medium. Transfer the eggs into a clean 35 mm petri dish containing only Whitten's 640 medium and evaluate them for normalcy. Keep the petri dish in a gassing chamber on a 37°C warming tray until ready for microinjection.
6. Remove 150–200 eggs at a time whose pronuclei have not yet fused from the petri dish and place in a microdrop of media in a microinjection dish. Also place two microdrops of BAC DNA, above and below the media. Cover all the three microdrops with 3 mL of equilibrated paraffin oil. Place the microinjection dish on the inverted microscope equipped with Leitz manipulators. Fill the microinjection

needle with the BAC DNA and start injecting into the eggs following standard microinjection procedure (see Notes 8 and 9). The concentration of the BAC DNA plays an important role in determining the ease with which it can be microinjected into an egg. 0.5–4 ng/µL DNA concentration should be tried..

3.5.4. Egg Transfer and Pseudopregnant Females

After microinjecting the BAC DNA into the eggs, culture the "healthy" looking eggs at 37°C overnight in Whitten's 640 media. Next day, transfer the embryos that have proceeded to the two-cell stage, into the oviduct of the pseudopregnant recipient females. Certain mutant/inbred lines do not culture well and need to be transferred directly after microinjection. To obtain recipient females setup mating of 6-wk-old females in natural estrus with vasectomized males a day before the embryo transfer. Females with a copulatory plug should be used as recipient mothers. Although the number of embryos transferred per female will depend upon the number of pseudopregnant recipient females available, avoid overcrowding. Ideally, transfer 8–10 embryos per oviduct that gives sufficient space per embryo and increases the survival rate.

3.5.5. Genotyping Founders and Examining BAC Rescue

If the eggs are collected from the intercross of homozygous mutant animals with a visible phenotype, then the presence of wild-type animals will suggest that the BAC rescues the mutant phenotype (*see* **Note 10**). Genotype the rescued animal to confirm that the rescue is linked to the presence of the BAC transgene. In the case of a nonvisible phenotype, genotype the mice to identify the homozygous mutant transgenic mice. Analyze these animals for functional complementation. When one or both the parents used for egg collection were wild-type then test for functional complementation in the F_1 or the F_2 generation.

Wean the pups and clip a quarter-inch portion of their tail to extract genomic DNA. Although PCR can be used for genotyping, Southern blotting technique is recommended. In addition to accurately identifying the transgenic mice, Southern blot analysis provides the approximate copy number of the transgene. The *Chloramphenicol* resistance gene (*see* **Note 11**) is an ideal probe to genotype BAC transgenic mice. To determine the number of copies of the BAC DNA integrated into the genome, rehybridize the blot with a genomic probe not present in the BAC insert. The intensity of the hybridization signal of this probe will serve as an internal control for quantitative analysis. Transgenic lines that contain 2–4 copies of the BAC should be preferred for further analysis. This is to avoid overexpressing the gene of interest as well as to increase the probability of having a few copies of the intact gene that may get disrupted during the integration of the BAC into the genome.

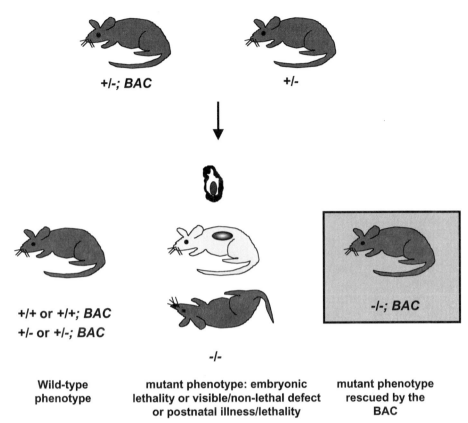

+/-; BAC   **+/-**

+/+ or +/+; BAC
+/- or +/-; BAC

-/-; BAC

-/-

| Wild-type phenotype | mutant phenotype: embryonic lethality or visible/non-lethal defect or postnatal illness/lethality | mutant phenotype rescued by the BAC |

Fig. 3. Genetics of BAC rescue. When heterozygous (+/–) mice are intercrossed, wild type (+/+), heterozygous (+/–) and homozygous (–/–) mutant offspring are obtained. The phenotypic effect of a recessive mutation is seen in homozygous animals. When the mutation results in embryonic lethality, no homozygous animal is obtained. To examine if a BAC clone is able to rescue the mutant phenotype, heterozygous mutant transgenic (+/–; *BAC*) mice are crossed with heterozygous mutant (+/–) mice. If homozygous mutant transgenic (–/–; BAC) mice are obtained (as shown in shaded box) that are indistinguishable from their wild type or heterozygous transgenic littermates then the BAC is able to complement the mutation.

As mentioned earlier, when BAC rescue cannot be performed in the founder animals, set up a mating between heterozygous mutant mice and six weeks old transgenic mice with 2–4 copies of BAC transgene that is heterozygous for the mutation (*see* **Fig. 3**). Homozygous mutant mice should be examined for rescue of the mutant phenotype. If the mutation is known to cause late-onset phenotype like tumorigenesis, behavioral phenotype, or other kind of illness, monitor mice for a longer period of time along with control littermates (homozygous mutant nontransgenic as well as heterozygous transgenic) to examine for the functional complementation by the BAC. In addition, intercross the rescued animals to test if they are fertile.

4. Notes

1. Use the vasectomized males every other week to obtain optimal performance. Use males until they are 10 mo old.
2. The DIC Nomarski attachment gives a much sharper image of the egg pronucleus, which speeds up the microinjection process. The microinjection is performed at 300× magnification (15× eyepiece and 20× objective).
3. In case, the 129/SV mouse strain is used, the Human Fetal Fluid media with glutamine and EDTA results in better survival of the embryos.
4. If the size of the gene is greater than the average insert size of the BAC library, it may be possible that the no BAC clone may contain the entire gene. Therefore, BAC complementation may not be feasible and other approaches, e.g., using yeast artificial chromosomes, may have to be considered.
5. Increasing the amount of BAC DNA to 2–5 μg and the length of sequencing primer to approx 33–35 bases (instead of 18–20 bases) improves the quality of the sequence.
6. Analyze the sequence for the presence of repetitive elements. PCR primer sequence should be selected from nonrepetitive regions.
7. The quality of the DNA plays a key role in the microinjection process. Make sure that the DNA is prepared fresh for microinjection. BAC DNA stored at 4°C for more than a week tends to gradually degrade and should not be used for microinjection. Do not precipitate the DNA after dialysis. Check the pH of the TE (10 mM Tris-HCl, pH 7.4; 0.1 mM EDTA) solution. Incorrect pH may result in lysis of the eggs. Although several laboratories use linear BAC DNA for microinjection, we prefer using supercoiled DNA. This prevents shearing of the DNA during microinjection. However, this may appear to increase the probability of disrupting the gene when the BAC integrates into the genomes. In contrast, others and we have found that at least one fully functional copy remains intact in transgenic lines that have 2–4 copies of the BAC DNA *(11,12)*.
8. The pipet puller is set to pull a long gradually tapering needle, which provides more control over the flow of DNA. The holding pipet opening is about one-third the size of the egg. The size of the opening of the needle for microinjecting the BAC DNA is larger than those used to inject smaller DNA fragments. BAC DNA is more viscous and tends to clog the needles with smaller opening. Prior to using the needle, etch the opening with 25% Hydrofluoric acid (HFl). This is done by attaching the needle to a piece of polyethylene tubing connected to a 10-cc syringe. Keeping constant pressure on the plunger of the syringe, slowly drag the tip of the needle across the surface of the acid, simultaneously.pushing air out of the syringe. The size of the air bubble produced increases with the size of the opening. The tip is etched until the desired size bubble is formed. Rinse the needle

twice with MilliQ water and then with methanol. The constant pressure exercised on the plunger of the syringe prevents any liquid from entering into the needle.

9. The microinjection needle is never reused. The holding pipet is changed when the microinjection dish is changed. On an average about 150–200 eggs are transferred to each microinjection dish and about 4–5 needles are used to inject all the eggs.

10. Rescue in founders may not occur owing to chimerism (i.e., not every cell of the founder animal may have the transgene) or chromatin reorganization. Sometimes, F_1 animals show rescue of the mutant phenotype, even if the founder transgenic animals do not show the rescue.

11. PCR primers to amplify chloramphenicol gene: forward primer 5′-CACTGGATAT ACCACCGTTG-3′ and reverse primer 5′-AGCATTCTGCCGACATGGAA-3′.

Acknowledgments

The authors would like to thank Srividya Swaminathan, Lino Tessarollo, and Scott Wilson for critical review of the manuscript. Research sponsored by the National Cancer Institute, DHHS.

References

1. Houdebinem, M. L., ed., (1997) Generation and use, in *Transgenic Animals*, Harwood Academic, New York.

2. Scholz, H., Bossone, S. A., Cohen, H. T., Akella, U., Strauss, W. M., and Sukhatme, V. P. (1997) A far upstream *cis*-element is required for Wilms' tumor-1 (WT1) gene expression in renal cell culture. *J. Biol. Chem.* **272,** 32,836–32,846.

3. Valarche, I., de Graaff, W., and Deschamps, J. (1997) A 3′ remote control region is a candidate to modulate Hoxb-8 expression boundaries. *Int. J. Dev. Biol.* **41,** 705–714.

4. Shizuya, H., Birren, B., Kim, U. J., et al. (1992) Cloning and stable maintenance of 300-kilobase-pair fragments of human DNA in *Escherichia coli* using an F-factor-based vector. *Proc. Natl. Acad. Sci. USA* **89,** 8794–8797.

5. Mejia, J. E. and Monaco, A. P. (1997) Retrofitting vectors for *Escherichia coli*-based artificial chromosomes (PACs and BACs) with markers for transfection studies. *Genome Res.* **7,** 179–186.

6. Jessen, J. R., Meng, A., McFarlane, R. J., et al. (1998) Modification of bacterial artificial chromosomes through chi-stimulated homologous recombination and its application in zebrafish transgenesis. *Proc. Natl. Acad. Sci. USA* **95,** 5121–5126.

7. Lee, E.-C., Yu, D., Martinez de Velasco, J., et al. (2001) A highly efficient *Escherichia coli*-based chromosome engineering system adapted for recombinogenic targeting and subcloning of BAC DNA. *Genomics* **73,** 56–65.

8. Farley, F. W., Soriano, P., Steffen, L. S., and Dymecki, S. M. (2000) Widespread recombinase expression using FLPeR (flipper) mice. *Genesis* **28,** 106–110.

9. Marangos,J. P. and Schmechel, D. E. (1987) Neuron Specific enolase, a clinically useful marker for neurons and neuroendocrine cells. *Annu. Rev. Neurosci.* **10,** 269–295.

10. Xu, C, F., Chambers, J. A., and Solomon, E. (1997) Complex regulation of the *BRCA1* gene. *J. Biol. Chem.* **272,** 20,994–20,997.

11. Chandler, J., Hohenstein, P., Swing, D. A., Tessarollo, L., and Sharan, S. K. (2001) Human BRCA1 gene rescues the embryonic lethality of *Brca1* mutant mice. *Genesis* **29,** 72–77.

12. Antoch, M. P., Song, E. J., Chang, A. M., et al. (1997) Functional identification of the mouse circadian *Clock* gene by transgenic BAC rescue. *Cell* **89,** 655–667.

13. Probst, F. J., Fridell, R. A., Raphael, Y., et al. (1998) Correction of deafness in *shaker-2* mice by an unconventional myosin in a BAC transgene. *Science* **280,** 1444–1447.

14. Wakabayashi, Y., Kikkawa, Y., Matsumoto, Y., et al. (1997) Genetic and physical delineation of the region of the mouse deafness mutation shaker-2. *Biochem. Biophys. Res. Commun.* **234,** 107–110.

15. Matesic, L. E., Yip, R., Reuss, A. E., et al. (2001) Mutations in Mlph, encoding a member of the Rab effector family, cause the melanosome transport defects observed in *leaden* mice. *Proc. Natl. Acad. Sci. USA* **98,** 10,238–10,243.

16. Wilson, S. M., Yip, R., Swing, D. A., et al. (2000) A mutation in *Rab27a* causes the vesicle transport defects observed in *ashen* mice. *Proc. Natl. Acad. Sci. USA* **97,** 7933–7938.

17. Means, G. D., Boyd ,Y., Willis, C. R., and Derry, J. M. (2001) Transgenic rescue of the *tattered* phenotype by using a BAC encoding Ebp. *Mamm. Genome* **12,** 323–325.

18. Kim, U. J., Birren, B. W., Slepak, T., et al. (1996) Construction and characterization of a human bacterial artificial chromosome library. *Genomics* **34,** 213–218.

19. Frengen, E., Weichenhan, D., Zhao, B., Osoegawa, K., van Geel, M., and de Jong, P.J. (1999) A modular, positive selection bacterial artificial chromosome vector with multiple cloning sites. *Genomics* **58,** 250–253.

20. Sinnett, D., Richer, C., and Baccichet, A. (1998) Isolation of stable bacterial artificial chromosome DNA using a modified alkaline lysis method. *Biotechniques* **24,** 752–754.

21. Hogan, B., Beddington, R., Constantini, F., and Lacy, E. (1994) *Manipulating the Mouse Embryo, A Laboratory Manual.* 2nd Edition, Cold Spring Harbor Laboratory Press, Cold Spring Harbor, NY.

14

Herpesviruses

A Brief Overview

Mathias Ackermann

1. Introduction

Herpesviruses have been detected in a vast variety of vertebrate species and in at least one invertebrate. It is anticipated that the approx 120 different herpesviruses known today represents only a fraction of the number that actually exists *(1)*. Each virus is closely associated with its main host species. This host-specific occurrence indicates that the herpesviruses have evolved with their hosts over long periods of time. Interestingly, many herpesviruses seem to be entirely avirulent within their original hosts. In contrast, upon infection of foreign hosts, i.e., those who did not participate during the process of coevolution, dramatic, often lethal diseases may occur *(2,3)*. However, many herpesviruses are associated with various degrees of disease in their original host. The potential of herpesviruses to infect a broad range of host cells and to either induce or distract immune reactions makes them interesting entities to study in the context of both, development of new vaccines and vectors for gene therapy *(1)*.

2. Structure

Members of the family Herpesviridae are classified on the basis of the architecture of the virion (*see* **Fig. 1**) *(1)*. The core of a mature virion contains the viral DNA, which is enclosed in a capsid. The capsid consists of 162 hexameric or pentameric capsomers and is approx 100 nm in diameter. The more or less amorphous, frequently asymmetrically distributed structures between capsid and the envelope are termed tegument. Finally, the herpesvirus envelope

From: *Methods in Molecular Biology, vol. 256:*
Bacterial Artificial Chromosomes, Volume 2: Functional Studies
Edited by: S. Zhao and M. Stodolsky © Humana Press Inc., Totowa, NJ

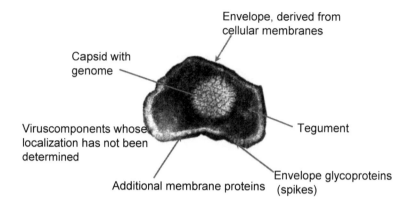

Fig. 1. Structural components of the herpesvirus particle.

is derived from cellular membranes and contains numerous spikes, consisting of glycoproteins.

Owing to variability in the thickness of the tegument and to the state of the envelope, the overall size of herpesvirions may vary between 120 and 300 nm. It is estimated that 30 to 35 different polypeptides are contained in one virus particle (*see* **Fig. 1**).

In contrast, the herpesvirus genomes may differ vastly, including molecular weight of the DNA, base composition, homogeneity of base distribution, and sequence arrangements, as well as some biological properties of the herpesviruses (reviewed in **ref. 4**). At least six classes of herpesvirus genomes are known, termed A through F (*see* **Fig. 2**). The differences are made up by the presence and locations of reiterations of terminal sequences greater than 100 basepairs.

1. In group A, for example, the channel catfish herpesvirus, a large sequence from one terminus is directly repeated at the other terminus.
2. In group B, the terminal sequence is directly repeated numerous times at both termini and the number of reiterations at both termini may vary. This group is exemplified by the herpes saimiri virus.
3. In group C, exemplified by the Epstein-Barr virus (EBV), the number of direct terminal reiterations is smaller, but there may be other sequences that are repeated and that subdivide the unique sequences of the genome into several delineated stretches.
4. In group D, the sequences from one terminus are repeated in an inverted orientation internally, which allows the unique sequences flanked by the repeats to invert relative to the remaining sequences. This group is exemplified by the Varizella-Zoster virus (VZV) as well as by the Bovine, Canine, Equine, Feline, and Porcine herpesviruses type 1.

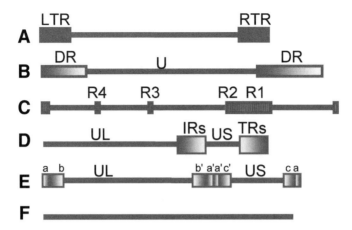

Fig. 2. Schematic overview of the herpesvirus genome arrangements. Boxed parts represent repeated sequences. The gradient within the boxes indicates the relative orientations of the repeats. For details, see text. *Abbreviations*: LTR: left terminal repeat; RTR: right terminal repeat; DR: direct repeat; U: unique sequence; R1 through R4: repeated sequences; UL: unique long; US: unique short; IRs: internal repeat sequences flanking US; TRs: terminal repeat sequences flanking US; a, b, c and a′, b, c′: inverted repeat regions among group E genomes, primes indicate inverted orientation.

5. In group E, sequences from both termini are repeated in an inverted orientation and juxtaposed internally, dividing the genomes into two components, which are each flanked by inverted repeats and which may invert relative to each other. This group is exemplified by the herpes simplex virus (HSV) and the human cytomegalovirus (HCMV).
6. Finally, in group F, exemplified by the tupaia herpesvirus, the sequences at the two termini are not identical and are not repeated, neither directly not in an inverted orientation.

The shared biological properties of the herpesviruses include: synthesis of viral DNA and assembly of capsids that occur in the nucleus, production of progeny virus accompanied by the destruction of the infected cell, and establishment of latency in the natural hosts.

Among the varying biological properties, it is important to note that some herpesviruses have a wide host cell range, whereas others have a narrow host range; some multiply rapidly, others more slowly; the cells and tissues, in which each virus remains latent, may vary with the virus species.

Based on those shared and varying biological properties, the herpesviruses have been classified into three subfamilies, i.e., *Alphaherpesvirinae*, *Betaherpesvirinae*, and *Gammaherpesvirinae*. The following properties are typical for members of the alphaherpesviruses: variable host range, relatively short replication cycle, rapid spread in cell culture, destruction of infected cells, and establishment of latency in sensory ganglia. The betaherpesviruses have a more restricted host range and a more slowly progressing reproductive cycle. Infected cells may become enlarged (cytomegalia) and may be persistently infected. Latency can be established in secretory glands, lymphoreticular tissue, and kidneys. The gammaherpesviruses, finally, are characterized by the most limited host range. Typically, viruses of this subfamily are specific for either T or B lymphocytes. All members replicate in lymphoblastoid cells, although some may also cause lytic infections in epitheloid and fibroblastic cells. Frequently, latency is established in lymphoid tissue.

3. Replication of the Herpesviruses
3.1. Attachment

First contacts between cells and herpesviruses are mediated by viral glycoproteins (glycoproteins gB and gC or their homologs), which bind to heparan sulfated proteoglycans on the cell surface. Subsequently, a viral receptor-binding glycoprotein binds to one of its cellular counterparts. Typically, alphaherpesviruses make use of at least one of the following entry mediators, i.e., either herpesvirus entry protein A (HveA), which belongs to the TNF receptor family, or HveC and HveB, which both belong to the immunoglobulin superfamily and are related to the poliovirus receptor CD155 (reviewed in **refs. 5,6**). In most cases, this secondary binding step is mediated by the viral glycoprotein gD. However, VZV, a member of the alphaherpesviruses, does not possess an obvious gD homolog. Therefore, this virus may not utilize any of the known herpesvirus entry mediators (reviewed in **ref. 7**).

Much less is known about the receptor(s) for the principal members of the betaherpesviruses, i.e., the cytomegaloviruses. An initial interaction of gB with heparan sulfate moieties is followed by a heparin-independent adsorption of gB to the putative receptor, a 30–36 kDa gB-binding cell surface protein, which is widely distributed (8). Interestingly, binding of CMV to cellular receptors is sufficient to initiate a signaling cascade, which appears to be part of the host cell response to viral infection (9). Human herpesviruses 6A and 6B (HHV-6A and HHV-6B), members of the roseolovirus genus of the betaherpesviruses, use CD46 as their cellular receptor (10). CD46, also known as the receptor for measles virus, is a member of the regulator of complement fixation family, which is expressed on the surface of all nucleated cells. However, expression of CD46 is not sufficient for infection, suggesting that a specific coreceptor may be needed. Human herpesvirus 7 (HHV-7), another member of the roseolovirus genus, uses CD4 as one of its receptors (11).

The adsorption processes of the gammaherpesviruses are even less well understood. EBV, the prototype of the genus lymphocryptovirus, has been studied most extensively. The most abundant EBV envelope glycoprotein, gp350/220, is known to use CD21 as its ligand *(12)*. CD21 is a member of the immunoglobulin superfamily, which is expressed to a high level on B lymphocytes. However, the EBV receptor on epithelial cells that lack CD21 has not been identified. Many gammaherpesviruses, which are members of the genus Rhadinovirus (also termed gamma 2 herpesviruses), for example, Human herpesvirus 8 (HHV-8) or Ovine herpesvirus 2 (OvHV-2) cannot replicate at all in cell cultures, which makes the identification of their receptors even more difficult *(13,14)*.

3.2. Penetration and Uncoating

Penetration, i.e., translocation of the virus from the outside to the inside of the cellular membrane, is difficult to observe because it is believed to occur very rapidly. Most alphaherpesviruses require the participation of gD, gB, and the heterodimer gH-gL, if entry should be followed by replication (reviewed in **ref. 15**). Generally, fusion of the viral envelope with the cellular membrane is believed to occur. This process is thought to release naked virus particles into the cytoplasm. However, studies employing cryotechniques combined with transmission and scanning electron microscopy gave a more-detailed insight into the entry pathway of Bovine alphaherpesviruses (BHV-1 and BHV-5) *(16)*. Apparently, the outer phospholipid layer of the viral envelope undergoes a unique fusion with the inner layer of the plasma membrane. Concomitantly, an invagination of the cellular membrane is formed close to the fusion site. Then, the virus particle enters the cytoplasm through the opened tip of the invagination, and the viral envelope defuses from the plasma membrane. Under these conditions, enveloped viral particles were even found close to the nuclear pores. Similarly, entry of the betaherpesvirus CMV involves the gH/gL complexes and occurs in a pH-independent manner *(17)*. Although entry by endocytosis does not result in productive infection of the alphaherpesviruses, entry of the roseolovirus members of the betaherpesvirinae appears to occur through an endocytic pathway rather than by fusion *(18)*.

On the other hand, gp85 and gp25, EBV-homologs to gH and gL, are important for penetration of this gammaherpesvirus. In addition, gp42, which is involved into engagement of human leukocyte antigen (HLA) class II, plays an important role for efficient B-lymphocyte infection by EBV. Together, these proteins are believed to mediate fusion of the viral membrane with cellular membranes (reviewed in **ref. 19**).

Following entry, the alphaherpesvirus particles are transported through the microtubular network to the nuclear pore. Interestingly, dyneins, which are responsible for ATP-dependent microtubular transportation, are capable to interact with the herpes membrane protein UL35. As evidenced by in vitro assays, the UL35 protein targets to the nuclear membrane *(20)*. Furthermore, the UL35 protein is a component of the intact virion *(21)*, which leaves room to the hypothesis that enveloped particles are transported to the nuclear pores. Very little is known concerning the fate of beta- and gammaherpesvirus particles throughout this stage.

4. Lytic Infection or Latency

After the cytoplasmic transport, the nucleocapsids associate with the nuclear pore complexes. At the pore, the nucleocapsids release their DNA into the nucleus. Notably, the typical herpesvirus DNA is infectious, i.e., upon transfection of viral genomic DNA, it is able to start to express genes and replicate without the help of other viral components. Yet, herpesvirus particles carry several proteins localized in the tegument that accomplish functions, including such that are associated to the regulation of viral and cellular gene expression. Upon disassembly of the virus particle some tegument proteins remain in the cytoplasm, whereas others also gain access to the nucleus. Both host and viral factors are considered to contribute to the decision, whether lytic replication or establishment of latency should primarily take place within an infected cell (*see* **Fig. 3**). The default pathway to be pursued seems to be characteristic and different for each the alpha-, beta-, or gammaherpesviruses (reviewed in **ref. 4**). Typically, most cells infected by either alpha- or betaherpesviruses support lytic replication, although only specific subsets of cells normally harbor latent virus. On the other extreme, EBV, as one prototype of the gammaherpesviruses, seems initially to favor the establishment of latency, whereas only a subset of cells supports lytic replication.

5. Genetic Content and Gene Expression

Herpesviruses encode between 70 and 200 known genes *(22–28)*. Most of the genes are preceded by a promoter sequence of between 51 and 200 nucleotides, located upstream of a TATA box. Downstream of the TATA box, a transcription initiation site of 20 to 25 nucleotides is followed by a nontranslated 5′ leader sequence of between 30 and 300 nucleotids, the open reading frame, a 10 to 30 nucleotide long 3′ nontranslated region, and polyadenylation signal (reviewed in **ref. 4**). Although most herpesvirus genes are not spliced, each herpesvirus encodes also for some spliced genes. Approx 26 genes are conserved among the three herpesvirus subfamilies. Their functions are associated to gene regulation, nucleotide metabolism, DNA replication, and virion formation and maturation. However, some gene clusters are conserved among the different virus subfamilies. For example, the alphaherpesviruses encode related latency-associated genes, glycoprotein D, a tegument-associated transactivator protein for immediate early (IE) genes, and an ICP4 homolog. The betaher-

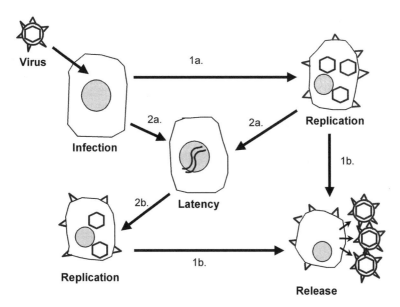

Fig. 3. Schematic representation of herpesvirus replication. Both host- and viral factors contribute to the decision, whether lytic replication (1a.,1b.) or establishment of latency (2a.) primarily takes place within the infected cell. Replication leads to the generation of virus progeny, which is released from the infected cell. Virus may be reactivated (2b.,1b.) at intervals, which also leads to the production of viral progeny.

pesviruses encode a block of so-called Beta-genes that have no counterparts in the other subfamilies. The gammaherpesviruses have acquired a number of genes that are homologs of cellular genes as well as a set of genes that is required for maintenance of the latent genomes in dividing cells.

Upon entry into the nucleus, the herpesvirus genomes are believed to circularize before the first set of genes, usually IE or latency-associated genes, is being expressed *(29)*. During productive infection, viral genes are expressed in a regulated cascade fashion *(30)*. The minimal sets of individual phases are termed IE (or α), early (E or β), and late (L or γ). In contrast, there is no shared pattern of gene expression among the herpesviruses, which is required for establishment, maintenance, and reactivation of latency. Some herpesviruses encode several proteins that are synthesized during latency. Others have apparently no requirement for gene expression throughout that stage.

5.1. Lytic Cycle

The promoters of the first set of genes to be expressed during lytic replication, the IE promoters, contain binding sites for cellular transcription factors but they can also be transactivated by virion tegument proteins. Among the functions of the IE proteins, at least three have to be mentioned at this stage: First, IE proteins are responsible for the activation of early gene expression as well as for autodownregulation of IE gene expression *(31)*. Second, IE proteins may interfere with cellular functions such as transcription of host genes and RNA splicing *(32,33)*. Third, they also may interfere with the antiviral response of the host. For example, ICP47 of HSV-1 is able to bind to a host protein that is associated with antigen processing (TAP)*(34–36)*. Thereby, it interferes with presentation of antigenic peptides by the class I major histocompatiblility complex (MHC-I). The products of the second set of genes expressed during lytic infection, the early proteins, are primarily involved into the viral nucleic acid metabolism. Their appearance signals the onset of viral DNA synthesis. Furthermore, the early proteins are involved into the onset of late gene expression. Simultaneous with the onset of viral DNA replication the transcription of late genes also increases. The main purpose of late gene expression is the production of large amounts of viral structural proteins for the assembly of progeny viral particles.

5.2. Latent State

As mentioned earlier, gene expression during latency differs vastly among herpesviruses. With EBV, even different types of latency can be distinguished, i.e., type I, type II, and type III, which differ with regard to latent gene expression. In type III, latently infected B lymphocytes, EBV expresses six nuclear proteins (the EBNAs) as well as two integral membrane proteins (the LMPs), two small nonpolyadenylated RNAs (EBERs), and Bam A rightward transcripts (BARTs) (reviewed in **ref. 19**). These viral gene products maintain the latent infection and cause the previously resting lymphocytes to continuously proliferate. Most EBV-infected B lymphocyte tumors have a latency I phenotype, which is characterized by expression of EBNA-1, EBERs, and BARTs. Type II latency, observed in vivo in Hodgkin's disease cells and in nasopharyngeal carcinoma, is characterized by an intermediate pattern of gene expression, including EBNA-1, the LMPs, EBERs, and BARTs. The lack of permissivity of in vitro-infected B lymphocytes for EBV replication may in part be related to the poor persistence of EBV episomes in such cells and the ability of the EBV DNA to be integrated into the cell genome, as lytic replication requires episomal viral DNA. The betaherpesvirus human CMV remains latent in lineage-committed myeloid cells (reviewed in **ref. 37**). Viral gene expression in latently infected progenitor cells is restricted to CMV latency-associated transcripts (CLTs). However, the pattern of latent gene expression changes as these cells mature into monocytes or dendritic cells. Remarkably, these cells may become productive as a result of differentiation under proinflammatory conditions.

Interestingly, the CLTs map to both DNA strands of the immediate early regions of CMV. Seven ORFs are predicted, four within sense CLTs and three within antisense CLTs. Infected individuals carry antibodies against the predicted latent proteins, indicating that they are synthesized during natural infection. The major viral gene products expressed during the latency of alphaherpesviruses are the latency-associated transcripts (LATs) *(38)*. LAT-negative mutants exhibit increased expression of lytic cycle genes in sensory neurons as compared with wild-type viruses or rescue mutants *(39)*. Thus, LATs may play a role in promoting latent infection by downregulating the lytic gene expression. Interestingly, the LATs have not been shown unambiguously to express their encoded ORFs.

6. DNA Amplification

Starting with early gene expression, several viral proteins begin to accumulate in the nucleus in order to form globular structures termed replication compartments, where viral DNA synthesis proceeds, starting from circularized templates *(40)*.

Several origins of DNA replication (ori) are known to serve as starting points for viral DNA synthesis. In HSV-1, one of the prototype alphaherpesviruses, they are termed oriS (located in the inverted repeats flanking the short unique sequence), oriL (mapping within the unique long sequence), and oriH (host-dependent ori) (reviewed in **ref. 15**). With cytomegaloviruses only a single ori, oriLyt, a positional homolog to oriL, has been identified (reviewed in **ref. 37**). From EVB, two ori's are known, i.e., ori lyt (used throughout lytic replication) and ori p (used for maintenance of the latent episomal DNA) (reviewed in **ref. 19**).

Seven viral proteins are essential for HSV-1 DNA replication. The ori-binding protein (OBP, encoded by UL9) is thought to unwind the DNA at one of the origins of DNA replication (ori) before recruiting ICP8 (UL29), an ssDNA-binding protein to the initiation site. These two proteins then recruit the DNA polymerase (UL31), its processivity factor (UL42), and the helicase-primase complex, consisting of the proteins encoded by UL5, UL8, and UL52, for initial rounds of Theta replication. Because the progeny DNA consists mainly of head-to-tail concatemers, it is assumed that Theta replication is rapidly replaced by a rolling-circle mechanism (reviewed in **ref. 41**)(*see* **Fig. 4**). Different requirements are known for lytic DNA replication of the prototype gammaherpesvirus EBV. No homolog to OBP was identified. Similar to HSV-1, the ssDNA binding protein (BALF2), the core DNA polymerase (BALF5), its processivity factor (BMRF1), and two components of the primase and helicase complex (BSLF1, BBLF4) are required. In addition, BBLF2/3 (a spliced primase helicase complex component), BKRF3, the uracil DNA glycosylase, and

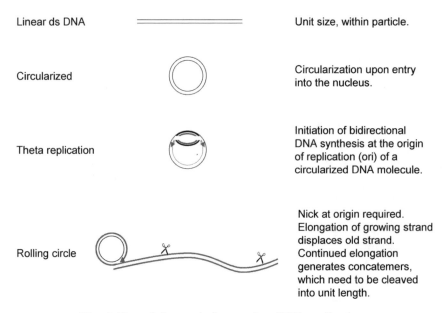

Fig. 4. Essential steps in herpesvirus DNA replication.

expression of the transactivators BZLF1, BRLF1, and BSMLF1 are essential for lytic DNA replication. DNA replication of the betaherpesviruses CMV is similar to the replication in the alphaherpesviruses, although proceeding considerably more slowly. Because no homolog to the OBP has been identified among CMVs, only six conserved proteins are found to be essential at the replication fork, i.e., the ssDNA binding protein ppUL57, the DNA polymerase (pUL54), its precessivity factor (ppUL45), and the three subunits of the helicase-primase complex (pUL70, pUL102, pUL105). Fascinatingly, the cells infected by CMVs have the appearance of transiting the S phase but are actually blocked from S phase progression when viral replication is supported. In contrast, the DNA replication system of other betaherpesviruses, such as HHV-6 and HHV-7, is more closely related to the system of the alphaherpesviruses than to the methods of CMV. For example, their oriLyts differ markedly from those of the CMVs. Furthermore, they have homologs to all the seven proteins needed for alphaherpesvirus DNA replication and they make use of even two different OBPs, OBP-1 and OBP-2.

7. Particle Formation and Genome Packaging

The newly produced capsid proteins are translocated to the nucleus, where particle assembly is initiated. A number of self interactions among major and minor capsid proteins leads to the formation of initially round procapsids, which mature then to become polyhedral (B capsids). It remains unclear, whether DNA is introduced into the round or the polyhedral capsids. However, encapsidation of viral DNA involves cleavage of the DNA concatemers into unit length monomers. Cleavage occurs when the capsids have been filled "head-full." The site, where cleavage takes place, is generally termed the "a-sequence," referring to the situation of HSV-1, where the cleavage and packaging signals have been mapped precisely near the genomic termini within the repeated sequences. There, the DNA packaging elements were fine mapped to the U_b and U_c domains of the "a" sequences and were designated *pac*1 and *pac*2 *(42,43)*. There seems to be little sequence specificity for such cleavage and packaging signals because the nonhomologous oligonucleotides of HSV-1, HSV-2, and CMV are interchangeable *(44)*. Very similar functional sequences have also been identified for EBV, HHV-6, and HHV-7 *(45–47)*. It is important to note that those cleavage/packaging sequences in combination with a functional origin of DNA replication are sufficient for the generation of amplicon systems that can be used for the expression of foreign genes in transduced cells.

8. Assembly and Egress

Following uptake of viral DNA, the virus particles have to acquire tegument proteins and an envelope before leaving the cell. At least two different pathways have been proposed for herpesvirus virion maturation. One pathway, which is favored by many scientists studying HSV, suggests that the envelope is acquired upon passage of the nascent virion from the nucleus to the inner nuclear membrane and that this envelope is retained until the virus leaves the cell (reviewed in **ref. *48***). According to this model, the virion envelope is processed by transit through Golgi stacks. Thus, nucleocapsids acquire a tegument layer upon budding through the inner nuclear membrane. Apparent changes of the tegument density are explained by maturation of the tegument layers. In contrast, the model favored by scientists studying VZV, EBV, CMV, and HHV-6, appears to involve a process of successive envelopment, deenvelopment, and reenvelopment at the trans-Golgi network as the virus particle moves from one membrane-bounded compartment to another. Some authors have described that tegument proteins are acquired from cytoplasmic pools of electron-dense material *(49)*. Others have observed membrane-bound nuclear structures, termed *tegusomes*, in which capsids may acquire their tegument *(50)*. Finally, the particles are released through secretory pathways. VZV is highly contagious in vivo, although in vitro mostly noninfectious particles are released from infected cells. This conundrum was recently resolved by the observation that VZV may be inactivated by acids and proteases in late endosomes. Epidermal cells, responsible for VZV shedding in vivo, lose their endosomal pathway during differentiation. Therefore, VZV will not be inactivated in these cells *(51)*.

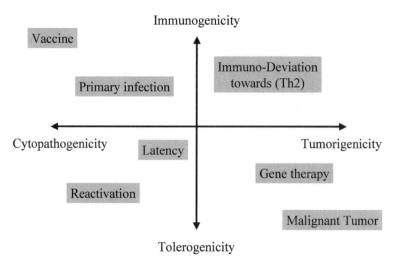

Fig. 5. A basic view of interactions between the herpesviruses and their hosts based on the dimensions of cytopathogenicity versus tumorigenicity and immunogenicity vs tolerogenicity. Subtle changes of the positions in this plain may determine overproductive replication and latency as well as overall health and disease.

Notably, the herpesviruses may not only spread from cell to cell through the extracellular space but also directly, probably through cell junctions.

Although the herpesviruses benefit from maintaining "their" cell alive for prolonged periods, productively infected cells do not survive. Cell death is partly a consequence of injury caused by viral replication and partly resulting from cellular responses to the infection. Viral injuries to the cell include shutoff of the synthesis of cellular proteins, degradation of cellular RNA, inhibition of splicing, shutoff of transcription, and selective degradation, stabilization or activation of cellular proteins.

9. Virus–Host Interactions

Although infection with herpesviruses never goes unnoticed by the infected organism, neither on a cellular or molecular level nor with respect to the native and adaptive immune system, the herpesviruses have found a variety of ways to successfully interact and cooperate with their hosts (reviewed in **ref. 52**). A more simplified view of these interactions and their consequences is presented in **Fig. 5**. Owing to their fine tuned adaptation to their original hosts, many of those virus–host interactions represent subtle changes in a balanced coexistence. Cytokines and chemokines may be induced and expression of cell surface markers may be altered. For example HHV-6A induces CD4 expression in infected natural killer (NK) cells, which, on one hand, may divert the immune response toward a Th2 type, but which, on the other hand, also renders those cells susceptible to superinfection by HIV-1 *(53,54)*. Others may downregulate MHC class I expression to reduce the effectiveness of CTL or modulate MHC class II expression to subtly alter the humoral immune response. Moreover, the gamma herpesviruses carry homologs of cellular gene products, for example CMV encodes UL18, an MHC-I homolog that binds β2 microblobulin to prevent the NK cell response *(55)*. Furthermore, virus-encoded chemokines may influence the behavior of leukocytes during the inflammatory response to infection. Thus, despite of a pronounced immune response, the herpesviruses may survive in their natural hosts, most of the time even without causing disease.

10. Significance of Cloning the Herpesvirus Genomes into BAC

BAC stands for the term *bacterial artificial chromosomes*, the overall theme of the present volume. The significance of the BAC technology arises from drawbacks that cannot be resolved by classical methods and from advantages that open the door for previously almost unthinkable experiments.

10.1. Technological Aspects and Drawbacks Inherent to the Pre-BAC Age

Early on, it was recognized that mutant herpesviruses may emerge on the basis of homologous recombination (reviewed in **ref. *56***). Originally, Morse et al. *(57,58)* constructed random intertypic recombinants between HSV-1 and HSV-2 by infecting cells at the same time with the two viruses. Obviously, there was enough sequence homology between these two viruses to allow recombination. However, the selection of viable recombinants was possible only under the strict precondition that the recombination events did not interfere with viral replication. Techniques to target specific viral genes improved gradually with time. Mocarski et al. *(59)* developed a technique to introduce alternative DNA sequences into the viral thymidine kinase (TK) locus by homologous recombination and to use this targeted deletion simultaneously as a means for the selection of wanted recombinants. For that purpose, they transfected plasmid cloned and genetically engineered mutant viral DNA, which was flanked by TK sequences into cells, which were subsequently infected with wild-type virus. Because the foreign sequences disrupted the viral *TK* gene upon integration by homologous recombination, recombinant viruses could be selected for by means of selection against TK activity in TK-negative cells infected with the recombinant progeny. Alternatively, starting from a TK-negative parent virus, the insertion of the viral *TK* gene into a targeted sequence followed by selection for TK-positive viral progeny, opened the door for the serial

construction of recombinant herpesviruses *(60)*. For a long time, the cotransfection of deproteinized intact herpes virus DNA with a molar excess of an
altered or foreign DNA fragment flanked by homologous sequences, was the
most successful way to produce recombinant progeny viruses for a wide variety of applications, including deletion of nonessential genes, introducing ran-
domized mutations, detection, and identification of so far unknown gene
products or functions *(60–63)*. In the same period, the system was modified to
allow also for the selection of viruses with mutations in sequences that are
essential for viral replication. De Luca et al. *(64)* were able to select a recombinant virus, which was deleted for both copies of the essential *ICP4* gene of
HSV-1, by growing the putative recombinant progeny in cell cultures, which
expressed the deleted gene product. The introduction of the genes such as
β-galactosidase or variations of enhanced green fluorescent protein (eGFP) for
the purpose of selection and tracing of viral mutants proved to be extremely
helpful in the recent years.

As exciting as the construction, selection, and characterization of those
recombinants was at the time, several great disadvantages could not be resolved:

1. **Time**. It took at least several months, often even years, to construct relevant viral
   mutants and to characterize them sufficiently before it was possible to address
   with their help biologically relevant questions.
2. **Inadvertant second mutations**. It is common knowledge that many mutations
   occur spontaneously during viral replication. In spite of this, numerous replication
   cycles are required for the selection and characterization of the desired mutants.
   Thus, each single step during the construction of specific viral mutants carried in
   it the inherent danger to create inadvertently further, second site mutations, which
   were able to spoil, obscure, complement the effect of the desired mutation
   (reviewed in **ref. 65**). Therefore, all mutations had to be rescued by generating a
   further mutant virus, which of course had to envisage the same dangers.
3. **Method not applicable**. It may sound trivial, but the methods described earlier
   are applicable only for a selected range of viruses. The efficiency with which the
   transfection of cells with deproteinized viral DNA led to the growth of a new
   generation of viruses depended both on the cells and the viruses used. For example, Vero cells are much more efficiently transfected than MDBK cells. HSV-1
   DNA is much more "infectious" than BHV-1 DNA (Ackermann, unpublished
   observation). Most importantly, all those techniques were not at all available for
   viruses that cannot be grown in conventional cell cultures or viruses that require
   rearrangements or deletions before the gain this ability. For example, clinical iso-
   lates of the human cytomegalovirus can be propagated in endothelial cell cultures, whereas laboratory strains cannot *(66)*. Both the human herpesvirus 8 and
   the Ovine herpesvirus 2 cannot be serially propagated in cell cultures *(13,14)*.
   Therefore, it was, thus far, impossible to create specific mutants of these viruses.

10.2. The BAC Age Opens the Door to the Previously Unthinkable

The advent of the technology to clone entire herpesvirus genomes into BACs opened essentially new, previously unthinkable possibilities (reviewed in **ref. 67**). The principles of BAC will be described in other sections of this book. However, the following advantages of the BAC technology over conventional virological and molecular biological approaches, as described earlier, need to be mentioned at this stage:

1. **Single copy**. The BACs are based on single copy plasmid vectors, which brings the advantage with it that intermolecular recombination events, an inherent problem throughout viral replication and selection of conventional viral recombinants, do not occur during bacterial propagation. The stability of BAC-cloned herpesviruses has been observed and reported with emphasis *(65,68–70)*.

2. **Biological fitness**. The construction of BAC-cloned viral mutants is essentially independent of the biological fitness of either the parent virus or the recombinant progeny *(65)*. Therefore, unlimited numbers of mutations and any mutation targeted into any region of the viral genome can be constructed.

3. **Technical aspects of mutagenesis**. Apart from conventional molecular biological procedures to mutagenize plasmid-cloned DNAs, a number of techniques, suitable for the modification of large genomes are available. Conservatively, the RecA-assisted restriction endonuclease cleavage technique may be applied as demonstrated by Fraefel et al. *(71)*. They used oligonucleotides designed to form regions of triple stranded DNA, which rendered them resistant to methylation by *Hha*I methylase. Later, these protected sites were used for cleavage by *HinP1*I. The packaging sequence of HSV-1 could be removed throughout this process, which represented the basis for the generation of the helpervirus-free amplicon packaging system. Alternatively, whole libraries of mutants can be generated by using the random transposon (Tn) mutagenesis as described by Brune and others *(66)*. Briefly, a Tn donor plasmid, carrying chloramphenicol resistance, and containing the enhanced *GFP* gene was introduced into bacteria containing the BAC (kanamycin resistance). Random mutagenesis was then induced by shifting the temperature from 31°C (permissive) to 44°C (nonpermissive), where the donor plasmid is lost unless it is integrated into the BAC. The library of mutants was then selected based on resistance to both kanamycin and chloramphenicol. Similarly, targeted mutations can be achieved in *Escherichia coli* (*E. coli*) by the RecA-mediated homologous recombination system. Saeki et al. *(70)* used this system to delete ICP27 from a BAC-cloned HSV-1. The same work also described the Cre-mediated site specific insertion technique, which was used to generate a definitively oversized BAC-cloned HSV-1 DNA, which was protected from being incorporated into amplicon particles.

4. **Prior characterization**. One simple but great advantage of the availability of mutagenized herpesvirus BAC clones lies in the possibility to characterize the desired mutants on the DNA level before they are used for time-consuming experiments.

5. **Applications**. HSV-1 in bacteria has led to the development of helper virus-free packaging systems for HSV-1 amplicons, which represent extraordinary tool for gene therapy. Recently, Wade-Martins et al. *(72)* were able to correct a genetic deficiency in human hypoxanthine phophoribosyltransferase (HPRT) by transferring the complete genetic locus with a size of 115 kbp to deficient cells by using this BAC-based helpervirus-free amplicon system. On the other hand, a new generation of DNA-based vaccines was developed on the basis of the BAC technology *(73)*. This application will be described in one of the following chapters.

Thus, BAC-cloned herpesviruses will play extremely important roles not only for the future understanding of the viruses themselves but also for their applications in gene therapy and immunology. Through BAC-technology, it is now possible to clone, propagate, and modify entire herpesvirus genomes independent of their ability to replicate in cell cultures *(13,14)*. By combining this with the properties of auxotrophic invasive bacteria *(74)*, it will also be possible to deliver those genomes back into living organisms in order regenerate viable virus and to study the pathogenesis of these viruses in their original host.

11. Conclusion

The herpesvirus-concept represents one of the most successful pathways for infection and survival of viruses in nature. A similar architecture of the virion and the ability to establish latency followed by periods of reactivation may be regarded as the unifying hallmarks of this enormously significant virus family. Yet, the herpesviruses have diverse host ranges. They make use of a variety of receptors and entry-mediators. They exploit different cellular mechanisms for penetration, uncoating, and traveling to the cell nucleus. They may or may not synthesize proteins to establish and/or maintain latency. They have established different pathways for gene expression, DNA replication, particle formation, egress, and even for manipulation of the host's response towards viral infection. The fate of infected cells may be destruction, survival, or even immortalization. Many of those events and their underlying molecular functions are still poorly understood. The advent of the technology to clone entire herpesvirus genomes into BACs, will help greatly to better understand the true nature of the herpesviruses and to take advantage of their great repertoire of abilities *(75,76)*. Because manipulations of the herpesvirus genomes in BACs allows mutagenesis and deletion of genes, which are normally essential for viral replication, it will be possible to construct rapidly a new generation of viral mutants, including such with multiple alterations in any combination of essential or nonessential genes.

References

1. Minson, A. C., Davison, A., Eberle, R., et al. (2000) Family *Herpesviridae*, in *Virus Taxonomy*, (Van Regenmortel, M. H. V., Fauquet, C. M., Bishop, D. H. L., et al., eds.), Academic, San Diego, CA, pp. 203–225.
2. Ackermann, M. (2001) Visions of the future of veterinary virology. *Vet. Sci. Tomorrow* Isssue 1, http://www.vetscite.org/cgi-bin/pw.exe/vst/reviews/index.htm.
3. Ackermann, M., Engels, M., Fraefel, C., et al. (2002) Herpesviruses: balance in power and powers imbalanced. *Vet. Microbiol.* **86,** 175–181.
4. Roizman, B. and Pellett, P. E. (2001) in *Fields Virology*, 4th edition, ch. 71, Lippincott Williams & Wilkins, Philadelphia, PA, pp. 2381–2397.
5. Shukla, D., Liu, J., Blaiklock, P., et al. (1999) A novel role for 3-O-sulfated heparan sulfate in herpes simplex virus 1 entry. *Cell* **99,** 13–22.
6. Shukla, D., Rowe, C. L., Dong, Y., Racaniello, V. R., and Spear, P. G. (1999) The murine homolog (Mph) of human herpesvirus entry protein B (HveB) mediates entry of pseudorabies virus but not herpes simplex virus types 1 and 2. *J. Virol.* **73,** 4493–4497.
7. Cohen, J. I. and Straus, S. E. (2001) in *Fields Virology*, 4th edition, ch. 78, Lippincott Williams & Wilkins, Philadelphia, PA, pp. 2787–2730.
8. Adlish, J. D., Lahijani, R. S., and St. Jeor, S. C. (1990) Identification of a putative cell receptor for human cytomegalovirus. *Virology* **176,** 337–3345.
9. Boyle, K. A., Pietropaolo, R. L., and Compton, T. (1999) Engagement of the cellular receptor for glycoprotein B of human cytomegalovirus activates the interferon-responsive pathway. *Mol. Cell. Biol.* **19,** 3607–3613.
10. Santoro, F., Kennedy, P. E., Locatelli, G., Malnati, M. S., Berger, E. A., and Lusso, P. (1999) CD46 is a cellular receptor for human herpesvirus 6. *Cell* **99,** 817–827.
11. Yasukawa, M., Inoue, Y., Ohminami, H., et al. (1997) Human herpesvirus 7 infection of lymphoid and myeloid cell lines transduced with an adenovirus vector containing the CD4 gene. *J. Virol.* **71,** 1708–1712.
12. Carel, J. C., Myones, B. L., Frazier, B., and Holers, V. M. (1990) Structural requirements for C3d,g/Epstein-Barr virus receptor (CR2/CD21) ligand binding, internalization, and viral infection. *J. Biol. Chem.* **265,** 12,293–12,299.
13. Hüssy, D., Stäuber, N., Leutenegger, C. M., Rieder, S., and Ackermann, M (2001) Quantitative fluorogenic PCR assay for measuring Ovine herpesvirus 2 replication in sheep. *Clin. Diagn. Lab. Immunol.* **8,** 123–128.
14. Moore, P. S. and Chang, Y. (2001) Kaposi's Sarcoma-associated herpesvirus, in *Fields Virology*, 4th edition, ch. 82, Lippincott Williams & Wilkins, Philadelphia, PA, pp. 2803–2833.
15. Roizman, B. and Knipe, D. M. (2001) in *Fields Virology* 4th edition, ch. 72, Lippincott Williams & Wilkins, Philadelphia, PA, pp. 2399–2459.
16. Wild, P., Schraner, E. M., Peter, J., Loepfe, E., and Engels, M. (1998) Novel entry pathway of Bovine herpesvirus 1 and 5. *J. Virol.* **72,** 9561–9566.

17. Compton, T., Nepomuceno, R. R., and Nowlin, D. M. (1992) Human cyto-megalovirus penetrates host cells by pH-independent fusion at the cell surface. *Virology* **191,** 387–395.
18. Cirone, M., Zompetta, C., Angeloni, A., et al. (1992) Infection by human herpesvirus 6 (HHV-6) of human lymphoid T cells occurs through an endocytic pathway. *AIDS Res. Hum. Retrov.* **8,** 2031–2037.
19. Kieff, E. and Rickinson, A. B. (2001) in *Fields Virology* 4th edition, ch. 74, Lippincott Williams & Wilkins, Philadelphia, PA, pp. 2511–2573.
20. Ye, G. J., Vaughan, K. T., Vallee, R. B., and Roizman, B. (2000) The herpes simplex virus 1 UL34 protein interacts with a cytoplasmic dynein intermediate chain and targets nuclear membrane. *J. Virol.* **74,** 1355–1363.
21. Purves, F. C., Spector, D., and Roizman, B. (1992) UL34, the target of the herpes simplex virus U(S)3 protein kinase, is a membrane protein which in its unphosphorylated state associates with novel phosphoproteins. *J. Virol.* **66,** 4295–4303.
22. Baer, R., Bankier, A. T., Biggin, M. D., et al. (1984) DNA sequence and expression of the B95-8 Epstein-Barr virus genome. *Nature* **310,** 207–211.
23. Chee, M. S., Bankier, A. T., Beck, S., et al. (1990) Analysis of the protein-coding content of the sequence of human cytomegalovirus strain AD169. *Curr. Top. Microbiol. Immunol.* **154,** 125–169.
24. Davison A. J. and Scott J. E. (1986) The complete DNA sequence of varicella-zoster virus. *J. Gen. Virol.* **67,** 1759–1816.
25. Ensser, A., Pflanz, R., and Fleckenstein, B. (1997) Primary structure of the Alcelaphine herpesvirus 1 genome. *J. Virol.* **71,** 6517–6525.
26. McGeoch, D. J., Dalrymple, M. A., Davison, A. J., et al. (1988) The complete DNA sequence of the long unique region in the genome of herpes simplex virus type 1. *J. Gen. Virol.* **69,** 1531–1574.
27. Russo, J. J., Bohenzky, R. A., Chien, M. C., et al. (1996) Nucleotide sequence of the Kaposi sarcoma-associated herpesvirus (HHV8). *Proc. Nat. Acad. Sci.* **93,** 14,862–14,867.
28. Virgin, H. W., Latreille, P., Wamsley, P., et al. (1997) Complete sequence and genomic analysis of murine gammaherpesvirus 68. *J. Virol.* **71,** 5894–5904.
29. Garber, D. A., Beverley, S. M., and Coen, D.M. (1993) Demonstration of circularization of herpes simplex virus DNA following infection using pulsed field gel electrophoresis. *J. Virol.* **197,** 459–462.
30. Honess, R. W. and Roizman, B. (1974) Regulation of herpesvirus macro-molecular synthesis. I. Cascade regulation of the synthesis of three groups of viral proteins. *J. Virol.* **14,** 8–19.
31. Honess, R. W. and Roizman, B. (1975) Regulation of herpesvirus macro-molecular synthesis. Sequential transition of polypeptide synthesis requires functional viral polypeptides. *Proc. Natl. Acad. Sci.* **72,** 1276–1280.
32. Hardwicke, M. A. and Sandri-Goldin, R. M. (1994) The herpes simplex virus regulatory protein ICP27 contributes to the decrease in cellular mRNA levels during infection. *J. Virol.* **68,** 4797–4810.

33. Hardy, W. R. and Sandri-Goldin, R. M. (1994) Herpes simplex virus inhibits host cell splicing, and regulatory protein ICP27 is required for this effect. *J. Virol.* **68**, 7790–7799.

34. Berger, C., Xuereb, S., Johnson, D. C., et al. (2000) Expression of herpes simplex virus ICP48 and human cytomegalovirus US11 prevents recognition of transgene products by CD8(+) cytotoxic T lymphocytes. *J. Virol.* **74**, 4465–4483.

35. Tomazin, R., van Schoot, N. E., Goldsmith, K., et al. (1998) Herpes simplex virus type 2 ICP47 inhibits human TAP but not mouse TAP. *J. Virol.* **72**, 2560–2563.

36. York, I. A., Roop, C., Andrews, D. W., Riddell, S. R., Graham, F. L., and Johnson, D. C. (1994) A cytosolic herpes simplex virus protein inhibits antigen presentation to CD8+ T lymphocytes. *Cell* **77**, 525–535.

37. Mocarski, E. S. and Courcelle, C. T. (2001) in *Fields Virology* 4th edition, ch. 76, Lippincott Williams & Wilkins, Philadelphia, PA, pp. 2629–2673.

38. Stevens, J. G., Wagner, E. K., Devi-Rao, G. B., Cook, M. L., and Feldman, L. T. (1987) RNA complementary to a herpesvirus alpha gene mRNA is prominent in latently infected neurons. *Science* **235**, 1056–1059.

39. Garber, D. A., Schaffer, P. A., and Knipe, D. M. (1997) LAT-associated function reduces productive-cycle gene expression during acute infection of murine sensory neurons with herpes simplex virus type 1. *J. Virol.* **71**, 5885–5893.

40. Quinlan, M. P., and Chen, L. B., and Knipe, D. M. (1984) The intranuclear location of a herpes simplex virus DNA-binding protein is determined by the status of viral DNA replication. *Cell* **36**, 857–868.

41. Lehman, I. R. and Boehmer, P. E. (1999) Replication of herpes simplex virus DNA. *J. Biol. Chem.* **274**, 28,059–28,062.

42. Deiss, L. P., Chou, J., and Frenkel, N. (1986) Functional domains within the a sequence involved in the cleavage-packaging of herpes simplex virus DNA. *J. Virol.* **59**, 605–618.

43. Deiss, L.P. and Frenkel, N. (1986) Herpes simplex virus amplicon: cleavage of concatemeric DNA is linked to packaging and involves amplification of the terminally reiterated a sequence. *J. Virol.* **57**, 933–941.

44. Spaete, R. R. and Frenkel, N. (1985) The herpes simplex virus amplicon: analyses of cis-acting replication functions. *Proc. Natl. Acad. Sci.* **82**, 694–698.

45. Deng, H. and Dewhurst, S. (1998) Functional identification and analysis of cis-acting sequences which mediate genome cleavage and packaging in human herpesvirus 6. *J. Virol.* **72**, 320–329.

46. Romi, H., Singer, O., Rapaport, D., and Frenkel, N. (1999) Tamplicon-7, a novel T-lymphotropic vector derived from human herpesvirus 7. *J. Virol.* **73**, 7001–7007.

47. Wang, F., Marchini, A., and Kieff, E. (1991) Epstein-Barr virus (EBV) recombinants: use of positive selection markers to rescue mutants in EBV-negative B-lymphoma cells. *J. Virol.* **65**, 1701–1709.

48. Enquist, L. W., Husak, P. J., Banfield, B. W., and Smith, G. A. (1998) Infection and spread of alphaherpesviruses in the nervous system. *Adv. Virus Res.* **51**, 237–347.

49. Torrisi, M. R., Gentile, M., Cardinali, G., et al. (1999) Intracellular transport and maturation pathway of human herpesvirus 6. *Virology* **257**, 460–471.

50. Roffman, E., Albert, J. P., Goff, J. P., and Frenkel, N. (1990) Putative site for the acquisition of human herpesvirus 6 virion tegument. *J. Virol.* **64,** 6308–6313.
51. Gershon, A. and Gershon, M., personal commun. 2001.
52. Alcami, A. and Koszinowski, H. U. (2000) Viral mechanisms of immune evasion. *Trends Microbiol.* **8,** 410–418.
53. Lusso, P., De Maria, A., Malnati, M., et al. (1991) Induction of CD4 and susceptibility to HIV-1 infection in human CD8+ T lymphocytes by human herpesvirus 6. *Nature* **349,** 533–535.
54. Lusso, P., Malnati, M. S., Garzino-Demo, A., Crowley, R. W., Long, E. O., and Gallo, R. C. (1993) Infection of natural killer cells by human herpesvirus 6. *Nature* **362,** 458–462.
55. Browne, H., Smith, G., Beck, S., and Minson, T. (1990) A complex between the MHC class I homologue encoded by human cytomegalovirus and β2 microglobulin. *Nature* **347,** 770–772.
56. Ackermann, M. (1988) The construction, characterization, and application of recombinant herpes viruses. *J. Vet. Med.* **B35,** 379–396.
57. Morse, L. S., Buchman, T. G., Roizman, B., and Schaffer, P. A. (1977) Anatomy of herpes simplex virus DNA. IX. Apparent exclusion of some parental DNA arrangements in the generation of intertypic (HSV-1 x HSV-2) recombinants. *J. Virol.* **24,** 231–248.
58. Morse, L. S., Pereira, L., Roizman, B., and Schaffer, P. A. (1978) Anatomy of herpes simplex virus DNA. XI. Mapping of viral genes by analysis of polypeptides and functions specified by HSV-1 x HSV-2 recombinants. *J. Virol.* **26,** 389–410.
59. Mocarski, E. S., Post, L. E., and Roizman, B. (1980) Molecular engineering of the herpes simplex virus genome: Insertion of a second L-S junction into the genome causes additional genome inversions. *Cell* **22,** 243–255.
60. Post, L. E. and Roizman, B. (1981) A generalized technique for deletion of specific genes in large genomes: ALPHA gene 22 of herpes simplex virus 1 is not essential for growth. *Cell* **25,** 227–232.
61. Ackermann, M., Longnecker, R., Roizman, B., and Pereira, L. (1986) Identification, properties, and gene location of a novel glycoprotein specified by herpes simplex virus 1. *Virology* **150,** 207–220.
62. Jenkins, F. J., Casadaban, M. J., and Roizman, B. (1985) Application of the mini-Mu-phage for target-sequence-specific insertional mutagenesis of the herpes simplex virus genome. *Proc. Natl. Acad. Sci.* **82,** 4773–4777.
63. Mavromara-Nazos, P., Ackermann, M., and Roizman, B. (1986) Construction and properties of a viable herpes simplex virus 1 recombinant lacking coding sequences of the ALPHA 47 gene. *J. Virol.* **60,** 807–812.
64. De Luca, N. A., McCarthy, A. M., and Schaffer, P. A. (1985) Isolation and characterization of deletion mutants of herpes simplex virus type 1 in the gene encoding immediate-early regulatory protein ICP4. *J. Virol.* **56,** 558–570.
65. Messerle, M., Crnkovic, I., Hammerschmidt, W., Ziegler, H., and Koszinowski, U.H. (1997) Cloning and mutagenesis of a herpesvirus genome as an infectious bacterial artificial chromosome. *Proc. Natl. Acad. Sci. USA* **94,** 14,759–14,763.

66. Brune, W., Ménard, C., Heesemann, J., and Koszinowski, U. H. (2001) A ribonucleotide reductase homolog of cytomegalovirus and endothelial cell tropism. *Science* **291,** 303–305.
67. Strathdee, C. A. (1999) Transposing BACs to the future. *Nat. Biotechnol.* **17,** 332–333.
68. DeFalco, J., Tomishima, M., Liu, H., et al. (2001) Virus-assisted mapping of neural inputs to a feeding center in the hypothalamus. *Science* **291,** 2608–2613.
69. Tobler, K., Fraefel, C., and Ackermann, M. (submitted) Amplicon-mediated *Cre*-ation of mutagenized BHV-1. Submitted to *J. Virol.*
70. Saeki, Y., Fraefel, C., Ichikawa, T., Breakefield, X. O., and Chiocca, E. A. (2001) Improved helper virus-free packaging sSystem for HSV amplicon vectors using an ICP27-deleted, oversized HSV-1 DNA in a bacterial artificial chromosome. *Mol. Ther.* **3,** 591–601.
71. Fraefel, C., Song, S., Lim, F., et al. (1996) Helper virus-free transfer of herpes simplex virus type 1 plasmid vectors into neuronal cells. *J. Virol.* **70,** 7190–7197.
72. Wade-Martins, R., Smith, E. R., Tyminski, E., Chiocca, E. R., and Saeki, Y. (2001) An infectious transfer and expression system for genomic DNA loci in human and mouse cells. *Nature Biotechnol.* **19,** 1067–1070.
73. Suter, M., Lew, A. M., Grob, P., et al. (1999) BAC-VAC, a novel generation of (DNA) vaccines: a bacterial artificial chromosome (BAC) containing a replication-competent, packaging-defective virus genome induces protective immunity against herpes simplex virus 1. *Proc. Natl. Acad. Sci.* **96,** 12,697–12,702.
74. Grillot-Courvalin, C., Goussard, S., Huetz, F., Ojcius, D. M, and Courvalin, P. (1998) Functional gene transfer from intracellular bacteria to mammalian cells. *Nature Biotechnol.* **16,** 862–866.
75. Britt, W. J. (2000) Infectious clones of herpesviruses: a new approach for understanding viral gene function. *Trends Microbiol.* **8,** 262–265.
76. Saeki, Y., Ichikawa, T., Saeki, A., et al. (1998) Herpes simplex virus type 1 DNA amplified as bacterial artificial chromosome in Escherichia coli: rescue of replication-competent virus progeny and packaging of amplicon vectors. *Hum. Gene Ther.* **9,** 2787–2794.

15

Cloning of β-Herpesvirus Genomes as Bacterial Artificial Chromosomes

Eva-Maria Borst, Irena Crnkovic-Mertens, and Martin Messerle

1. Introduction

Herpesviruses form a family of DNA viruses that are of considerable medical importance *(1)*. Although the genomes of many members of the herpesviruses have been completely sequenced, our knowledge on the function of the majority of herpesvirus genes is still quite insufficient. This is especially true for the members of the β-herpesviruses, i.e., the cytomegaloviruses (CMVs), because their slow replication kinetics, cell association, and large genome size (230 kb) made the construction of viral mutants a difficult and tedious procedure *(2)*. Conventional mutagenesis protocols for herpesviruses are based on the insertion of marker genes into the viral genome, which allows to disrupt or delete viral genes *(3–5)*. Unfortunately, the method has certain limitations. The protocols rely on recombination events in eukaryotic cells that are relatively rare and difficult to control. Accordingly, adventitious deletions, rearrangements, and illegitimate recombination events in the viral genomes have frequently been observed. It is especially cumbersome that the verification of the mutant genome structure can only be done at the very end of the lengthy isolation procedures of the mutant virus. Generation of viral mutants requires an obligatory selection process against the wild-type virus. In the end, the recombinant virus has to be plaque-purified and separated from the wild-type virus. This makes the isolation of viral mutants with growth disadvantages a difficult task. The introduction of mutations into essential viral genes is even more complicated because besides the required selection against wild-type virus it also depends on the quality of a complementing cell line that provides the essential gene product in *trans*.

From: *Methods in Molecular Biology, vol. 256:*
Bacterial Artificial Chromosomes, Volume 2: Functional Studies
Edited by: S. Zhao and M. Stodolsky © Humana Press Inc., Totowa, NJ

In order to circumvent these obstacles, we have developed a completely new approach for the construction of herpesvirus mutants that is based on cloning of the virus genome as a bacterial artificial chromosome (BAC) in *Escherichia coli* (*E. coli*) *(6)*. Here, we describe the procedures for the insertion of the BAC vector sequences into the viral genome (*see* **Fig. 1A**), for isolation of circular genome intermediates from infected cells and for transformation and propagation of the BACs in *E. coli*. Furthermore, we provide protocols for the isolation of BACs from bacteria, for characterization of BACs by restriction enzyme digestion and for reconstitution of recombinant viruses by transfection of BACs into permissive cells (*see* **Fig. 1C**).

Once the viral genome is cloned as a BAC, mutagenesis can be done by homologous recombination in *E. coli* (*see* **Fig. 1B**) *(6–8)*. This approach has clear advantages *(9)*. The powerful methods of bacterial genetics allow the rapid and efficient introduction of any kind of mutation (deletion, insertion, and point mutation) into the cloned viral genome. Even mutagenesis of essential viral genes now becomes feasible because introduction of the mutation into the viral genome is initially independent of the phenotype of the mutant virus *(10)*. Manipulated genomes can be characterized prior to the reconstitution of the recombinant virus and several rounds of mutagenesis can be performed on the BAC-cloned genomes. Because transfection of the mutated BAC in permissive cells leads to reconstitution of mutant virus only (*see* **Fig. 1C**), no selection against wild-type virus and no tedious purification of the mutant virus is required. The method has now been adopted by a number of laboratories for cloning of various herpesvirus genomes *(11–17)* and the cloning of other herpesvirus genomes is under way. In summary, the genetic approach considerably

---

Fig.1. *(see facing page)* Cloning and mutagenesis of the CMV genome as a bacterial artificial chromosome. (**A**) The BAC vector sequences are integrated into the CMV genome by homologous recombination in eukaryotic cells and recombinant viruses are selected. Circular genome intermediates are then isolated from infected cells and electroporated into *E. coli*. (**B**) Mutagenesis of the BAC cloned CMV genome is done by homologous recombination in *E. coli*. The mutant allele (mut) plus flanking sequences homologous to the integration site of the mutant allele (**A** and **B**) are provided by a shuttle plasmid. Following recombination with the CMV BAC the shuttle plasmid is integrated into the viral BAC genome, leading to a cointegrate. The cointegrate can resolve either via region **A** or region **B**, giving rise to a BAC with either the mutant (mut) or the wildtype (wt) allele. (**C**) The mutated CMV BAC is isolated from the bacteria and transfected into permissive cells, leading to the reconstitution of mutant CMV. (Reference Messerle et al., 1997. *Proc. Natl. Acad. Sci. USA*, Vol. 94, pp. 14759–14763. Copyright 1997 National Academy of Sciences, USA.)

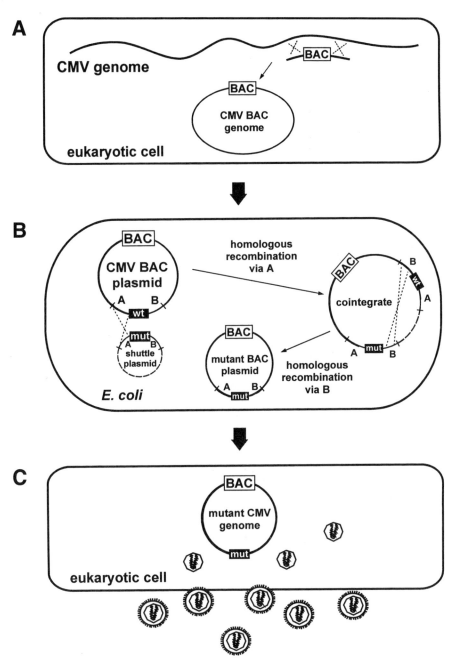

speeds up the construction of herpesvirus mutants and facilitates the analysis of herpesvirus gene functions *(9)*.

2. Materials

Standard equipment and materials for molecular biology and cell culture work is required (e.g., a table top centrifuge, incubators for bacteria and eukaryotic cells, a refrigerator or cold room, a freezer (–20°C), pipets, microtest tubes, cell culture dishes, and flasks, 6-well dishes, and so on).

2.1. Construction of the Recombination Plasmid

1. Restriction endonucleases (e.g., NEB).
2. Ligase and ligation buffer (e.g., Roche).
3. Thermomixer (e.g., Eppendorf Thermomixer 5436).
4. Sterile double-distilled water (ddH$_2$0).
5. *E. coli* strain DH10B (or any other *E. coli* strain useful for plasmid cloning purposes). Store at –70°C as a glycerol stock.
6. Competent *E. coli* DH10B prepared according to standard protocols *(18)*. Competent *E. coli* bacteria ready for transformation are also available from commercial suppliers (e.g., GIBCO BRL, Stratagene).
7. LB medium: 10 g tryptone peptone (Difco 0123-17), 5 g yeast extract (Difco 0127-17-9), 8 g NaCl and 1 L ddH$_2$0. Autoclave.
8. LB agar plates: 500 mL LB medium plus 9 g Bacto Agar (Difco 0140-01). Autoclave. Cool down to 50°C before adding the desired antibiotic (e.g., chloramphenicol [17 µg/mL]).
9. Chloramphenicol (Appligene Oncor 130112). Stock solution: 34 mg/mL in 70% ethanol. Store at –20°C.
10. Optional: Oligonucleotides (the oligonucleotides are custom-made by commercial suppliers; they contain appropriate restriction enzyme recognition sites to support individual cloning of the required DNA fragments).
11. Materials for polymerase chain reaction (PCR): thermostable DNA-Polymerase, deoxynucleotide 5′-triphosphates (dNTPs), primers.

2.2. Generation of a Recombinant Virus

1. Restriction endonucleases (*see* **Subheading 2.1.**).
2. 1 *M* Tris-HCl stock solution pH 8.0. Autoclave.
3. 0.5 *M* ethylenediaminetetraacetic acid (EDTA) pH 8.0. Autoclave.
4. TE buffer: 10 m*M* Tris-HCl pH 8.0, 0.1 m*M* EDTA pH 8.0. Prepare from stock solutions using sterile ddH$_2$0.
5. 3 *M* sodium acetate pH 5.2.
6. 100% ethanol p.a. (store at –20°C).
7. 70% ethanol.
8. Cell culture media (GIBCO-BRL).
9. Trypsin for cell culture (0.25% trypsin plus 1 m*M* EDTA, GIBCO-BRL 25200-056).

10. Phosphate-buffered saline (PBS) for cell culture (GIBCO-BRL 14190-094).
11. OPTI-MEM1 medium (GIBCO-BRL 51985-018). Store at 4°C in the dark.
12. 10% sodium dodecyl sulfate (SDS) stock solution in ddH$_2$O.
13. 100 mM Tris-HCl pH 8.0, 20 mM EDTA pH 8.0. Prepare from stock solution with sterile ddH$_2$0.
14. 1% SDS. Prepare from 10% SDS stock solution with sterile ddH$_2$0.
15. Proteinase K. Stock solution: 20 mg/mL in 10 mM Tris-HCl pH 7.5, 10 mM NaCl, 1 mM EDTA, 0.5% SDS, store at –20°C.
16. Phenol-Chloroform (1:1) solution, store at 4°C in the dark.
17. Tissue culture dishes and flasks.
18. Pasteur pipets.
19. Overhead mixer REAX2 (Heidolph, Schwabach, Germany).
20. Electroporator for bacteria and eukaryotic cells (e.g., Bio-Rad Gene Pulser or EasyjecT, peqLab Biotechnologie GmbH, Erlangen, Germany).
21. Electroporation cuvets 0.2 and 0.4 cm (Bio-Rad gene pulser cuvettes 1652086 and 1652088).
22. Incubator for cell culture (37°C, 5% CO$_2$).
23. Table top centrifuge [e.g., Eppendorf centrifuge 5415C or 5417R (4°C)].

2.3. Selection of Recombinant Virus

1. Mycophenolic acid (Sigma M5255).
2. Xanthine (Sigma X4002).
3. 0.1 M NaOH solution.
4. 0.22 µm sterile syringe filters PVDF (P666.1; Roth, Karlsruhe, Germany).

2.4. Hirt Extraction

1. 20 mM EDTA pH 8.0. Prepare from stock solution with sterile ddH$_2$O.
2. 1.2% SDS. Prepare from stock solution with sterile ddH$_2$O.
3. 5 M NaCl. Autoclave.
4. Microtest tubes (2 mL; e.g., Eppendorf).
5. Phenol-Chloroform (*see* **Subheading 2.2.**).
6. Glycogen solution 35 mg/mL (e.g., 37-1810; peqLab, Erlangen, Germany). Store at –20°C.
7. Isopropanol.
8. 70% ethanol.

2.5. Electroporation of Circular Viral Genomes into E. coli

1. Dialysis membrane 0.025 µm (Millipore VSWP04700).
2. Electrocompetent *E. coli* DH10B prepared according to standard protocols (*18*). Please note that highly competent *E. coli* bacteria are also commercially available from several suppliers (GIBCO BRL, Stratagene).
3. Electroporator (*see* **Subheading 2.2.**).
4. LB medium and LB agar plates.

2.6. Isolation of BACs From E. coli Clones and Characterization of BACs by Restriction Enzyme Digestion and Agarose Gel Electrophoresis

1. Tubes with snap caps (13 mL) for culture of bacteria.
2. Shaker for bacteria.
3. Polypropylene conical tubes (15 mL; Falcon 352096).
4. Underbench Centrifuge (e.g., Kendro Varifuge 3.0R).
5. Sol I: 50 mM glucose, 10 mM EDTA pH 8.0, 25 mM Tris-HCl pH 8.0. Autoclave.
6. Sol II: 0.2 M NaOH, 1% SDS (prepare freshly).
7. Sol III: 3 M potassium acetate (58.8 g potassium acetate in 120 mL ddH$_2$O, add 23 mL glacial acid, add ddH$_2$0 to a final volume of 200 mL). Store at 4°C.
8. RNase A, stock solution 10 mg/mL in 10 mM Tris-HCl pH 7.5, 15 mM NaCl. Boil for 20 min, store at –20°C.
9. Restriction endonucleases (*see* **Subheading 2.1.**).
10. Agarose electrophoresis grade (GIBCO-BRL 15510-027).
11. 10X TBE stock solution: 0.9 M Tris, 0.9 M boric acid, 20 mM EDTA disodium salt, autoclave.
12. Ethidium bromide stock solution (Sigma tablets E2515, dissolve 1 tablet [100 mg] in 10 mL ddH$_2$0), store at 4°C in the dark. Please note that ethidium bromide is a mutagen.
13. Gel electrophoresis units with power supply (e.g., Hoefer HE 33 mini horizontal submarine unit, Hoefer HE 99X max. submarine unit, electrophoresis power supply EPS 1001, Amersham Pharmacia).
14. Ultraviolet (UV) transilluminator.

2.7. Storage of BAC Clones as Glycerol Stocks

1. 50% glycerol in ddH$_2$0. Autoclave.
2. Cryotube vials 1.8 mL and 4.5 mL (Nunc).
3. Freezer (–70°C).

2.8. Isolation of BACs From Large E. coli Cultures

1. Erlenmeyer beaker (2 L).
2. Nucleobond PC 500 Kit (Machery-Nagel, Düren, Germany) or Large Construct Kit (Qiagen, Hilden, Germany).
3. High-performance centrifuge (e.g., Beckman J2 or Sorvall RC-3B with rotor SM-24)
4. Centrifuge tubes (13 mL; e.g., Sarstedt 55518, 95 × 16 mm)

2.9 Reconstitution of Recombinant Viruses by Transfection of BACs into Permissive Cells

1. Opti-MEM (*see* **Subheading 2.2.**).
2. SuperFect transfection reagent (Qiagen, Hilden, Germany).

3. Methods

3.1. Construction of a Recombination Plasmid for Generation of a Virus Genome Containing the BAC Vector Sequences

For generation of the CMV BAC, the viral genome and the BAC vector have to be joined. We decided to integrate the BAC vector sequences into the virus genome by homologous recombination in eukaryotic cells (*see* **Fig. 1A**), i.e., by generating a recombinant virus, because conventional cloning by ligation of the linear viral genome to the BAC vector was considered to be difficult due to the large size of the CMV genome. Also, insertion of the BAC vector between the genomic termini might subsequently interfere with the packaging process of the viral genomes. After enrichment of the recombinant virus, circular viral DNA containing the BAC vector sequences is isolated from the cells and transferred to *E. coli* (*see* **Fig. 1B**) *(6,7)*.

In order to insert the BAC vector sequences into the CMV genome, a recombination plasmid has to be constructed (*see* **Fig. 2**). This recombination plasmid contains the BAC vector flanked by sequences homologous to the desired integration site in the viral genome (left and right side homologies) plus a selection or screening marker (e.g., *gpt* [guanosine phosphoribosyl transferase] or *gfp* [green fluorescent protein]) (*see* **Fig. 2**, step 4). As a backbone for construction of the recombination plasmid, we use a multi-copy plasmid vector, e.g., pUC19. Because we have hints for a size constraint of the CMV genome with regard to its packaging into capsids *(6,19)*, one has to take care that the upper limit of the CMV genome size is not exceeded by insertion of the BAC vector (*see* **Note 1**). Therefore, we recommend to replace a nonessential region of the viral genome by the BAC vector sequences and the *gpt* (guanosine phosphoribosyl transferase) selection marker *(6,7)*. Because there is no size constraint for the BAC in bacteria, the missing sequences can be easily reinserted by homologous recombination in *E. coli* once the viral genome is cloned as an infectious BAC *(19,20)*.

1. Choose a suitable plasmid vector (e.g., the multicopy plasmid vectors of the pUC or pBluescript series) providing appropriate restriction enzyme recognition sites for construction of a recombination plasmid. Please note that several cloning steps have to be carried out (*see* **Fig. 2**; right side homology, left side homology, BAC vector including a selection or screening marker). In most cases, it is convenient to clone in a first step an oligonucleotide linker containing the required restriction enzyme recognition sites into the plasmid vector (*see* **Fig. 2**, step 1). The BAC vectors used in our laboratory *(6–8)* are derivatives of the BAC plasmid pBAC108L *(21)* and can be linearized and inserted via a unique *Pac*I site (*see* **Fig. 2**, step 4). Because it is desirable to excise the recombination fragment from the plasmid backbone prior to transfection into eukaryotic cells in order to facil-

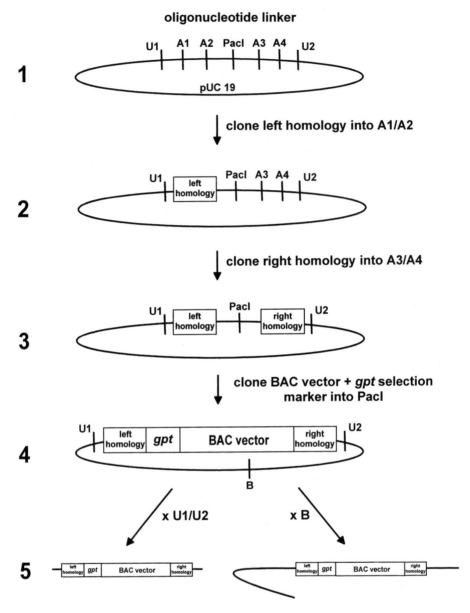

Fig. 2. Construction of a recombination plasmid for generation of a CMV BAC genome. Step 1. An oligonucleotide linker providing the restriction enzyme sites needed is cloned into a high-copy plasmid. Steps 2 and 3. The regions homologous to the intended integration site of the BAC vector sequences in the viral genome are inserted into the plasmid (restriction enzyme sites A1/A2 are used for the left side homology and A3/A4 for the right side homology). Step 4. The BAC vector sequences plus a suitable selection marker (e.g., the *gpt* gene) are cloned into the *Pac*I restriction site. Step 5. Prior to transfection the recombination fragment is excised from the plasmid using the restriction enzyme sites U1/U2. Alternatively, the plasmid is linearized via a unique restriction site in the plasmid backbone (site B).

itate recombination into the viral genome (*see* **Fig. 2**, step 5), one should flank the homology arms with unique restriction enzyme sites (sites U1 and U2 in **Fig. 2**). Alternatively, one should leave one unique site (site B in **Fig. 2**) in the plasmid backbone to allow linearization of the recombination plasmid.

2. Clone the viral sequences homologous to the intended integration site of the BAC vector in the viral genome (left and right side homology) into the plasmid (*see* **Fig. 2**, steps 2 and 3). We recommend using homologies of at least 500 bp in size. The viral sequences may be obtained by PCR.

3. Clone the BAC vector sequences with the selection or screening marker into the *Pac*I restriction site between the homologies (*see* **Fig. 2**, step 4; *see* **Note 2**).

3.2. Generation of a Recombinant Virus by Cotransfection of the Recombination Plasmid and Viral DNA

3.2.1. Preparation of the Recombination Plasmid

1. Excise the recombination fragment containing the BAC vector, the selection marker and the flanking homologies with suitable restriction enzymes from the plasmid backbone (*see* **Fig. 2**, step 5). Alternatively, linearize the recombination plasmid by digestion with a restriction enzyme that cuts in the vector backbone (*see* **Fig. 2**, step 5). Digest 40 µg of the recombination plasmid with appropriate restriction enzymes in a total volume of 50 µL.

2. Add 150 µL TE buffer to the reaction tube (final volume 200 µL). Add 20 µL (1/10 vol) 3 *M* sodium acetate pH 5.2 and mix carefully.

3. Add 550 µL (2.5 vol) ice-cold ethanol and mix carefully.

4. Pellet the precipitated DNA by centrifugation for 10 min at 14,000 rpm and 4°C in a tabletop centrifuge.

5. Wash the DNA with 1 mL 70% ethanol and centrifuge for 5 min as in **step 4**.

6. Air-dry the pellet and resuspend in 20 µL sterile water.

7. Check the integrity and quantity of the plasmid DNA by gel electrophoresis.

3.2.2. Preparation of the Viral DNA

1. Infect about 3×10^6 permissive cells at a multiplicity of infection (moi) of 0.1 PFU/cell.

2. When the cytopathic effect (CPE) is complete, harvest the cell supernatant and trypsinize the cells. Collect the cells and the supernatant and spin down the cells at 300*g* for 5 min.

3. Wash the cells with 5 mL PBS and centrifuge at 300*g* for 5 min.

4. Resuspend the cell pellet in 500 µL 100 m*M* Tris-HCl pH 8.0/20 m*M* EDTA pH 8.0 and add 500 µL of 1% SDS. Mix carefully.

5. Add 40 µL proteinase K (20 mg/mL stock) and incubate for 3 h (or overnight) at 56°C.

6. Add 1 mL phenol-chloroform to the DNA solution. Mix carefully (e.g., by incubating the micro test tubes on a slowly rolling overhead mixer for 10 min). Separate the phases by centrifugation at 14,000 rpm in a tabletop centrifuge for

5 min. Transfer the upper (aqueous) phase to new vials using cut pipet tips (*see* **Note 3**).

7. Add 1/10 vol 3 *M* sodium acetate pH 5.2 and 2.5 vol ethanol. Mix carefully.
8. Spool out the DNA with the tip of a Pasteur pipet and wash with 2 mL 70% ethanol
9. Blot dry on the wall of a microtest tube.
10. Dissolve the DNA in 100 µL TE buffer by incubation at 37°C for several hours without shaking followed by incubation at 4°C overnight.
11. Digest 10 µL of the viral DNA with a suitable restriction enzyme and check by gel electrophoresis (*see* **Subheading 3.7.**) for quantity and quality (*see* **Note 4**).

3.2.3. Transfection of the Recombination Plasmid by Electroporation

1. One day before transfection split 4×10^6 cells so that they will be about 50% confluent the next day.
2. On the day of transfection, harvest the cells by trypsin treatment. Spin down the cells at 300*g* for 5 min and wash the pellet with 5 mL PBS. Pellet the cells at 300*g* for 5 min.
3. Resuspend the cell pellet in 800 µL Opti-MEM.
4. Add 40 µg of the linearized recombination plasmid and 2–5 µg of viral DNA, mix carefully.
5. Transfer to a 4-mm electroporation cuvet and electroporate at room temperature at 280 V and 1500 µF (*see* **Note 5**).
6. Immediately after electroporation, seed the cells in a T25 flask containing pre-warmed culture medium, incubate at 37°C and 5% CO_2.

Plaques will usually appear within 1 wk. If the cells become confluent, it is helpful to split the cells to new tissue culture dishes. Wait until complete CPE appears, then save the supernatant and proceed with **Subheading 3.4.** to select and enrich recombinant viruses.

3.3. Generation of the CMV BAC: Transfection of the Recombination Construct and Superinfection With Virus

As an alternative to cotransfection, the recombination construct can be transfected into permissive cells followed by superinfection with the virus.

1. Prepare 40 µg of the linearized recombination construct and electroporate into permissive cells as described above. The cells should be at least 50% confluent the next day.
2. One day after transfection, wash the cells once with culture medium and infect with virus at an multiplicity of infection (moi) of 5 PFU/cell in a small volume of culture medium (*see* **Note 6**). Add cell culture medium after infection.
3. Incubate the cells at 37°C until complete cytopathic effect occurs (several days), collect the supernatant and spin at 300*g* for 5 min to remove cells and cell debris.

Store an aliquot (1 mL) of the supernatant at –70°C and use the remaining supernatant to select recombinant virus (continue with protocol in **Subheading 3.4.**).

3.4. Selection of Recombinant Virus

Recombinant viruses are enriched by selection with mycophenolic acid and xanthine utilizing the *E. coli* enzyme guanosine phosphoribosyl transferase (*gpt*). Selection is performed essentially according to published protocols *(5,22)*. The *E. coli* enzyme *gpt* is able to synthesize and to provide purine precursors from xanthine in order to support replication of recombinant viruses under conditions in which the cellular purine synthesis is blocked by mycophenolic acid. Please note that proliferation of cells is impaired under these conditions.

3.4.1. Preparation of Selection Medium

1. Prepare stock solutions of 50 mM mycophenolic acid and 50 mM xanthine, both dissolved in 0.1 M NaOH. Store at –20°C.
2. Add mycophenolic acid and xanthine to the cell culture medium to a final concentration of 100 μM for mycophenolic acid and 25 μM for xanthine. Filter with sterile filter (0.22 μm) and store at 4°C.

3.4.2. Enrichment of Recombinant Viruses

1. Add the supernatant obtained by the procedures described in **Subheading 3.2.** or **3.3.** to a T25 flask containing a confluent monolayer of cells. Incubate for 4 h, wash twice with PBS and apply the selection medium (*see* **Note 7**).
2. Incubate until the cells show complete CPE (takes several days). Take the supernatant, clarify by centrifugation as described in **Subheading 3.3., step 3**, freeze an aliquot and transfer the remaining medium to a T25 flask with confluent cells. Incubate, wash, and apply selection as described in **step 1**.
3. Repeat **step 2**.
4. After the third round of selection (*see* **step 3**) DNA can be isolated from the infected cells following the procedure of Hirt (*see* protocol in **Subheading 3.5.**). However, it is advisable to continue in parallel with the selection procedure for some more rounds to generate a second Hirt extract if the first attempt is not successful (*see* **Note 8**).

3.5. Isolation of Circular Viral Genomes From Infected Cells: Generation of a Hirt Extract

For the isolation of circular viral genomes, we applied the protocol originally described by Hirt *(23)* for extraction of polyoma virus DNA from infected cells. This procedure allows the selective enrichment and isolation of circular, supercoiled and low molecular weight DNA, whereas large chromosomal and linear DNA molecules are excluded.

1. Harvest the cells by trypsin treatment. Collect the cells by centrifugation for 5 min at 300g.
2. Wash the cell pellet with 5 mL culture medium, pellet cells by centrifugation for 5 min at 300g.
3. Wash the pellet with 5 mL sterile PBS, collect the cells by centrifugation (*see* **step 2**).
4. Resuspend the cells in 500 µL 20 mM EDTA pH 8.0 and transfer to a 2-mL microtest tube.
5. Add 500 µL 1.2% SDS and mix carefully by inverting the tubes to lyse the cells.
6. Add 0.66 mL 5 M NaCl, mix carefully by inverting the tubes. A white precipitate should become visible at this step.
7. Incubate overnight or at least for 4 h at 4°C (*see* **Note 9**).
8. Centrifuge for 30 min at 4°C and 14 000 rpm in a table top centrifuge (*see* **Note 10**).
9. Transfer the supernatant with a cut blue tip to a new 2-mL microtest tube. Place half of the sample into a second 2-mL microtest tube. Extract both samples with 1 vol phenol-chloroform for 10 min at 4°C, e.g., by slowly rolling the tubes on an overhead mixer.
10. Centrifuge for 5 min at 4°C and 14 000 rpm in a table top centrifuge.
11. Transfer the upper phase with a cut pipet tip to two new 2 mL tubes (*see* **Note 11**).
12. Add 20 µg of glycogen as a carrier for precipitation.
13. Precipitate the DNA with 0.8 vol isopropanol, centrifuge immediately for 20 min at 4°C and 14 000 rpm in a tabletop centrifuge.
14. Wash the pellet with 1 mL 70% ethanol (*see* **Note 12**).
15. Air-dry the pellet and dissolve in 100 µL TE buffer.

3.6. Electroporation of Circular Genomes into E. coli

1. Fill a 10-cm Petri dish with TE buffer.
2. Put a 0.025 µm dialysis membrane onto the surface of the TE buffer.
3. Apply 20 µL of the Hirt extract (from **step 15** of method in **Subheading 3.5.**) with a cut pipet tip onto the dialysis membrane.
4. Dialyze for 30 min at room temperature.
5. Meanwhile, thaw a 50-µL aliquot of electrocompetent *E. coli* strain DH10B on ice and cool down a 0.2-cm electroporation cuvet on ice (*see* **Note 13**).
6. Add 5 µL of the dialyzed Hirt extract to the electrocompetent *E. coli* and mix briefly by flicking the tube. Transfer the whole content to the cold 0.2 cm electroporation cuvet.
7. Electroporate at 400 Ω, 25 µF and 2.5 kV (*see* **Note 14**)
8. Add 500 µL prewarmed LB medium to the cuvet and transfer the content to a 1.5-mL microtest tube.
9. Incubate at 37°C for 1 h with shaking (900 rpm) on a thermomixer.
10. Plate the bacteria on two LB-Agar plates containing 17 µg/mL chloramphenicol (*see* **Note 15**).
11. Incubate overnight at 37°C (*see* **Note 16**).

3.7. Isolation of BACs From E. coli Clones and Characterization of BACs by Restriction Enzyme Digestion and Agarose Gel Electrophoresis

1. Pick single bacterial clones into tubes containing 10 mL LB with 17 µg/mL chloramphenicol (*see* **Note 17**).
2. Incubate at 37°C and 180 rpm on a shaker for at least 18 h (*see* **Note 18**).
3. Transfer 9 mL of the culture to a new tube (15 mL) and centrifuge at 2500*g* for 5 min in an underbench centrifuge.
4. Resuspend the bacterial pellet with a pipet tip in 200 µL Sol I and transfer to a 2-mL microtest tube.
5. Add 300 µL Sol II to lyse the cells and mix carefully by inverting the tubes.
6. Incubate 5 min at room temperature (not longer!) (*see* **Note 19**).
7. Add 300 µL ice-cold Sol III, mix by inverting the tubes, and incubate on ice for 5 min.
8. Centrifuge for 5 min at 4°C and 14,000 rpm in a table top centrifuge.
9. Decant the supernatant into new 2 mL microtest tubes.
10. Add 800 µL isopropanol and mix carefully by inverting the tubes to precipitate the DNA (*see* **Note 20**).
11. Centrifuge 20 min at 4°C and 14 000rpm in a tabletop centrifuge.
12. Wash the DNA pellets with 1 mL 70% ethanol and centrifuge 5 min as in **step 11**.
13. Air-dry the pellets at room temperature and dissolve the DNA in 100 µL TE buffer plus 20 µg/mL RNaseA for 20 min at 37°C without shaking the tubes (*see* **Note 21**).
14. Use 45 µL of each sample for restriction enzyme digestion in a total volume of 50 µL with about 20 U of a suitable restriction enzyme per sample. Incubate for about 4 h. In parallel, digest an aliquot of DNA of the parental virus as a control.
15. Dissolve agarose electrophoresis grade in 250 mL 0.5X TBE buffer to a final concentration of 0.5–0.8% depending on the size of the expected fragments.
16. Cool down the agarose by magnetic stirring for 30 min at room temperature.
17. Add 20 µL ethidium bromide stock solution.
18. Pour a 15 × 20-cm gel and allow solidifying for at least 1 h at room temperature.
19. Load the 50 µL restriction enzyme digested samples onto the gel and run in 0.5X TBE buffer at 2.5 V/cm for 14–18 h (depending on the size of the expected fragments).
20. Visualize the DNA bands in the gel with a UV transilluminator, take a photo and identify mutant BACs by the DNA restriction pattern.

3.8. Storage of BAC Clones As Glycerol Stocks

1. Grow a clone containing the desired BAC in 5 mL LB medium with 17 µg/mL chloramphenicol for at least 18 h.
2. Transfer 800 µL of the culture together with 800 µL sterile 50% glycerol solution into a cryotube and store at –70°C.
3. Always go back to this master tube in order to grow up the clone.

3.9. Isolation of BACs From Large E. coli Cultures

1. Inoculate 500 mL LB medium containing 17 µg/mL chloramphenicol with 1 mL culture volume of a bacterial clone containing the BAC (usually the culture volume left in **Subheading 3.7., step 3**). Use a 2 L Erlenmeyer beaker.
2. Incubate for 22–24 h at 37°C on a shaker with 180 rpm.
3. Purify BAC DNA on silica columns by ion exchange chromatography, e.g., by using the NucleoBond PC 500 Kit (Macherey-Nagel, Düren, Germany) or the Qiagen Large Construct Kit (Qiagen, Hilden, Germany) following the instructions of the manufacturers (*see* **Note 22**).
4. Dissolve the BAC DNA in 200 µL TE buffer for at least 45 min at 37°C without shaking (*see* **Note 21**).
5. Check 5 µL of the DNA by restriction enzyme digest and gel electrophoresis as described in **Subheading 3.7.** Use at least three different restriction enzymes to assure the integrity of the BAC DNA. For HCMV and MCMV restriction enzymes like *Eco*RI, *Hin*dIII, *Xho*I, *Bgl*II, and *Eco*RV (*see* **Fig. 3**) have proven to be useful for characterization of the BAC cloned genome (*see* **Note 23**).

3.10. Transfection of BACs into Permissive Cells for Reconstitution of Recombinant Viruses

For reconstitution of recombinant virus the BAC DNA is transfected into permissive cells. The following protocol is adjusted for transfection in 6-well plates.

1. Split the cells the day before transfection into a 6-well plate so that they will be about 80% confluent the next day (*see* **Note 24**).
2. Dilute 10 µL (i.e., about 2 µg) of the purified BAC DNA with 80 µL Opti-MEM (final volume is 100 µL).
3. Add 10 µL of SuperFect transfection reagent (Qiagen, Düren, Germany) to the DNA solution, mix carefully by flicking the tube, and incubate for 10 min at room temperature.
4. Wash the cells once with PBS.
5. Add 600 µL of complete cell growth medium to the reaction tube, mix shortly by flicking the tube, and transfer the content to one well of the 6-well plate with a 2-mL plastic pipet.
6. Incubate for 3 h at 37°C and 5% CO_2.
7. Remove the transfection complexes and wash cells three times with culture medium.
8. Incubate until viral plaques appear in the cell monolayer (*see* **Note 25**).

When plaque formation occurs, wait until CPE is complete. Freeze the supernatant at –70°C and isolate total DNA from the cells (*see* **Subheading 3.2.**) in order to check the DNA pattern of the recombinant virus genome as described earlier. If the expected DNA pattern of the recombinant virus can be

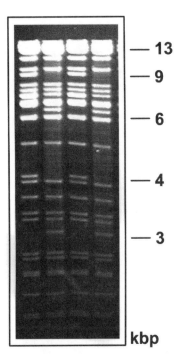

Fig. 3. Characterization of CMV BACs by restriction enzyme analysis. BAC DNA was isolated from 500 mL bacterial culture each and finally dissolved in 200 μL TE buffer. Five microliters of the DNA was digested with *Eco*RV, separated on a 0.6% agarose gel for 15 h at 2.5 V/cm and visualized by ethidium bromide staining and UV transillumination Lanes 1, 3: parental CMV BAC. Lanes 2, 4: CMV BAC carrying a mutuation of the SCP gene.

verified, prepare a virus stock and check the recombinant for its biological properties in vitro and (if possible) in vivo.

4. Notes

1. The size constraints may be less critical for other herpesvirus genomes. In this case, the BAC vector sequences can be inserted without replacement of nonessential sequences.
2. Note that bacterial clones containing the desired recombination construct can be identified by double selection using the antibiotics whose resistances are encoded by the BAC vector and the plasmid backbone, respectively.
3. The DNA extraction may be repeated if the aqueous phase appears milky or still contains some material of the interphase. Whenever working with high-molecular weight DNA like viral, chromosomal or BAC DNA, always cut pipet tips with a scissor prior to pipeting and never vortex the samples, instead mix gently by

inverting the tubes. Otherwise shearing of the DNA molecules can occur. Store the high-molecular weight DNA in TE buffer at 4°C, do not freeze.

4. The procedure can be upscaled in order to prepare larger amounts of viral DNA.

5. The electroporation conditions are given for primary cells like human foreskin fibroblasts. For permanent cell lines use 250 V and 1500 µF, for sensitive cells use 200 V and 1500 µF.

6. Efficiency of infection of transfected cells is rather poor. Therefore, a multiplicity of infection (moi) of 5–10 PFU/cell should be used. The viral infectious dose can be increased 10- to 20-fold (or even more) by centrifugation of the cells for 30 min at 800*g* immediately after addition of the viral inoculate (centrifugal enhancement of the infection *[24]*).

7. It is important that the cells are confluent when the selection is applied, because the selection medium inhibits cell proliferation. We observed that cells detach form the tissue culture dishes if they are not confluent at this step. This may resemble a CPE.

8. In general, chances to isolate a sufficient amount of recombinant viral DNA by the Hirt protocol are good when at least 10% of plaques contain the recombinant virus. It might, therefore, be helpful to insert a marker like the gene for the green fluorescent protein into the BAC genome (i.e., provide the *gfp* gene on the recombination plasmid) in order to simplify detection of recombinant virus *(8)*.

9. During this step, proteins, large chromosomal and linear DNA molecules precipitate, whereas circular, supercoiled, and low molecular weight DNA will remain in the supernatant.

10. It is essential to do the centrifugation step at 4°C.

11. It may happen that the upper phase is not clear after centrifugation. This can be avoided by increasing the time of phenol-chloroform extraction to 15–20 min and extend the centrifugation step to 10 min.

12. As the samples are already rich in salt it is not necessary to add 1/10 vol 3 *M* sodium acetate for precipitation. Also, because of the high-salt concentration, DNA precipitation will occur immediately. The samples should not be incubated at –20 or –70°C in order to avoid extensive precipitation of salt. Washing with 70% ethanol should be extensive (leave the samples in 70% ethanol at room temperature overnight).

13. Preparation of electrocompetent *E. coli* cells is done following standard procedures *(18)*. Preferentially, we use the *E. coli* strain DH10B because in our hands it proved to be the most competent strain for electroporation of large DNA molecules like BACs.

14. For optimizing the electroporation conditions for large BAC molecules, please see the publication of Birren et al. *(25)*. We found, however, that the electroporation conditions are not critical at this step.

15. The BAC vector encodes chloramphenicol resistance.

16. The CMV BAC is a low copy plasmid with 1–2 copies per bacterial cell. It might, therefore, take up to 24 h until clones appear on the agar plates.

17. It is important to grow the bacteria with good aeration. Please use appropriate incubation tubes.

18. As the BAC is low copy the bacteria can be grown for up to 24 h in order to increase the bacterial mass and thereby the DNA yield.
19. After lysis of the cells always use cut pipet tips when handling the samples and never vortex—instead, mix carefully by inverting the tubes in order to avoid shearing of the high molecular weight DNA. Do not incubate longer than 5 min because this will result in excessive release of chromosomal DNA.
20. When working with *E. coli* strain DH10B it is not necessary to perform a phenol-chloroform extraction.
21. Large BAC DNA molecules are sometimes difficult to bring into solution.
22. For the NucleoBond PC 500 kit do the optional filtration step after addition of buffer S3 instead of centrifugation and load the sample twice onto the column as recommended. The final DNA pellet is extremely small and sometimes hard to detect on the bottom of the tube. We prefer using plastic tubes and a Sorvall SM24 rotor for centrifugation of the precipitated DNA instead of a Sorvall SS34 rotor with Corex tubes. Immediately after centrifugation decant the supernatant and make sure not to lose the tiny pellet.
23. Measuring the concentration of BAC DNA in a spectrophotometer often gives inconsistent results, because the DNA solution may be inhomogeneous. The amount of BAC DNA purified from 500 mL of a bacterial culture is usually about 50 µg in total. We routinely check the quality and quantity of the DNA as described in **Subheading 3.9., step 5** without measuring the optical density. Usually, 5 µL of digested BAC DNA can be easily detected on an agarose gel by UV transillumination.
24. For the HCMV BAC transfection of MRC-5 cells gives better results than transfection of human foreskin fibroblasts (HFF). However, because the virus productivity on MRC-5 cells is about one order of magnitude lower than on HFF cells, the reconstituted virus should be grown on HFF for high titer virus production. For the MCMV BAC transfection of murine embryonic fibroblasts (MEF) or NIH 3T3 is used for reconstitution of recombinant viruses.
25. It is advisable to split the cells into 6-cm dishes when they reach confluence because CMV replicates better in growing cells. Do not aspirate the (virus containing) supernatant from the transfected cells. Instead, transfer it to the new 6-cm dish together with fresh medium.

Acknowledgment

This work was supported by the Wilhelm-Roux-Program of the Medical Faculty, Martin-Luther-University of the Halle-Wittenberg and the NBL3-Program of the German Ministry for Education and Research (BMBF).

References

1. Roizman, B. (1996) Herpesviridae in *Fields Virology* (Fields, B. N., Knipe, D. M., and Howley, P. M., eds.), Lippincott-Raven, Philadelphia, PA, pp. 2221–2230.
2. Mocarski, E. S. and Kemble, G. W. (1996) Recombinant cytomegaloviruses for study of replication and pathogenesis. *Intervirology* **39,** 320–330.

3. Post, L. E. and Roizman, B. (1981) A generalized technique for deletion of specific genes in large genomes: alpha gene 22 of herpes simplex virus 1 is not essential for growth. *Cell* **25,** 227–232.
4. Spaete, R. R. and Mocarski, E. S. (1987) Insertion and deletion mutagenesis of the human cytomegalovirus genome. *Proc. Natl. Acad. Sci. USA* **84,** 7213–7217.
5. Greaves, R. F., Brown, J. M., Vieira, J., and Mocarski, E. S. (1995) Selectable insertion and deletion mutagenesis of the human cytomegalovirus genome using the Escherichia coli guanosine phosphoribosyl transferase (gpt) gene. *J. Gen. Virol.* **76,** 2151–2160.
6. Messerle, M., Crnkovic, I., Hammerschmidt, W., Ziegler, H., and Koszinowski, U. H. (1997) Cloning and mutagenesis of a herpesvirus genome as an infectious bacterial artificial chromosome. *Proc. Natl. Acad. Sci. USA* **94,** 14,759–14,763.
7. Borst, E. M., Hahn, G., Koszinowski, U. H., and Messerle, M. (1999) Cloning of the human cytomegalovirus (HCMV) genome as an infectious bacterial artificial chromosome in Escherichia coli: a new approach for construction of HCMV mutants. *J. Virol.* **73,** 8320–8329.
8. Adler, H., Messerle, M., Wagner, M., and Koszinowski, U. H. (2000) Cloning and mutagenesis of the murine gammaherpesvirus 68 genome as an infectious bacterial artificial chromosome. *J. Virol.* **74,** 6964–6974.
9. Brune, W., Messerle, M., and Koszinowski, U. H. (2000) Forward with BACs: new tools for herpesvirus genomics. *Trends Genet.* **16,** 254–259.
10. Borst, E. M., Mathys, S., Wagner, M., Muranyi, W., and Messerle, M. (2001) Genetic evidence of an essential role for cytomegalovirus small capsid protein in viral growth. *J. Virol.* **75,** 1450–1458.
11. Delecluse, H. J., Hilsendegen, T., Pich, D., Zeidler, R., and Hammerschmidt, W. (1998) Propagation and recovery of intact, infectious Epstein-Barr virus from prokaryotic to human cells. *Proc. Natl. Acad. Sci. USA* **95,** 8245–8250.
12. Saeki, Y., Ichikawa, T., Saeki, A., et al. (1998) Herpes simplex virus type 1 DNA amplified as bacterial artificial chromosome in Escherichia coli: rescue of replication-competent virus progeny and packaging of amplicon vectors. *Hum. Gene. Ther.* **9,** 2787–2794.
13. Stavropoulos, T. A. and Strathdee, C. A. (1998) An enhanced packaging system for helper-dependent herpes simplex virus vectors. *J. Virol.* **72,** 7137–7143.
14. Smith, G. A. and Enquist, L. W. (1999) Construction and transposon mutagenesis in Escherichia coli of a full-length infectious clone of pseudorabies virus, an alphaherpesvirus. *J. Virol.* **73,** 6405–6414.
15. Horsburgh, B. C., Hubinette, M. M., Qiang, D., MacDonald, M. L., and Tufaro, F. (1999) Allele replacement: an application that permits rapid manipulation of herpes simplex virus type 1 genomes. *Gene Ther.* **6,** 922–930.
16. Schumacher, D., Tischer, B. K., Fuchs, W., and Osterrieder, N. (2000) Reconstitution of Marek's disease virus serotype 1 (MDV-1) from DNA cloned as a bacterial artificial chromosome and characterization of a glycoprotein B-negative MDV-1 mutant. *J. Virol.* **74,** 11,088–11,098.

17. Delecluse, H. J., Kost, M., Feederle, R., Wilson, L., and Hammerschmidt, W. (2001) Spontaneous activation of the lytic cycle in cells infected with a recombinant Kaposi's sarcoma-associated virus. *J. Virol.* **75,** 2921–2928.

18. Sambrook, J. F. and Russell, D. W. (2000) *Molecular Cloning: A Laboratory Manual.* Cold Spring Harbor Laboratory Press, Cold Spring Harbor, NY.

19. Wagner, M., Jonjic, S., Koszinowski, U. H., and Messerle, M. (1999) Systematic excision of vector sequences from the BAC-cloned herpesvirus genome during virus reconstitution. *J. Virol.* **73,** 7056–7060.

20. Hobom, U., Brune, W., Messerle, M., Hahn, G., and Koszinowski, U. H. (2000) Fast screening procedures for random transposon libraries of cloned herpesvirus genomes: mutational analysis of human cytomegalovirus envelope glycoprotein genes. *J. Virol.* **74,** 7720–7729.

21. Shizuya, H., Birren, B., Kim, U. J., et al. (1992) Cloning and stable maintenance of 300-kilobase-pair fragments of human DNA in Escherichia coli using an F-factor-based vector. *Proc. Natl. Acad. Sci. USA* **89,** 8794–8797.

22. Vieira, J., Farrell, H. E., Rawlinson W. D., and Mocarski, E. S. (1994) Genes in the HindIII J fragment of the murine cytomegalovirus genome are dispensable for growth in cultured cells: insertion mutagenesis with a lacZ/gpt cassette. *J. Virol.* **68,** 4837–4846.

23. Hirt, B. (1967) Selective extraction of polyoma DNA from infected mouse cell cultures. *J. Mol. Biol.* **26,** 365–369.

24. Osborn, J. E. and Walker, D. L. (1968) Enhancement of infectivity of murine cytomegalovirus in vitro by centrifugal inoculation. *J. Virol.* **2,** 853–858.

25. Sheng, Y., Mancino, V., and Birren, B. (1995) Transformation of Escherichia coli with large DNA molecules by electroporation. *Nucl. Acids Res.* **23,** 1990–1996.

16

Construction of HSV-1 BACs and Their Use for Packaging of HSV-1-Based Amplicon Vectors

Thomas G. H. Heister, Andrea Vögtlin, Lars Müller, Irma Heid, and Cornel Fraefel

1. Introduction

Viruses are parasites that are strictly dependent on the host cell for replication. In general, the more complex a virus, the more complicated the network of interactions between the virus and the host. Many of these interactions are still unrevealed or not understood in detail. The generation of mutant viruses is an important strategy to obtain insights into the functional role of individual virus genes. A classical genetic approach used chemical mutagenesis to produce mutant viruses, however, identifying the mutation responsible for an observed phenotype and excluding second-site mutations is difficult. Development of site-directed mutagenesis of the herpes simplex virus type 1 (HSV-1) genome by homologous recombination allowed targeted mutation of individual HSV-1 genes *(1,2)*. This technique remains laborious and, in some cases, impossible, because null mutants of genes that are essential for HSV-1 replication can only be produced on complementing cell lines *(3)*. A more-efficient strategy for mutating HSV-1 was achieved by cloning the entire virus genome as a set of overlapping clones in cosmids. This allows the manipulation of specific virus genes on individual cosmid clones in *Escherichia coli (E. coli)* and their subsequent characterization. Cotransfection of this modified cosmid set into permissive cells leads to the reconstitution of recombinant HSV-1 genomes by homologous recombination *(4)*. However, one problem of the cosmids is their instability, in particular those clones that contain repeated sequences. Therefore, alternative cloning vectors, which permit the amplification, manipulation, and characterization of large DNA fragments are used *(5)*.

From: *Methods in Molecular Biology, vol. 256:*
Bacterial Artificial Chromosomes, Volume 2: Functional Studies
Edited by: S. Zhao and M. Stodolsky © Humana Press Inc., Totowa, NJ

One of these vectors is the bacterial artificial chromosome (BAC). The first herpesvirus genome cloned as a BAC was the 230-kb mouse cytomegalovirus (CMV) genome *(6)*. When transfected into permissive cells, this CMV-BAC supported the production of virus progeny. Several other herpes virus genomes have been cloned recently as BACs, including Epstein-Barr virus *(7)*, HSV-1 *(8–10)*, pseudorabies virus *(11)*, and human CMV *(12)*. These BACs facilitate the manipulation of the virus genomes in *E. coli* and subsequent analysis of the introduced genetic alterations in eukaryotic cells. Despite the low copy number, sufficient BAC DNA can be obtained from a few milliliters of bacterial culture for restriction endonuclease analysis, hybridization, or polymerase chain reaction (PCR). Because BAC DNA is supercoiled, it is resistant to shear-induced breakage during isolation and, hence, even BACs as large as 300 kb require no extraordinary measures in handling the DNA *(5)*. This chapter describes the construction of two HSV-BACs, fHSVΔpac, and fHSVpac+ (*see* **Subheading 3.1.**), as well as the use of fHSVΔpac as helper HSV-1 genome for the packaging of HSV-1 amplicon vectors into virus particles (*see* **Subheading 3.2.**). HSV-1 amplicon vectors are bacterial plasmids that contain: (i) a transgene cassette; (ii) sequences from *E. coli*, including an origin of DNA replication and an antibiotic resistance gene, for propagation in bacteria; and (iii) sequences from HSV-1, including an origin of DNA replication *(ori)* and a DNA cleavage/packaging signal *(pac)*, which support replication and packaging of the amplicon DNA into virions in the presence of helper functions *(13)*. Because the HSV-1 *cis*-elements, *ori* and *pac*, required for replication and packaging are smaller than 1 kb, the amplicon has the potential to accommodate up to 150 kb of foreign DNA. This large transgene capacity, combined with the broad host range, makes the HSV-1 amplicon an excellent candidate as a gene delivery vehicle.

2. Materials

2.1. Cloning of the 152-kb HSV-1 Genome into a BAC: fHSVΔpac and fHSVpac+

2.1.1. Construction of HSV-1 Transfer Plasmid fH98-102pac

1. pBluescriptSK+ (pBsSK+; Stratagene, La Jolla, CA).
2. Synthetic oligonucleotide duplex 5′ tcgagggcccttaattaagatcttaattaagggccc 3′.
3. pHSVPrPUC (H. Federoff, University of Rochester, NY).
4. pBeloBAC11 (Research Genetics, Huntsville, AL).
5. cos56 of HSV-1 cosmid set C *(4)* (requests for cosmid set C should be addressed to Drs. Cunningham and Davison; MRC Virology Unit, Institute of Biomedical and Life Sciences, University of Glasgow, Glasgow G11 5JR, U.K.).
6. Restriction endonucleases: *Xho*I, *Xho*II, *Bgl*II, *Hinc*II, *Eco*RV, *Bam*HI (New England Biolabs, Beverly, MA).

2.1.2. Construction of rHSVf

1. Cosmid set C6Δa48Δa *(4,14)* (Cornel Fraefel, Institute of Virology, University of Zurich, 8057 Zurich, Switzerland. E-mail:Cornelf@vetvir.unizh.ch).
2. Restriction endonuclease *Pac*I (New England Biolabs).
3. Phenol (Merck, Dietikon, Switzerland).
4. Chloroform (Merck, Dietikon, Switzerland).
5. VERO 2-2 cells *(15)* (requests for VERO 2-2 cells should be addressed to Dr. Sandri-Goldin; Department of Microbiology and Molecular Genetics, College of Medicine, Medical Sciences I, B240, University of California, Irvine, Irvine, CA 92697-4025).
6. 60-mm-diameter tissue-culture dishes (Falcon, Zurich, Switzerland).
7. Dulbecco's modified Eagle medium (DMEM) containing 1% penicillin/streptomycin (P/S), 10% fetal bovine serum (FBS), and 500 µg/mL G418 (GIBCO-BRL, Basel, Switzerland).
8. Humified 37°C, 5% CO_2 incubator (Heraeus Instruments, Zurich, Switzerland).
9. LipofectAMINE (GIBCO, BRL).
10. Cell scraper, 1.8-cm blade (Falcon).
11. Dry ice-ethanol bath.
12. 37°C water bath.

2.1.3. Construction of fHSVΔpac and fHSVpac+

1. African green monkey kidney cells (VERO, ECACC).
2. 175 cm² tissue-culture flasks (Falcon).
3. Sorvall RC-5B centrifuge (Kendro Laboratory Products AG, Zurich, Switzerland).
4. Sorvall SS34 rotor, Beckmann SW 28 rotor (or equivalent).
5. Balanced salt solution (BSS): 137 m*M* NaCl, 5.4 m*M* KCl, 10 m*M* Tris-HCl, 10 m*M* ethylenediaminetetraacetic acid (EDTA), pH 7.5.
6. TE buffer: 10 m*M* Tris-HCl, 10 m*M* EDTA, pH 7.5.
7. Proteinase K solution (20 mg/mL in H_2O).
8. Sodium dodecyl sulfate (SDS) solution (10%) in H_2O.
9. Ethanol (Merck).
10. Low-melting temperature agarose (GIBCO, BRL).
11. β-agaraseI (New England BioLabs).
12. T4 DNA ligase (New England Biolabs).
13. Bio-Rad gene pulser version 10-90 (or equivalent) (Bio-Rad, Munich, Germany).
14. Electrocompetent *E. coli* DH10B (GIBCO, BRL) (*see* **Note 1**).
15. LB medium (DIFCO, Sparks, USA).
16. Chloramphenicol (100 mg/mL in ethanol).
17. Calf intestinal phophatase (CIP; New England Biolabs).

2.1.4. Isolation and Purification of HSV-BAC DNA From Bacteria

1. LB medium containing 15 µg/mL chloramphenicol.
2. Lab-Shaker at 37°C (Adolf Kühner AG, Birsfelden, Switzerland).

3. Dimethyl sulfoxide (DMSO) (Merck, Dietikon, Zurich).
4. Sterile 2-mL cryogenic vials (Corning, Cambridge, Canada).
5. GSA bottles.
6. Sorvall GSA and SS-34 rotors (or equivalent).
7. Paper towels.
8. Plasmid Maxi Kit, which includes Qiagen-tip 500 columns and buffers P1, P2, P3, QBT, QC, and QF (Qiagen, Basel, Switzerland).
9. Folded filter (∅ 120 mm; Schleicher & Schüll, Dassel, Germany).
10. Isopropanol (Merck).
11. 70% (v/v) ethanol.
12. TE buffer: 10 m*M* Tris-HCl, 10 m*M* EDTA, pH 7.4.
13. 30-mL centrifuge tubes (Beckman, Munich, Germany).
14. Ultraclear centrifuge tubes 13 × 51 mm (Beckman, Munich, Germany).
15. Cesium chloride (CsCl; Fluka, Buchs SG, Switzerland).
16. 10 mg/mL ethidium bromide in H_2O (*see* **Note 2**).
17. Paraffin oil (Merck).
18. TV 865 rotor.
19. 1-mL disposable syringe.
20. 21-gage hypodermic needle.
21. UV-lamp (366 nm).
22. *N*-butanol (Merck).
23. Dialysis cassettes, Slide-A-Layzer 10K (10,000 MWCO; Pierce, Rockford, IL).
24. 1-mL Tuberculin syringe.
25. 36-gage hypodermic needle.

2.1.5. Characterization of Purified HSV-BAC DNA

1. Spectrophotometer with quartz cuvets for DNA measuring (Pharmacia, Freiburg, Germany) (*see* **Note 3**).
2. Restriction endonucleases (New England Biolabs).

2.2. Packaging of HSV-1 Amplicon Vectors into HSV-1 Virions Using fHSVΔpac As the Replication-Competent, Packaging-Defective HSV-1 Helper DNA

1. VERO 2-2 Cells *(15)*.
2. 60-mm-diameter tissue-culture dishes.
3. DMEM with 10, 6, or 2% FBS. Supplement medium with 500 µg/mL G418 unless otherwise stated (Gibco, BRL).
4. Humidified 37°C/34°C, 5% CO_2 incubator (Heraeus Instruments, Zurich, Switzerland).
5. OptiMEM I reduced-serum medium (Gibco, BRL).
6. 15-mL conical centrifuge tubes (Falcon).
7. fHSVΔpac (*9*; *see* **Subheading 3.1.**) or fHSVΔpacΔ27ΔKn and pEBHICP27 *(16)*.
8. HSV-1 amplicon DNA (e.g., pHSVGFP; *17*).
9. LipofectAMINE (Gibco, BRL).

10. Cell scraper 1.8-cm blade (Falcon).
11. Skan-sonifier (Skan AG, Basel-Allschwil, Switzerland).
12. 30-mL centrifuge tube (Beckman, Munich, Germany).
13. Sorvall SS-34 rotor.
14. Ultraclear centrifuge tube 25 × 89 mm and 14 × 95 mm (Beckman, Munich, Switzerland).
15. Centrikon T2060 Ultracentrifuge (Kontron, Zurich, Switzerland).
16. Phosphate-buffered saline (PBS).
17. 10, 30, and 60% (w/v) sucrose in PBS.
18. Beckman SW28 and SW40 rotor (Beckman, Munich, Switzerland).
19. Dry ice-ethanol bath.
20. 24-well tissue-culture plates (Greiner, Nürtingen, Germany).
21. Paraformaldehyde 4% in H_2O (Sigma, Munich, Germany).

3. Methods

3.1. Cloning of the 152-kb HSV-1 Genome into a BAC: fHSVΔpac and fHSVpac+

3.1.1. Construction of HSV-1 Transfer Plasmid fH98-102pac

HSV-1 transfer plasmid fH98-102pac (*see* **Fig. 1A**) is constructed to facilitate integration of *E. coli* F-factor-derived sequences, as well as the HSV-1 DNA cleavage/packaging signal *(pac)* into a *pac* signal-deleted HSV-1 genome cloned as a set of five overlapping cosmids (cosmid set C6Δa48Δa; *14*). Construction of fH98-102pac includes several cloning steps. In the first step, the *Xho*I site of pBsSK+ is substituted for the polylinker *Xho*I-*Pac*I-*Bgl*II-*Pac*I-*Xho*I to generate pBXPBPX. Then, the 1.4-kb *Xho*II fragment containing the HSV-1 *pac* signal from plasmid pHSVPrPUC is inserted into the single *Bgl*II restriction site of pBXPBPX. The resulting plasmid pBXPpacPX allows the recovery of the HSV-1 *pac* signal by *Xho*I digestion and insertion of the 1.4-kb fragment into the single *Xho*I site of pBeloBAC11 to produce pBeloBACpac. In pBeloBACpac, the *pac* signal is flanked by *Pac*I sites; of note, wild-type HSV-1 DNA does not contain *Pac*I sites. To facilitate homologous recombination with the HSV -1 genome, an approx 4-kb *Hinc*II fragment (nucleotides 98,742-102,732) of the HSV-1 genome is isolated from cos56 of HSV-1 cosmid set C *(4)*, subcloned into the *Eco*RV site of pBsSK+, recovered as an approx 4-kb *Bam*HI fragment and introduced into the single *Bam*HI site of pBeloBACpac to form fH98-102pac (*see* **Note 4**).

1. Digest plasmid pBsSK+ with *Xho*I for 1–2 h at 37°C in a water bath and ligate overnight at 16°C in a water bath to the synthetic oligonucleotide duplex 5′ tcgagggcccttaattaagatcttaattaagggccc 3′ to construct pBXPBPX.
2. Digest pHSVPrPUC with *Xho*II for 1–2 h at 37°C in a water bath and insert the 1.4-kb fragment into the *Bgl*II site of pBXPBPX by ligation overnight at 16°C.

A

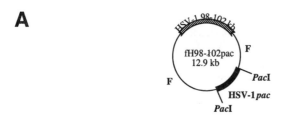

B

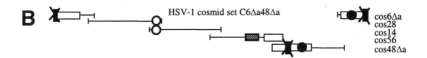

HSV-1 cosmid set C6Δa48Δa

cos6Δa
cos28
cos14
cos56
cos48Δa

C

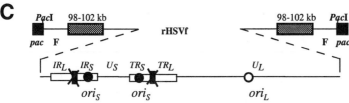

D

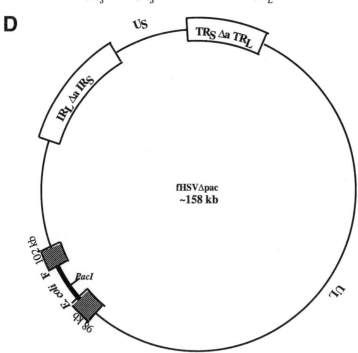

3. Digest the resulting plasmid, pBXPpacPX, with *Xho*I for 1–2 h at 37°C in a water bath and insert the 1.4-kb fragment into the unique *Xho*I site of pBeloBAC11 by ligation overnight at 16°C.
4. Digest cos56 of HSV-1 cosmid set C *(4)* with *Hinc*II for 1–2 h at 37°C in a water bath and insert the 4-kb fragment into the *Eco*RV site of pBsSK+ by ligation overnight at 16°C. The resulting clone is designated pBH98-102.
5. Digest pBH98-102 with *Bam*HI for 1–2 h at 37°C in a water bath and insert the 4-kb fragment into the unique *Bam*HI site of pBeloBACpac to form transfer plasmid fH98-102pac (*see* **Fig. 1A**).

3.1.2. Construction of rHSVf

The rHSVf DNA is generated by homologous recombination of the previously described plasmid fH98-102pac and DNA from HSV-1 cosmid set C6aΔ 48aΔ (*14; see* **Fig. 1B**). Cosmid set C6Δa48Δa represents the HSV-1 genome, with *pac* signals deleted, as a set of five overlapping clones. Recombination of the five cosmid clones leads to the generation of a replication-competent but packaging-defective HSV-1 genome. A replication- and packaging-competent HSV-1 genome is generated after an additional recombination event with fH98-102pac. The resulting rHSVf virus (*see* **Fig. 1C**) can be propagated and amplified in VERO cells, viral DNA can be isolated and used for further manipulations (*see* **Note 4**).

1. Digest the five cosmids of set C6Δa48Δa with *Pac*I for 1–2 h at 37°C to release the HSV-1 inserts and purify the DNA by phenol and chloroform extraction. (Cornel Fraefel, Institute of Virology, University of Zurich, 8057 Zurich, Switzerland. E-mail: Cornelf@vetvir.unizh.ch).

Fig. 1. (*see opposite page*) Cloning of the 152-kb HSV-1 genome in a bacterial artificial chromosome. (**A**) Transfer plasmid fH98-102pac, which is based on the single-copy F-plasmid from *E. coli*, contains the HSV-1 *pac* signal flanked by *Pac*I restriction sites and sequences between nucleotides 98,968 and 102,732 of the HSV-1 genome to facilitate homologous recombination with DNA from cosmid set C6Δa48Δa shown in B. (**B**) Cosmid set C6Δa48Δa represents the HSV-1 genome deleted for the *pac* signals (X) and includes cos6Δa, cos14, cos28, cos48Δa, and cos56. (**C**) Structure of the replication-competent rHSVf DNA produced after cotransfection of and recombination between fH98-102pac and cosmid set C6Δa48Δa in VERO 2-2 cells. (**D**) Structure of fHSVΔpac, which was constructed by deleting the *pac* signals of rHSVf with restriction endonuclease *Pac*I. Components of the 152-kb HSV-1 genome: U_L, unique long segment; U_S, unique short segment; IR_S, internal repeat of the short segment; TR_S, terminal repeat of the short segment; IR_L, internal repeat of the long segment; TR_L, terminal repeat of the long segment; ori_S, origin of DNA replication of the short segment; ori_L, origin of DNA replication of the long segment; *pac*, DNA cleavage/packaging signal.

2. Plate VERO 2-2 cells at a density of 1×10^6 cells per 60-mm-diameter tissue-culture dish. Grow in DMEM containing 10% FBS, 1% P/S, and 500 μg/mL G418 in a humidified 37°C, 5% CO_2 incubator.
3. On the following day, transfect the cells with 0.5 μg of fH98-102pac DNA and 0.4 μg of each of the five cosmid DNAs from set C6Δa48Δa by the Lipofect-AMINE procedure as described by the manufacturer.
4. After 2½ d, scrape the cells into the medium, freeze, and thaw the suspension three times using a dry ice-ethanol bath and a 37°C water bath.
5. Remove cell debris by centrifugation at 1400g for 10 min. The supernatant contains the putative recombinant HSV-1, rHSVf.

3.1.3. Construction of fHSVΔpac and fHSVpac+

For the generation of the final constructs, fHSVΔpac and fHSVpac+, the rHSVf virus (*see* **Fig. 1C**) is amplified on VERO cells, and viral DNA is isolated. To produce fHSVΔpac (*see* **Fig. 1D**), virion DNA from rHSVf is digested with *Pac*I restriction endonuclease to delete the *pac* signal. The large fragment containing the rest of the rHSVf genome is purified by gel-electrophoresis and then self-ligated before electroporation into electrocompetent *E. coli* DH10B cells. The resulting clones are characterized by restriction enzyme analysis to identify fHSVΔpac (*see* **Fig. 1D**). To construct fHSVpac+, fHSVΔpac DNA is linearized with *Pac*I, treated with alkaline phosphatase to prevent religation, and then ligated to the 1.4-kb *Pac*I fragment isolated from pBXPpacPX. The ligation products are electroporated into electrocompetent *E. coli* DH10B cells and resulting clones are characterized by restriction enzyme analysis to identify fHSVpac+.

1. Supernatant from **step 5** in **Subheading 3.1.2.** is amplified over three passages on VERO cells. When virus induced cytopathic effects (CPE) are visible, the cell supernatant is used to inoculate $12 \times 175 \text{ cm}^2$ tissue-culture flasks.
2. When CPE is complete, scrape the cells into the medium, freeze, and thaw the suspension (240 mL) three times in a dry ice-ethanol bath and 37°C water bath.
3. Remove the cell debris by centrifugation at 1400g for 10 min. Then, pellet the virus in the supernatant at 28,000g at 4°C for 1 h and discard the supernatant.
4. Resuspend the pellet in 80 mL of BSS.
5. Centrifuge at 28,000g at 4°C for 30 min and resuspend the pellet in 5.5 mL TE buffer containing 150 μL of a 20 mg/mL proteinase K solution.
6. Incubate the solution at 37°C for 1 h.
7. Add 300 μL of a 10% SDS solution and incubate overnight at 37°C.
8. Extract the viral DNA with phenol and chloroform, precipitate with ethanol, and resuspend in TE buffer.
9. Digest the DNA with *Pac*I and fractionate the fragments by gel electrophoresis through 0.4% low-melting temperature agarose at 40 V overnight.

10. Excise the 160-kb band, treat with β-agaraseI, and self-ligate the purified DNA with T4 DNA ligase overnight at 16°C in a water bath.
11. Electroporate the ligation products into electrocompetent *E. coli* DH10B cells according to instructions provided by the supplier (*see* **Note 1**).
12. Plate the cells on LB plates containing 12.5 µg/mL chloramphenicol and incubate overnight at 37°C.
13. Prepare plasmid DNA from the clones by alkaline lysis and characterize the DNA by restriction enzyme analysis. Digest fHSVΔpac DNA with *Kpn*I, *Dra*I, *Hin*dIII, *Bg*LII, *Eco*RI, or *Pac*I for 1–2 h at 37°C. For analysis of the restriction endonuclease patterns of the BAC DNA load the reaction mixtures on a 0.4% agarose gel and separate the fragments overnight at 40 V in TAE electrophoresis buffer. Then stain the gel with ethidium bromide (*see* **Note 2**). *Kpn*I digestion produces several DNA fragments (the exact size of fragments could be calculated from the corresponding HSV-1 genome) and a 1.9-kb fragment that contains the HSV-1 ori_L. *Dra*I, *Hin*dIII, *Bgl*II, or *Eco*RI digestion results in a specific DNA restriction pattern corresponding to the used HSV-1 genome. *Pac*I digestion leads to linearization of fHSVΔpac DNA, because of the previously deleted *Pac*I-flanked HSV-1 *pac* signal (*see* **step 9**) and subsequent self-ligation (*see* **step 10**). For restriction endonuclease patterns of fHSVΔpac (*Kpn*I, *Dra*I, *Hin*dIII, *Bgl*II, *Eco*RI, *Pac*I) and fHSVpac+ (*Eco*RI, *Pac*I), *see* **ref. 9**, **Fig. 2**.
14. To generate fHSVpac+, digest fHSVΔpac DNA with *Pac*I, treat with alkaline phosphatase and ligate overnight at 16°C to the 1.4-kb *Pac*I fragment from pBXPpacPX.
15. Electroporate the ligation products into electrocompetent *E. coli* DH10B and select clones on LB plates containing 12.5 µg/mL chloramphenicol. Identify correct clones by restriction enzyme digestion. Digest fHSVpac+ DNA with *Eco*RI or *Pac*I, respectively, for 1–2 h at 37°C. Separate the reaction mixture in a 0.4% agaraose gel as described above. *Pac*I digestion produces two fragments, the 1.4-kb fragment that contains the HSV-1 *pac* signal and the approx 156-kb HSV-BAC backbone *(9)*.

3.1.4. Isolation and Purification of HSV-BAC DNA From Bacteria

The following method is a modified Qiagen-tip 500 protocol.

1. Inoculate 6 mL of LB medium containing 12.5 µg/mL chloramphenicol with a single colony harbouring the HSV-BAC DNA.
2. Grow for 8 h at 37°C with shaking at 225 rpm (*see* **Note 5**).
3. The next day, inoculate 4 × 1 L of LB medium containing 12.5 µg/mL chloramphenicol with 1.5 mL of the preculture from **step 2** per 1 L.
4. Grow for 12–16 h at 37°C with shaking at 225 rpm (*see* **Note 5**).
5. Place 1-mL aliqouts of bacterial culture in cryogenic storage vials and add 70 µL DMSO. Mix well and freeze at –80°C for long-term storage (up to several years).
6. Distribute the 4 L overnight culture from **step 4** to 6 GSA bottles and pellet the bacteria for 5 min at 4000*g* (5000 rpm in a Sorvall GSA rotor) and 4°C.

7. Decant medium and invert each bottle on a paper towel for 1 to 2 min to remove all liquid. Resuspend each of the six bacterial pellets in 5 mL buffer P1 (*see* **Notes 6, 7**) and combine the six aliquots (total 30 mL). Add 130 mL of buffer P1 and distribute to four new GSA bottles (40 mL per bottle).

8. Add 40 mL of buffer P2 to each bottle and mix gently by inverting the bottles four to six times. Incubate for 5 min at room temperature (*see* **Note 8**).

9. Add 40 mL of buffer P3 and mix immediately by inverting the bottles six times. Incubate the bottles for 30 min on ice; invert the bottles occasionally. Centrifuge 30 min at 17,000*g* (11,000 rpm in a Sorvall GSA rotor) and 4°C.

10. Filter the supernatants through a folded filter (Ø 120 mm) into four fresh GSA bottles.

11. Precipitate the DNA with 0.7 volumes (84 mL per bottle) of isopropanol, mix gently, and centrifuge immediately for 30 min at 17,000*g* (11,000 rpm in a Sorvall GSA rotor) and 4°C.

12. Carefully remove the supernatants and mark the locations of the pellet. Wash the DNA pellet by adding 20 mL of cold 70% ethanol to each and centrifuge for 30 min at 16,000*g* (10,000 rpm in a Sorvall GSA rotor) and 4°C.

13. Resuspend each of the four pellets in 2 mL of TE buffer, pH 7.4. Pool the four pellets (total volume 8 mL) and add 52 mL of QBT buffer (final volume 60 mL).

14. Equilibrate 2 Qiagen-tip 500 columns with 10 mL of QBT buffer each, and allow the column to empty by gravity flow.

15. Transfer the solution from **step 13** through a folded filter (Ø 120 mm) onto the equilibrated Qiagen-tip 500 columns (30 mL per column) and allow it to enter the resin by gravity flow.

16. Wash the columns with 2 × 30 mL of QC buffer.

17. Elute the DNA by adding 15 mL of preheated (65°C) QF buffer to each column and collect the eluate in 30-mL centrifuge tubes.

18. To precipitate the DNA, add 0.7 volumes (10.5 mL) of isopropanol, mix, and centrifuge for 30 min at 20,000*g* (13,000 rpm in a SS34 rotor) and 4°C.

19. Resuspend each of the DNA pellets in 3 mL TE buffer at 65°C. This can take several hours.

20. Prepare two Beckman Ultra Clear centrifuge tubes (13 × 51 mm) with 3 g CsCl and add the DNA solution from **step 19** (3 mL per tube) *(18)*. Mix the solution gently until the salt is dissolved. Add 300 µL of ethidium bromide (10 mg/mL in water) (*see* **Note 2**) to the DNA/CsCl solution. Immediately mix the ethidium bromide solution (which floats on the surface) with the DNA/CsCl solution. Then overlay the solution with 300 µL light paraffin oil and seal the tubes.

21. Centrifuge at 48,000 rpm for 17 h at 20°C in a TV 865 rotor (or equivalent).

22. Two bands of DNA, located in the center of the gradient, should be visible in normal light. The upper band, which usually contains less material, consists of linear and nicked circular HSV-BAC DNA. The lower band consists of closed circular HSV-BAC DNA.

23. Harvest the lower band using a disposable 1-mL syringe fitted with a 21-gage hypodermic needle under UV-light and transfer it to a microfuge tube.

24. Remove ethidium bromide from the DNA solution by adding an equal volume of n-butanol in TE/CsCl (3g CsCl dissolved in 3 ml TE, pH 7.4) *(19)*.
25. Mix the two phases by vortexing and centrifuge at 1500 rpm for 3 min at room temperature in a bench centrifuge.
26. Carefully transfer the lower, aqueous phase to a fresh microfuge tube. Repeat **steps 24–26** four to six times until all the pink color disappears from both the aqueous phase and the organic phase.
27. Add an equal volume of isopropanol, mix and centrifuge at 1500 rpm for 3 min at room temperature.
28. Tranfer the aqueous phase to a fresh microfuge tube.
29. To remove the CsCl from the DNA solution, dialyze 6 h against TE, pH 7.4 at 4°C. Then, change the TE buffer and dialyse over night. For dialysis, the DNA solution is injected into a dialysis cassette, Slide-A-Layzer 10K (10.000 MWCO) using a 1-mL Tuberculin syringe fitted with a 36-gage hypodermic needle. After dialysis, the solution is recovered from the dialysis cassette by using a fresh 1 mL Tuberculin syringe fitted with a 36-gage hypodermic needle. The DNA solution is then transferred to a clean microfuge tube.

3.1.5. Characterization of Purified HSV-BAC DNA

Recombination events and also deletion of sequences from HSV-BAC clones may occur during amplification in bacteria. Therefore, isolated DNA should be analyzed before used in further experiments. The complete sequence of the HSV-1 genome is known, so that restriction patterns can be predicted and used for characterization of the HSV-BAC DNA. The functionality of HSV-BAC DNA can be analyzed by transfection into permissive cells and a subsequent standard plaque assay. For example, transfection of fHSVpac+ DNA into permissive cells should result in the generation of virus plaques within the cell monolayer after a few days. Transfection of fHSVΔpac DNA into permissive cells does not result in the generation of virus progeny because of the absence of *pac* signals.

1. Determine the absorbance of the DNA solution from **step 29** at 260 nm (A_{260}) and 280 nm (A_{280}) using a UV spectrophotometer (*see* **Note 3**). Store the DNA at 4°C (*see* **Note 9**).
2. Verify the fHSVΔpac and fHSVpac+ DNA by restriction endonuclease analysis (*see* **Subheading 3.1.3.**). For restriction endonuclease patterns of fHSVΔpac (*Kpn*I, *Dra*I, *Hin*dIII, *Bgl*II, *Eco*RI, *Pac*I) and fHSVpac+ (*Eco*RI, *Pac*I) *see* **ref. 9**, **Fig. 2**.
3. For a functional analysis of fHSVΔpac and fHSVpac+, transfect DNA into VERO or 2-2 cells (*see* **Subheading 3.2.**). fHSVΔpac is packaging-defective and does not give rise to progeny virus, whereas fHSVpac+ is packaging-competent and should give rise to plaques and progeny virus.

3.2. Packaging of HSV-1 Amplicon Vectors into HSV-1 Virions Using fHSVΔpac As the Replication-Competent, Packaging-Defective HSV-1 Helper DNA

In this section, we outline the protocol for the generation of helper virus-free amplicon vectors. Packaging of amplicons depends on HSV-1 helper-functions that were conventionally provided by helper viruses, resulting in the inevitable coexistance of helper virus and amplicons in the vector stocks *(13,20)*. A significant improvement in this regard was the development of a helper virus-free packaging system. Initially, a system based on five overlapping cosmids that span the HSV-1 genome with the *pac* signals deleted was used to provide the helper functions *(4,14)*. The limitations of this system include: (i) the genetic instability of the cosmid clones; (ii) complicated procedure in preparing five HSV-1 DNA fragments from cosmids; and (iii) difficulty in large-scale production of vector stocks. To overcome these problems, the entire 152-kb HSV-1 genome, without *pac* signals, was cloned as a BAC in *E. coli* *(9,16; see* **Subheading 3.1.**). Here, we provide a detailed protocol for amplicon vector packaging using fHSVΔpac as the replication-competent, packaging-defective HSV-1 helper genome.

1. On the day before transfection, plate 1.2×10^6 VERO 2-2 cells per 60-mm tissue-culture dish (*see* **Note 10**). Grow the cells in DMEM containing 10% FBS, 1% P/S, and 500 µg/mL G418 at 34°C in a humidified 5% CO_2 incubator.
2. For each 60-mm dish to be transfected, place 250 µL OptiMEM I into each of two 15-mL conical tubes. Add 2 µg fHSVΔpac DNA and 0.6 µg amplicon DNA to one tube (*see* **Note 11**). To the other tube, add 16 µL LipofectAMINE (*see* **Note 12**). Mix the contents of the tubes gently (do not vortex). Combine the contents of the two tubes (*see* **Note 13**). Mix well (do not vortex) and incubate 45 min at room temperature.
3. Approx 10 min prior to the end of the incubation time, wash the cultures prepared the day before (**step 1**) once with 2 mL OptiMEM I. Add 0.9 mL OptiMEM I to the tube from **step 3** containing the DNA-LipofectAMINE transfection mixture (1.4 mL total volume). Aspirate all medium from the culture, add the transfection mixture, and incubate 5.5 h in a humidified 34°C, 5% CO_2 incubator.
4. Aspirate the transfection mixture and wash the cells three times with 2 mL OptiMEM I. After aspirating the last wash, add 3.5 mL DMEM, 6% FBS and incubate 2 to 3 d in a humidified 34°C, 5% CO_2 incubator (*see* **Note 14**).
5. Scrape the cells from **step 5** into the medium, then transfer the suspension to a 15-mL conical tube.
6. Sonicate the suspension for 20 s with 20% output energy. Keep the sample on ice at all times.
7. Remove the cell debris by centrifugation for 10 min at 1400g and 4°C. Remove a sample for titration, then proceed to concentration and purification of the vector stock.

8. Transfer the supernatant from **step 7** to a 30-mL centrifuge tube and spin 2 h at 20,000*g* (13,000 rpm in Sorvall SS-34 rotor) and 4°C. Alternatively, purify and concentrate the vector stock using a discontinuous sucrose gradient (*see* **step 9**). Resuspend the pellet in a small volume (e.g., 300 µL) of 10% sucrose. Remove a sample for titration, then divide into aliquots (e.g., 30 µL), freeze in a dry ice-ethanol bath and store at –80°C.

9. For purification of HSV-1 amplicon stocks, prepare a sucrose gradient in a Beckman Ultra-Clear centrifuge tube 25 × 89 mm by adding the following solutions into the tube: 7 mL 60% sucrose, 7 mL 30% sucrose, and 3 mL 10% sucrose. Carefully add the vector stock from **step 8** (up to 20 mL) on top of the gradient and centrifuge 2 h at 100,000*g* and 4°C, using a Beckman SW28 rotor. Aspirate the 10 and 30% sucrose layers from the top and collect the virus band at the interface between the 30% and the 60% layers (*see* **Note 15**).

10. Transfer the interphase to a Beckman Ultra-Clear centrifuge tube 14 × 95 mm, add approx 15 mL PBS, and pellet virus particles for 1 h at 100,000*g* and 4°C, using a Beckman SW40 rotor.

11. Resuspend the pellet in a small volume (e.g., 300 µL) of 10% sucrose. Divide into aliquots (e.g., 30 µL) and freeze in a dry ice-ethanol bath. Store at –80°C. Before freezing, retain a sample of the stock for titration.

12. To determine the concentration of infectious vector particles in the HSV-1 amplicon stock, cells are infected with samples collected from stocks either before or after purification and/or concentration (*see* **steps 8**, **11**, and **13**). Cells expressing the transgene are counted 24–48 h after infection to calculate the titer (transducing units per milliliter) (*see* **Note 16**).

13. Plate cells, e.g., VERO, BHK 21, or 293 cells at a density of 1.0×10^5 per well of a 24-well tissue-culture plate in 0.5 mL DMEM with 10% FBS. Incubate at 37°C overnight.

14. Aspirate the medium and wash each well once with PBS. Remove PBS and add 0.1, 1, and 5 µL samples collected from vector stocks, diluted to 250 µL each in DMEM with 2% FBS. Incubate 1 to 2 d at 37°C.

15. Remove the inoculum and fix the cells for 20 min at room temperature with 250 µL of 4% paraformaldehyde, pH 7.0. Wash the fixed cells three times with PBS, then proceed with the appropriate detection protocol, depending on the transgene (*see* **Note 14**).

4. Notes

1. We recommend the use of *E. coli* strain DH10B. Standard electroporation gives high-efficiency transformation (approx 10^6 transformants per µg of DNA).

2. Caution: Ethidium bromide (EtBr) is a powerful mutagen and is moderately toxic. Gloves should be worn when working with solutions that contain this dye. EtBr solution should be made up in a fume hood or other confined space. Personnel preparing EtBr solutions should wear adequate protective clothing. Stock solutions of 10 mg/mL EtBr should be stored in a light-protected glass container at 4°C. Such solutions are stable for several years.

3. A A_{260} value of 1.0 is equivalent to 50 µg/mL of double-stranded DNA. Additionally, the ratio between A_{260} and A_{280} provides information about DNA purity. Typically, pure DNA preparations have a A_{260}/A_{280} value of 1.8; do not use DNA preparations with a ratio below this value.

4. As fHSVΔpac was constructed specifically to provide a packaging-defective HSV-1 helper genome for the packaging of HSV-1 amplicon vectors, the cloning of the wild-type HSV-1 genome requires some modifications to the protocol presented here: (i) The transfer plasmid used to introduce BAC sequences into the HSV-1 genome contains HSV-1 sequences that facilitate homologous recombination with the viral DNA, but no *pac* signal. (ii) DNA from "rHSVf" is not isolated from virions, but from the nuclei of infected cells, and then electroporated into *E. coli*. The transfer plasmid is cotransfected with wild-type HSV-1 DNA into VERO cells. The resulting recombinant HSV-1, rHSVf, is amplified on VERO cells.

5. It is convenient to grow the 6-mL preculture during the day and the larger culture overnight. Do not exceed the stated incubation times, as this may increase the risk of introducing modifications into the HSV-BACs.

6. The lyophilized RNase A provided in the Qiagen Plasmid Maxi Kit should be dissolved in P1 buffer. The dissolved RNase A will be stable for 6 mo when stored at 4°C.

7. The bacteria should be resuspended completely, leaving no cell clumps. Clumps will decrease the lysis of bacteria and reduce the DNA yield.

8. Do not allow the lysis reaction to proceed for more than 5 min, as it will strongly reduce the DNA yield.

9. The HSV-BAC DNA can be stored at 4°C for several weeks.

10. The cells should be confluent on the day of transfection.

11. If fHSVΔpacΔ27ΔKn *(16)* is used as the replication-competent packaging-defective HSV-1 genome, add 0.2 µg of plasmid pEBHICP27 *(16)*, which complements for the *IE2* gene deleted in the fHSVΔpacΔ27ΔKn clone, to the transfection mix.

12. Per set of 15-mL conical tubes, the transfection mixture for up to six 60-mm dishes can be combined. In the author's experience, six dishes can conveniently be manipulated at once.

13. The contens of the LipofectAMINE tube should be added slowly to the tube containing the DNA. This is best accomplished by using a 1000-µL pipet. If the contents of the two tubes are combined too rapidly, formation of an insoluble complex between the large BAC-DNA and LipofectAMINE can occur, decreasing the transfection efficiency. The (insoluble) complex can be seen as violette flakes floating in the tube.

14. Two days after transfection, at least 50% of the cells should show cytopathic effects (the cells should round up but remain attached to the plate). The use of an amplicon vector that expresses the gene for GFP, such as pHSVGFP *(17)*, can be an invaluable tool to monitor efficiency and spread of the vector during the course of the packaging experiment. Detection of GFP-fluorescence requires no fixation of cells. The cell suspension can be shock-frozen using a dry ice-ethanol bath and

stored at –80°C for an extended period of time in case an interruption of the protocol is desired. When ready to continue the protocol, thaw the suspension using a 37°C water bath. Do not leave the suspension at 37°C any longer than necessary to thaw it.

15. The interface between the 30 and 60% sucrose layers appears as a cloudy band when viewed with a fiber-optic illuminator.

16. The titers expressed as transducing units per milliliter (t.u./mL) are relative and do not necessarily reflect numbers of infectious particles per milliliter. Factors influencing relative transduction efficiencies include: (a) the cells used for titration; (b) the promoter regulating the expression of the transgene; (c) the transgene; and (d) the sensitivity of the detection method. Strategies to detect transgene expression may include coexpression—as a fusion protein or via an internal ribosome entry site—of a reporter gene that is easily detectable (such as the gene for green fluorescent protein (GFP) or the *E. coli lacZ* gene), or immunocytochmical staining using an antibody that specifically recognizes the transgene product.

References

1. Mocarski, E. S., Post, L. E., and Roizman, B. (1980) Molecular engineering of the herpes simplex virus genome: insertion of a second L-S junction into the genome causes additional genome inversions. *Cell* **22,** 243–255.
2. Post, L. E. and Roizman, B. (1981) A generalized technique for deletion of specific genes in large genomes: alpha gene 22 of herpes simplex virus 1 is not essential for growth. *Cell* **25,** 227–232.
3. DeLuca, N. A., McCarthy, A. M., and Schaffer, P. A. (1985) Isolation and chracterization of deletion mutants of herpes simplex virus type 1 in the gene encoding immediate–early regulatory protein ICP4. *J. Virol.* **56,** 558–570.
4. Cunningham, C. and Davison, A. J. (1993) A cosmid-based system for constructing mutants of herpes simplex virus type 1. *Virology* **197,** 116–124.
5. Shizuya, H., Birren, B., Kim, U. J., et al. (1992) Cloning and stable maintenance of 300-kilobase-pair fragments of human DNA in *Escherichia coli* using an F-plasmid-based vector. *Proc. Natl. Acad. Sci. USA* **89,** 8794–8797.
6. Messerle, M., Crnkovic, I., Hammerschmidt, W., Ziegler, H., and Koszinowski, U. H. (1997) Cloning and mutagenesis of a herpesvirus genome as an infectious bacterial artificial chromosome. *Proc. Natl. Acad. Sci. USA* **94,** 14,759–14,763.
7. Delecluse, H. J., Hilsendegen, T., Pich, D., Zeidler, R., and Hammerschmidt, W. (1998) Propagation and recovery of intact, infectious Epstein–Barr virus from prokaryotic to human cells. *Proc. Natl. Acad. Sci. USA* **95,** 8245–8250.
8. Stavropoulos, T. A. and Strathdee, C. A. (1998) An enhanced packaging system for helper-dependent herpes simplex virus vectors. *J. Virol.* **72,** 7137–7143.
9. Saeki, Y., Ichikawa, T., Saeki, A., et al. (1998) Herpes simplex virus type 1 DNA amplified as bacterial artificial chromosome in *Escherichia coli*: rescue of replication-competent virus progeny and packaging of amplicon vectors. *Hum. Gene Ther.* **9,** 2787–2794.

10. Horsburgh, B. C., Hubinette, M. M., Qiang, D., Mc Donald, M. L. E., and Tufaro, F. (1999) Allele replacement: an application that permits rapid manipulation of herpes simplex virus type 1 genomes. *Gene Ther.* **6,** 922–930.

11. Smith, G. A. and Enquist, L. W. (1999) Construction and transposon mutagenesis in *Escherichia coli* of a full-length infectious clone of pseudorabies virus, an alpha herpesvirus. *J. Virol.* **73,** 6405–6414.

12. Borst, E. M., Hahn, G., Koszinowski, U. H., and Messerle, M. (1999) Cloning of the human cytomegalovirus (HCMV) genome as an infectious bacterial artificial chromosome in *Escherichia coli*: a new approach for construction of HCMV mutants. *J. Virol.* **73,** 8320–8329.

13. Spaete, R. R. and Frenkel, N. (1982) The herpes simplex virus amplicon: A new eucaryotic defective-virus cloning-amplifying vector. *Cell* **30,** 305–310.

14. Fraefel, C., Song, S., Lim, F., et al. (1996) Helper virus-free transfer of herpes simplex virus type 1 plasmid vectors into neural cells. *J. Virol.* **70,** 7190–7197.

15. Smith, I. L., Hardwicke, M. A., and Sandri-Goldin, R. M. (1992) Evidence that the herpes simplex virus immediate-early protein ICP27 acts post-transcriptionally during infection to regulate gene expression. *Virology* **186,** 74–86.

16. Saeki, Y., Fraefel, C., Ichikawa, T., Breakefield, X. O., and Chiocca, A. (2001) Improved helper virus-free packaging system for HSV amplicon vectors using an ICP27-deleted, oversized HSV-1 DNA in a bacterial artificial chromosome. *Mol. Ther.* **3,** 591–601.

17. Aboody-Guterman, K. S., Pechan, P. A., Rainov, N. G., et al. (1997) Green fluorescent protein as a reporter for retrovirus and helper virus-free HSV-1 amplicon vector-mediated gene transfer into neural cells in culture and *in vivo*. *Neuroreport* **8,** 3801–3808.

18. Maniatis, T., Fritsch, E. F., and Sambrook, J. (1989) in *Molecular Cloning: A Laboratory Manual*, Cold Spring Harbor Laboratory Press, Cold Spring Harbor, NY, pp. 1.42–1.48.

19. Maniatis, T., Fritsch, E. F., and Sambrook, J. (1989) in *Molecular Cloning: A Laboratory Manual*, Cold Spring Harbor Laboratory Press, Cold Spring Harbor, NY, pp. 1.46.

20. Geller, A. I., and Breakefield, X. O. (1988) A defective HSV-1 vector expresses *Escherichia coli* beta-galactosidase in cultured peripheral neurons. *Science* **241,** 1667–1669.

17

Mutagenesis of Viral BACs With Linear PCR Fragments (ET Recombination)

Markus Wagner and Ulrich H. Koszinowski

1. Introduction

In the last few years, mutagenesis of viruses with large DNA genomes (like the herpesviruses) was simplified by cloning the viral genomes as bacterial artificial chromosomes (BACs) and their subsequent transfer into *Escherichia coli (E. coli.)*. Owing to the high frequency of restriction sites, classical cloning methods for site-directed mutagenesis are not applicable to these large viral BACs. One possibility for mutagenesis is allele replacement by a two-step recombination procedure (*see* Chapter 18). A much more rapid one-step procedure for introduction of mutations into the viral BACs is homologous recombination between a linear DNA fragment and the viral BAC by double crossing-over (*see* principle in **Fig. 1**). Recombination was originally performed with the recombination functions *RecE* and *RecT* and, therefore, termed *ET recombination* or *ET mutagenesis (1)*. Meanwhile, the recombination functions *redα (exo)* and *redβ (bet)* from bacteriophage λ have been shown to be a good alternative for *RecE* and *RecT* because they are slightly more efficient for double crossing-over. In addition to *redα/RecE* and *redβ/RecT*, expression of the exonuclease inhibitor *redγ (gam)* is necessary to allow mutagenesis in bacteria because the gam protein inhibits bacterial exonucleases and protects the linear recombination fragment from degradation. We found that the viral BACs remained more stable when using the *red* recombination functions, in contrast to *RecE* and *RecT*. Compared to the *RecA* mediated two-step recombination procedure, ET recombination is much faster because the *red* recombinases used allow efficient homologous recombination with very short homology arms (below 50 nts), which can be provided by synthetic oligonucleotides. Therefore,

From: *Methods in Molecular Biology, vol. 256:*
Bacterial Artificial Chromosomes, Volume 2: Functional Studies
Edited by: S. Zhao and M. Stodolsky © Humana Press Inc., Totowa, NJ

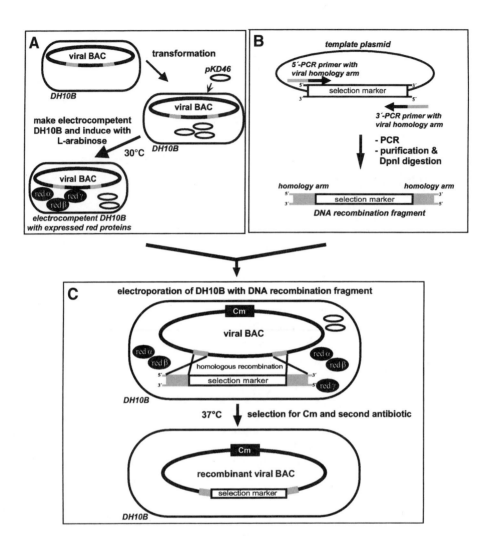

Fig. 1. Principle steps of mutagenesis. (A) Preparation of electrocompetent bacterial cells for mutagenesis. Prior to mutagenesis the plasmid pKD46, encoding the recombination functions, is introduced into DH10B containing the viral BAC. In the next step, electrocompetent bacterial cells are prepared by growing the bacteria at 30°C. At the same time, expression of recombinases is induced by addition of 0.1% L-arabinose to the growth medium. (B) Generation and preparation of the DNA recombination fragment. The DNA recombination fragment that contains a selection marker flanked by homologies to the viral genome (gray boxes) is generated by PCR. The resulting DNA fragment is purified and digested with *Dpn*I to remove residual template DNA. (C) Mutagenesis of the viral BAC. For homologous recombination, the DNA recombination fragment is electroporated into the DH10B containing the viral BAC and the expressed recombinases redα, -β, and -γ. After selection with chloramphenicol (Cm) for the BAC and the appropriate antibiotic for the introduced selection marker over night at 37°C, bacteria with the recombinant viral BAC are obtained.

any classical cloning procedures including construction and amplification of plasmids are obviated. In addition, mutagenesis is independent of the availibility of appropriate restriction enzyme sites, because the homologies in the oligonucleotides can freely be choosen. For a comprehensive overview on ET recombination, see the review of Muyrers et al. *(2)*. A specific problem concerning ET mutagenesis of viral BACs may represent the occurrence of repeated sequences within the viral genomes, which can result in undesired recombination events and in consequence to potential instabilities of the viral BACs. Instability is mainly caused by introducing the viral BACs into bacterial strains that constitutively express recombination functions like *RecA* or *RecE/RecT*, or by high expression levels of recombinases. This problem was observed for different viral BACs (e.g., MCMV, especially with EBV and MHV-68) and in part solved by using the inducible BAD-promotor for controlled expression of the recombinases on the high-copy plasmid pBADαβγ *(1,3)*. Using pBADαβγ, the expression of recombinases can be controlled by the concentration of L-arabinose in the growth medium and by the induction time. Once the right conditions have been defined, a correct site-specific homologous recombination for mutagenesis is achievable, without occurrence of unwanted recombination events elsewhere in the viral genome. However, definition of the optimal L-arabinose concentration and induction time for mutagensis is crucial and has to be determined for each individual viral BAC. This procedure, which sometimes is quite laborious and frustrating, can now be obviated by inducible expression of all three *red* functions from a low copy plasmid backbone, like pKD46 *(4)*. Now even proper mutagenesis of formerly instable viral BACs of EBV and MHV-68 can be achieved using a standard L-arabinose induction procedure. Therefore, we describe mutagenesis using plasmid pKD46 in this protocol. A major technical concern when using this protocol is the preparation of highly electrocompetent bacterial cells, because inefficient transformation rates are the main source of failure (*see* **Note 14**).

Despite all advantages of ET recombination, a main disadvantage remains the necessity to introduce a selection marker along with the mutation. Thus, the resistance marker remains present in the mutated BAC. This track of the mutagenesis can be minimized by using a selection marker that is flanked with FRT (FLP recognition target site) or *loxP* sequences. These sites allow excision of the selection marker after mutagenesis by the site specific recombinases FLP *(5)* and Cre *(6)*, respectively. A simple method for excision of the selection marker using FLP recombinase is described in **Subheading 3.4.** of this protocol. It allows elimination of the selection marker, but still leaves one 34 bp FRT site as a track of the mutagenesis within the viral BAC.

Recently, there have been several publications which describe modified procedures for ET recombination that avoid any tracks of mutagenesis and allow even introduction of point mutations *(7,8)*. However, the efficiacy of these mutagenesis protocols is still quite low and, therefore, they require laborious screening procedures for detection of positive clones. Because viral BACs generally tend to be more instable, it still has to be shown whether these procedures also work with viral BACs. Here, we describe the basic usage of this method for introduction of a selection marker in place of a viral sequence. Because recombination between the viral BAC and the recombination fragment was found to be successful with homologies that can be apart from few basepairs (bp) to more than 180 kbp, single and multiple deletions of any size can be introduced into viral genomes. We confirmed this by the mutagenesis of our MCMV BAC genome pSM3fr *(9)*, in which we easily introduced deletions from 3 bp to 183 kbp in size (unpublished data). The generation of viral deletion mutants will probably be the most used application of this method for virologists. The described protocol can also be applied on nonviral BACs.

2. Materials

All chemicals should be of molecular biology grade. All solutions should be made with double-distilled water (ddH$_2$O). For information on reagents of kits used, please refer to the provided information from the manufacturer.

2.1. Preparation of Electrocompetent Bacterial Cells for Mutagenesis

1. Propagate the viral BAC in bacterial strain DH10B (Life Technologies, Inc.) (*see* **Note 1**).
2. Recombination plasmid pKD46 *(4)* (*see* **Note 2**).
3. LB medium according to Sambrook and Russel *(10)* (*see* **Note 3**).
4. Agar plates with standard concentrations of agar and LB medium: Add 15 to 25 µg/mL chloramphenicol (for selection of the BAC) and 50 µg/mL ampicillin (for selection of plasmid pKD46).
5. Make a fresh sterile 10% (w/v) stock solution of L-arabinose (Sigma-Aldrich, Inc. A3256) in standard LB medium.
6. Prepare 750 mL 10% glycerol in ddH$_2$O and precool it on ice at least for 3 h.
7. Additional tools needed: Sorvall centrifuge, sterile centrifuge tubes, and 1.5-mL Eppendorf vials, Kleenex tissues, glas pipets, liquid nitrogen (N$_2$).

2.2. Generation and Preparation of the DNA Recombination Fragment

1. For PCR:
   a. 1.5 µL 20–50 µ*M* 5′- and 3′-PCR-primers (*see* **Subheading 2.5.**).
   b. 1 µL (3–10 ng) template DNA with selection marker (e.g., plasmid pACYC177 for kanamycin resistance gene).
   c. 2 µL 10 m*M* dNTPs.

    d. 10 μL 10X PCR buffer.

    e. 0.75 μL Expand High Fidelity polymerase (~2.5 U), 83.25 μL ddH$_2$O (*see* **Note 4**).

2. PCR purification kit (Qiagen, #28104).
3. *Dpn*I restriction enzyme.
4. 100% EtOH.
5. 70% EtOH.
6. 3 *M* sodium acetate, pH 5.2.
7. Thermocycler for PCR.

2.3. Electroporation of Bacterial Cells With DNA Recombination Fragment for Mutagenesis

1. Electrocompetent bacterial cells (*see* **Subheading 3.1.**).
2. DNA recombination fragment (*see* **Subheading 3.2.**).
3. LB medium according to Sambrook and Russel *(10)*.
4. 0.2-cm cuvets for electroporation.
5. Bio-Rad Gene Pulser or equivalent electroporator.
6. Agar plates with standard concentrations of agar and LB medium: Add 15 to 25 μg/mL chloramphenicol and appropriate concentration of antibiotics for the selection marker on the DNA recombination fragment (*see* **Note 5**).

2.4. Excision of the Selection Marker Using FLP Mediated Recombination

1. Prepare transformation-competent DH10B containing the mutated viral BAC, which carries the selection marker flanked by two FRT-sites (*see* **Note 6**).
2. Plasmid pCP20 *(5)*.
3. Agar plates with standard concentrations of agar and LB medium: Add 15 to 25 μg/mL chloramphenicol and 50 μg/mL ampicillin.
4. Agar plates with standard concentrations of agar and LB medium: Add 15 to 25 μg/mL chloramphenicol.

2.5. Design and Synthesis of Oligos for PCR

1. Synthesize 5′- and 3′-PCR-primers that possess a homology arm to the target DNA at the 5′-end and a 3′-priming region to an antibiotic resistance marker at the 3′-end (*see* **Fig. 2**).
2. The sequences of the homology arms are determined by the sequence of the target DNA flanking the site in which the PCR fragment should be inserted. The recommended length of homology is between 35 and 50 nts (*see* **Note 7**).
3. The sequence of the 3′-priming region to the antibiotic resistance marker should be between 18 and 25 nts. It is selected as for any other PCR using standard software programs (*see* **Note 8**).
4. Additional sequences or mutations (e.g., restriction sites, FRT sites or epitope tags) can be included into one or both oligonucleotides (between the homology

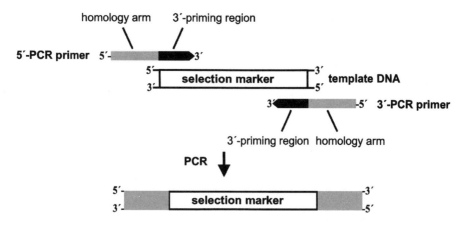

Fig. 2. Primer design for generation of the DNA recombination fragment by PCR. A 5'-PCR primer and a 3'-PCR primer, both containing a 3'-priming region to the ends of a selection marker and to the flanking regions of the viral target sequence (homology arm), have to be designed. Additional sequences like FRT sites or epitope tags can be introduced into the PCR primers (optional insertion of sequences). After PCR, the DNA recombination fragment contains the selection marker with flanking homology arms that are complementary to the target DNA and allow recombination between the DNA recombination fragment and the viral BAC.

arm and the 3'-priming region—*see* **Fig. 2**). The size of the inserted sequence is only restricted by the limits of faithful oligonucleotide synthesis (*see* **Note 9**).

5. Since the synthesized oligonucleotides are more than 60 nts in length, their quality is a major concern and HPLC or PAGE purification is recommended (*see* **Note 10**).

3. Methods

3.1. Preparation of Electrocompetent Bacterial Cells for Mutagenesis

For linear fragment mutagenesis, the plasmid pKD46 has to be introduced into DH10B containing the viral BAC. In addition, these bacterial cells have to be prepared for electroporation and the expression of the recombination functions redα, -β, and -γ need to be induced by addition of L-arabinose to the growth medium. All preparation steps should be done under conditions as cold as possible and permanently on ice. All used materials should be precooled to at least 0°C, including the rotor for centrifugation. Precool the 5 mL glass pipets before use by pipetting cold 10% glycerol up and down several times.

1. Transform the DH10B bacteria containing the viral BAC with 5 to 10 ng of plasmid pKD46 using standard procedures. Plate 1/10 of the transformed bacterial cells onto an agar plate containing 15 to 25 μg/mL chloramphenicol and 50 μg/mL ampicilin and incubate overnight at 30°C (*see* **Note 11**).

2. Pick a single colony and grow in 5 mL LB medium containing 15 to 25 µg/mL chloramphenicol and 50 µg/mL ampicilin on a shaker at 180 rpm overnight at 30°C.
3. Prepare 200 mL LB medium containing 15 to 25 µg/mL chloramphenicol, 50 µg/mL ampicilin, and 0.1% (w/v) κ-arabinose (*see* **Note 12**).
4. Inoculate the prepared LB medium with 2 mL of overnight bacteria culture and shake at 30°C until bacterial cells reach an OD_{600} of 0.3 to 0.5 (*see* **Note 13**).
5. Put on ice for 15 min and centrifuge for 10 min at 7000 rpm (Sorvall centrifuge) and −4°C to pellet the bacterial cells.
6. Discard supernatant, put tube on ice, and resuspend bacterial cells in 10 mL ice-cold 10% glycerol with a precooled 5-mL pipet. Add 200 mL ice-cold 10% glycerol.
7. Centrifuge for 10 min at 7000 rpm and −4°C and repeat **step 6**.
8. Centrifuge for 10 min at 7000 rpm and −4°C, repeat **step 6** and centrifuge again for 10 min at 7000 rpm and −4°C. Discard supernatant and remove residual liquid within the tube with a Kleenex tissue, but avoid to touch the pellet.
9. Resuspend bacterial cells to a total volume of 900 µL ice-cold 10% glycerol. Transfer 60 µL aliquots into 1.5-mL Eppendorf vials that have been precooled on dried ice and freeze immediately in liquid N_2. Store at −80°C (*see* **Note 14**).

3.2. Generation and Preparation of the DNA Recombination Fragment

1. Prepare PCR in a total volume of 100 µL (*see* **Note 4**):
   a. 1.5 µL 5′-oligonucleotide primer (concentration 20–50 µ*M*).
   b. 1.5 µL 3′-oligonucleotide primer (concentration 20–50 µ*M*).
   c. 1 µL template DNA with selection marker (3–10 ng in total).
   d. 2 µL 10 m*M* dNTPs.
   e. 10 µL 10X PCR buffer.
   f. 0.75 µL Expand High Fidelity polymerase (≈2.5 U).
   g. Add ddH$_2$O to final volume of 100 µL.
2. Perform 30 PCR cycles using the following conditions (*see* **Note 15**):
   a. 5 min 95°C
   b. 30 s 94°C/2 min 62°C to 45°C (decrease 1°C per cycle)/2 min 68°C for 18 cycles (*see* **Note 16**).
   c. 30 s at 94°C/2 min at 45°C/2 min at 68°C for 12 cycles.
3. Purify PCR product using the Qiagen PCR purification kit and finally elute the DNA in 50 µL ddH$_2$O.
4. Add 6 µL 10X *Dpn*I reaction buffer and 2 µL *Dpn*I restriction enzyme (10 U/µL) to the eluted DNA and incubate for 1.5 h at 37°C to remove template DNA (*see* **Note 17**).
5. Precipitate the DNA by adding 180 µL 100% EtOH and 6 µL 3 *M* sodium acetate. Incubate at −80°C for 15–30 min and centrifuge 15 min at 13,000 rpm in a microfuge at 4°C.
6. Wash DNA pellet with 70% EtOH and redissolve in 10 µL ddH$_2$O.

3.3. Electroporation of Bacterial Cells With DNA Recombination Fragment for Mutagenesis

1. Precool 0.2-cm cuvets on ice for at least 5 min (*see* **Note 18**).
2. Thaw 60 μL of prepared electrocompetent bacterial cells on ice and add 5 μL of prepared DNA recombination fragment (*see* **Note 19**).
3. Immediately electroporate bacterial cells at 2.5 kV, 200 Ω, and 25 μF (Bio-Rad Gene Pulser), add 1 mL LB medium (without antibiotics) and transfer back to Eppendorf vial.
4. Shake at 37°C for 1.5 to 2 h.
5. Pellet the bacterial cells at 6000 rpm in a microfuge for 30 s and redissolve pellet in 200 μL LB medium.
6. Plate half of the resuspended bacteria onto an agar plate containing 15 to 25 μg/mL chloramphenicol and the second appropriate antibiotic corresponding to the used selection marker and incubate overnight at 37°C (*see* **Note 20**).

3.4. Excision of the Selection Marker Using FLP Mediated Recombination

1. Transform bacterial cells containing the mutated viral BAC (that possesses the introduced antibiotic marker flanked by FRT sites) with 2 to 10 ng of plasmid pCP20 using standard procedures (*see* **Note 21**).
2. Shake at 30°C for 1 h and plate 1/10 of the transformed bacterial cells onto an agar plate containing 25 μg/mL chloramphenicol and 50 μg/mL ampicilin.
3. Incubate overnight at 30°C (*see* **Note 22**).
4. Pick single colonies and streak on agar plates containing 25 μg/mL chloramphenicol only. Incubate overnight at 43°C.
5. Pick single colonies on agar plates containing 25 μg/mL chloramphenicol alone and in parallel on agar plates containing chloramphenicol plus the antibiotic of the introduced selection marker, respectively. Incubate overnight at 37°C and check colonies for loss of the selection marker.

4. Notes

1. DH10B is the most suitable *E. coli* strain for propagation of viral BACs because even large BACs can be easily introduced into these bacterial cells by electroporation and stably maintained. Genotype: F⁻ mcrA Δ(mrr-hsdRMS-mcrBC) φ80d*lac*ZΔM15 Δ*deo*R recA1 endA1 araD139 Δ(*ara, leu*)7697 *gal*U *gal*K λ⁻ rpsL nupG. Mutagenesis with linear PCR fragments works with quite high efficiacy in DH10B. Other bacterial strains (like HB101, DK1, and NS3145) can also be used, as long as the viral BAC can be introduced and stably propagated. The most commonly used BACs contain the chloramphenicol resistance gene. We advise using 15 to 25 μg/mL chloramphenicol for BAC selection. If a different selection marker is on the BAC, the protocol has to be adjusted accordingly.
2. As an alternative to plasmid pKD46, other plasmids or bacterial strains *(11)* that express the genes *red*α, -β, and -γ or RecE, RecT, and *red*γ, respectively, can be

used for mutagenesis. Possible alternative plasmids are pBADETγ *(1)*, pBADαβγ *(1)*, and pGETrec *(12)*. Owing to the higher expression level of the recombinases from these high-copy plasmids compared to pKD46, the stability of the viral BAC may be affected. Therefore, we recommend to use the low-copy plasmid pKD46. If other plasmids than pKD46 are used, changes in the protocols are required (refer to **refs.** *1* and *12*).

3. Because expression of the recombinases from plasmid pKD46 can be repressed by glucose, it is important that only LB medium is used and no glucose is added.

4. We recommend using the Expand High Fidelity PCR System (Roche Diagnostics, 1732641). *See* **Note 8** for choosing an appropriate selection marker as template DNA.

5. After insertion of the DNA recombination fragment, the BAC contains a second selection marker for selection. When choosing the appropriate antibiotic concentration, remember not to use standard concentrations for high-copy plasmids. BACs are low-copy plasmids and require lower antibiotic concentrations (e.g., for kanamycin 25 μg/mL).

6. For introducing plasmid pCP20 into DH10B, chemical transformation according to standard procedures is sufficient.

7. The mutagenesis efficacy increases with the length of the two homology arms from 25 to 60 nts tested so far. A length of around 40 nts is a good compromise between efficacy and costs. So far, we did not observe that certain sequence contents work better than others. To circumvent unwanted recombination events, the linear DNA recombination fragment should be targeted to unique sequences within the BAC DNA.

8. The length of the 3′-priming region to the selection marker is less critical, 18 to 25 nts are recommended. But because the first annealing step during PCR starts via these priming regions, efficient annealing of the 3′-priming region is important for good amplification of the DNA recombination fragment. It is recommended to test different possible priming sequences. In principle, a variety of selection markers can be used (kanamycin, tetracyclin, zeocin, blasticidin, hygromycin). Ampicillin and chloramphenicol are already used for selection of the plasmid pKD46 and the viral BAC, respectively. Some selection markers, especially zeocin, tend to give rise to pseudocolonies (colonies which are not growing in liquid culture). We recommend to use the kanamycin resistance gene from Tn903, which is for instance contained in plasmid pACYC177 (New England Biolabs).

9. To excise the selection marker from the target DNA after mutagenesis, the marker has to be flanked by FRT or *lox*P-sites, respectively. This can be achieved by including these sequences between the homology arm and the 3′-priming region of each oligonucleotide. Subsequently, the selection marker can be excised by FLP or Cre recombinase. The minimal FRT sequence is: 5′-GAAGTTCC TATTCTCTAGAAAG TATAGGAACTTC (*see* **Subheading 3.4.** for use of FLP).

10. We recommend HPLC or PAGE purification of the oligonucleotides that is offered by most companies synthesizing oligonucleotides, but it is not essential. If you do not HPLC purify the oligonucleotides, extract them once with phenol:chloroform (Briefly: Take 100 μL of oligonucleotide solution, add 12 μL 3 *M* sodium acetate,

pH 7.5 and 120 μL phenol:chloroform. Vortex for 30 s, separate the phases by centrifugation for 5 min and transfer the DNA containing fraction to a new Eppendorf vial. Precipitate with 360 μL 100% ethanol, wash with 70% ethanol, dry the DNA pellet and redissolve it in 100 μL ddH$_2$O).

11. Plasmid pKD46 contains the temperature sensitive origin of replication pSC101. This origin maintains a low copy number and replicates at 30°C. The plasmid is lost from bacterial cells when incubated at temperatures above 37°C. Selection should be done with 50 μg/mL ampicillin.

12. Preparation of LB medium with 0.1% (w/v) L-arabinose: add 2 mL of a freshly prepared 10% L-arabinose/LB medium stock solution to 198 mL LB medium. Viral BACs with repeat sequences may tend to be instable after expression of recombinases owing to unwanted recombination events via these repeats. Lowering the amount of recombinases expressed in the bacterial cells decreases the risk of BAC instability, but at the same time efficiency of insertion of the DNA recombination fragment decreases. Therefore, the integrity of the viral BAC after preparation of the electrocompetent cells should be controlled. In case of BAC instability the time of induction with 0.1% L-arabinose should be shortened. BAC stability and efficiency of mutagenesis may be tested for different induction times.

13. We obtained efficient recombination with electrocompetent bacterial cells that have been grown to an OD$_{600}$ between 0.25 and 0.6. In our experiments it seemed that there was a tendency to a sligthly higher efficiency of recombination with bacterial cells grown to an OD$_{600}$ <0.4.

14. It is critical to get highly electrocompetent bacterial cells. Standard protocols for preparation of chemically competent cells yield lower competences and do not work. One should reach a competence of at least 10^8 colonies/μg DNA after transformation with a standard high copy plasmid. This should be checked before mutagenesis.

15. The annealing temperature of 62°C for performing PCR with oligonucleotides of 60 to 90 nts in length is working in most cases. But sometimes touchdown PCR *(10)* is the only successful method. Therefore, we generally perform touchdown PCR as described in this protocol.

16. An elongation temperature of 68°C is advised when using the Expand High Fidelity PCR System (*see* **Note 4**). Otherwise, 72°C may be favourable. For amplification of DNA fragments longer than 1 kb, the elongation time should be increased.

17. Residual PCR template plasmid in the preparation of DNA recombination fragment is the main source of false-positive background colonies during mutagenesis. Therefore, a complete digestion of template plasmid after PCR is essential. Because PCR derived DNA is not methylated, the residual methylated template DNA is removed by *Dpn*I digestion, which only affects methylated DNA. To make sure that the template plasmid is methylated, it should be propagated in a dam-positiv bacterial strain (e.g., DH5α), which methylates at GATC sites.

18. If the same cuvets are to be used several times, they need to be washed at least 10 times with ddH$_2$O before electroporation.

19. The amount of DNA recombination fragment for transformation should be between 0.5 and 1.5 μg.
20. The second antibiotic depends on the template DNA used for generation of the DNA recombination fragment. Incubation of the bacterial cells at 37°C on LB agar plates containing chloramphenicol and the second antibiotic for the selection marker of the DNA recombination fragment allows selection for mutant BACs. At the same time, plasmid pDK46 is lost from the culture.
21. Plasmid pCP20, expressing the FLP recombinase, contains the temperature sensitive origin of replication pSC101. This origin maintains a low copy number and replicates at 30°C. The plasmid is lost from bacterial cells when incubated at temperatures above 37°C. Selection should be applied with 50 μg/mL ampicillin.
22. During incubation at 30°C, FLP recombinase is expressed from plasmid pCP20 and recombines the two FRT sites. This results in excision of the sequences between the two FRT sites. Only a single FRT site remains at the original insertion site of the DNA recombination fragment within the target DNA.

Acknowledgment

This work was supported by the Deutsche Forschungsgemeinschaft through SFB 455.

References

1. Zhang, Y., Buchholz, R., Muyrers, J. P., and Stewart, A. F. (1998) A new logic for DNA engineering using recombination in Escherichia coli. *Nat. Genet.* **20**, 123–128.
2. Muyrers, J. P., Zhang, Y., and Stewart, A. F. (2001) Techniques: Recombinogenic engineering—new options for cloning and manipulating DNA. *TIBS* **26**, 325–331.
3. Muyrers, J. P., Zhang, Y., Testa, G., and Stewart, A. F. (1999) Rapid modification of bacterial artificial chromosomes by ET-recombination. *Nucl. Acids Res.* **27**, 1555–1557.
4. Datsenko, K. A. and Wanner, B. L. (2000) One-step inactivation of chromosomal genes in Escherichia coli K-12 using PCR products. *PNAS* **97**, 6640–6645.
5. Cherepanov, P. P. and Wackernagel, W. (1995) Gene disruption in Escherichia coli: TcR and KmR cassettes with the option of FLP-catalyzed excision of the antibiotic-resistance determinant. *Gene* **158**, 9–14.
6. Adler, H., Messerle, M., Wagner, M., and Koszinowski, U. H. (2000) Cloning and mutagenesis of the murine gammaherpesvirus 68 genome as an infectious bacterial artificial chromosome. *J. Virol.* **74**, 6964–6974.
7. Muyrers, J. P., Zhang, Y., Benes, V., Testa, G., Ansorge, W., and Stewart, A. F. (2000) Point mutation of bacterial artificial chromosomes by ET recombination. *EMBO Rep.* **1**, 239–243.
8. Swaminathan, S., Ellis, H. M., Waters, L. S., et al. (2001) Rapid engineering of bacterial artificial chromosomes using oligonucleotides. *Genesis* **29**, 14–21.

9. Wagner, M., Jonjic, S., Koszinowski, U. H., and Messerle, M. (1999) Systematic excision of vector sequences from the BAC-cloned herpesvirus genome during virus reconstitution. *J. Virol.* **8,** 7056–7060.
10. Sambrook, J. and Russel, D. (2001) *Molecular Cloning: A Laboratory Manual.* Cold Spring Harbor Laboratory Press, Cold Spring Harbor, NY.
11. Yu, D., Ellis, H. M., Lee, E.-C., Jenkins, N. A., Copeland, N. G., and Court, D. L. (2000) An efficient recombination system for chromosome engineering in Escherichia coli. *PNAS* **97,** 5978–5983.
12. Nefedov, M., Williamson, R., and Ioannou, P. A. (2000) Insertion of disease-causing mutations in BACs by homologous recombination in Escherichia coli. *Nucl. Acids Res.* **28,** e79i–e79iv.

18

Mutagenesis of Herpesvirus BACs by Allele Replacement

Eva-Maria Borst, György Pósfai, Frank Pogoda, and Martin Messerle

1. Introduction

For mutagenesis of BAC-cloned herpesvirus genomes, we have adapted a two-step replacement procedure originally described by O'Connor et al. *(1)* for the manipulation of large DNA segments in *Escherichia coli (E. coli)*. In principle, the mutant allele to be introduced is provided on a so-called shuttle plasmid. The mutant allele has to be flanked by regions homologous to the desired integration site of the mutation in the BAC (regions A and B, *see* **Fig. 1**, step 1). By homologous recombination in *E. coli*, the shuttle plasmid will completely insert into the BAC, leading to a cointegrate (*see* **Fig. 1**, step 2). As a consequence, the cointegrate carries the wild-type, as well as a mutant allele, and a duplication of the homologous sequences flanking the mutation and the wild-type locus. The cointegrate can spontaneously resolve by homologous recombination via regions A or B giving rise to either the wild-type herpesvirus BAC plasmid or a mutant BAC, depending on which of the homologous regions (A or B) is used for recombination (*see* **Fig. 1**, step 3). The propensity for resolution of the cointegrate goes up with increasing sizes of the homologous sequences. However, even if homologous sequences of 2 or 3 kbp are provided, resolution of the cointegrates remains a rather rare event. Therefore, we have introduced a negative selection marker into the shuttle plasmid (the *sacB* gene) that allows to select against the bacteria still containing a nonresolved cointegrate (*see* **Fig. 1**, step 4) and to identify bacterial clones harboring the mutant or the parental BAC. The desired mutant BAC is finally identified

From: *Methods in Molecular Biology, vol. 256:*
Bacterial Artificial Chromosomes, Volume 2: Functional Studies
Edited by: S. Zhao and M. Stodolsky © Humana Press Inc., Totowa, NJ

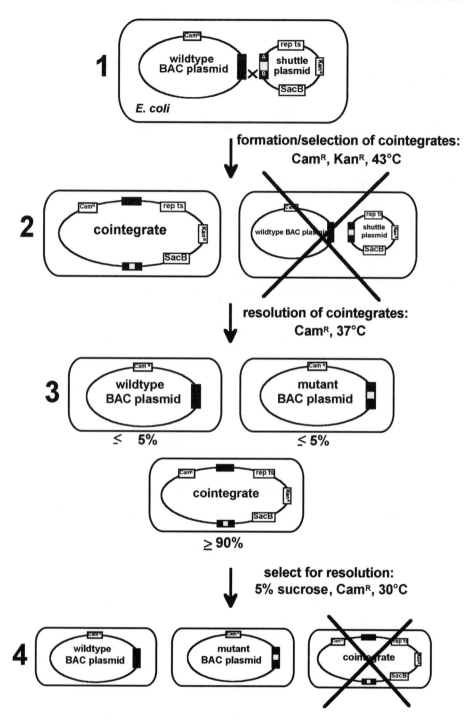

by restriction enzyme digestion or PCR. The mutant BAC can be transfected into permissive cells in order to reconstitute mutant virus (*see* Chapter 15).

Two-step replacement procedures with various modifications have been independently developed by a number of other laboratories for mutagenesis of BACs and other plasmids *(1,2,7,9–14)*. The two-step replacement method has several advantages and also some limitations in comparison to other mutagenesis techniques, e.g., the recently developed protocols that utilize recombinases of bacteriophages and linear recombination fragments *(15–18)*. Construction of the shuttle plasmid requires several cloning steps that depend on suitable restriction enzyme sites present in the viral genome or on the generation of appropriate DNA fragments by PCR. Thus, the procedure takes more time than ET-mutagenesis *(15)* that uses PCR-amplified linear DNA fragments for recombination. However, once the homology arms are cloned, the shuttle plasmid can be used for targeting various gene insertions to the same position in the viral genome. The mutagenesis procedure itself takes about 7 d and works highly reliable in our hands. A major advantage of the two-step replacement procedure is the fact that no remnants of the mutagenesis procedure (like selection markers or any other foreign sequences) remain in the BAC genomes after mutagenesis. This is in contrast to ET-mutagenesis that usually requires the insertion of an antibiotic resistance gene *(15)*. Although FRT- or loxP-flanked selection markers can subsequently be removed with the help of site-specific recombinases *(15,17,19)*, there still remain some foreign sequences inserted in the BAC that may interfere with gene expression. Nevertheless, ET-mutagenesis is certainly the fastest technique to introduce deletions into BACs. We believe, however, that the two-step replacement procedure is still the method of choice to bring in point mutations as well as gene insertions into BAC-cloned herpesvirus genomes. The recently described protocols for insertion of point mutations utilizing the bacteriophage λ recombination functions either represent also a two-step replacement strategy *(20)* or are quite inefficient and, therefore,

---

Fig. 1. *(see opposite page)* Two-step replacement procedure for mutagenesis of the BAC-cloned CMV genome by homologous recombination in *E. coli*. Step 1: The shuttle plasmid carrying the desired mutation plus flanking homologies (A and B) is transformed into bacteria that already contain the BAC. Step 2: Through homologous recombination via region A or B the shuttle plasmid is completely integrated into the viral BAC genome, leading to a cointegrate. Bacteria containing the cointegrate are selected by incubation at 43°C. Step 3: Resolution of the cointegrate via A or B leads to either the wildtype or the mutant BAC. Step 4: Bacteria harboring a resolved cointegrate are selected at 30°C on agar plates containing 5% sucrose.

require the screening of hundreds of colonies to identify a positive clone *(21)*. Another aspect concerns the stability of the BACs during mutagenesis. Viral genomes often contain direct sequence repeats that are optimal substrates for recombination enzymes. Thus, in recombination-proficient bacteria there is a risk for the occurrence of undesired deletions or rearrangements in the BAC-cloned viral genomes *(22)*. We made the observation that the recombinogenic activity provided by the *rec*A gene transiently expressed from the backbone of the shuttle plasmid is much weaker than the activity of the bacteriophage recombinases used in the ET-mutagenesis procedure *(15)*. This is underpinned by the fact that the two-step replacement procedure requires homologous sequences of at least 2.5 kbp in size in order to achieve efficient recombination whereas homology arms as small as 40 nucleotides in length are sufficient for ET-mutagenesis *(15)*. Accordingly, instability of the BAC-cloned virus genomes was never a problem when the two-step replacement procedure was applied. In contrast, it was sometimes difficult to properly adjust the recombinogenic activity of the bacteria when mutagenesis was performed with bacteriophage recombinases (Messerle, M. and Wagner, M., unpublished results). Altogether, we think that the two-step replacement procedure is highly useful for mutagenesis of BAC-cloned herpesvirus genomes.

2. Materials

1. Competent *E. coli* strain DH10B bacteria for plasmid cloning.
2. High copy cloning vector (e.g., pBluescript or pUC-series).
3. Shuttle plasmid of the pST76-family *(2)* (e.g., pST76-KSR *[3]*).
4. Electrocompetent *E. coli* strain DH10B containing the BAC cloned virus genome.
5. Incubators for bacteria at 30, 37, and 43°C.
6. Restriction enzymes.
7. 50% sucrose solution: 100 g sucrose in a final volume of 200 mL ddH$_2$O. Sterilize by filtration (0.25 μm) and store at 4°C. For preparation of 5% sucrose agar plates take 450 mL LB agar and add 50 mL of 50% sucrose solution.
8. Sterile tooth picks.
9. Inoculation needle.
10. LB medium: 10 g tryptone peptone (Difco 0123-17), 5 g yeast extract (Difco 0127-17-9), 8 g NaCl add 1 L ddH$_2$O. Autoclave.
11. LB agar plates: 500 mL LB medium + 9 g Bacto Agar (Difco 0140-01). Autoclave. Cool down to 50°C before adding the desired antibiotic.
12. LB-agar plates containing 17 μg/mL chloramphenicol plus 25 μg/mL kanamycin (15–30 plates).
13. LB-agar plates containing 17 μg/mL chloramphenicol (20 plates).
14. LB-agar plates containing 17 μg/mL chloramphenicol plus 5% sucrose (10 plates).
15. LB-agar plates containing 25 μg/mL kanamycin (10 plates).

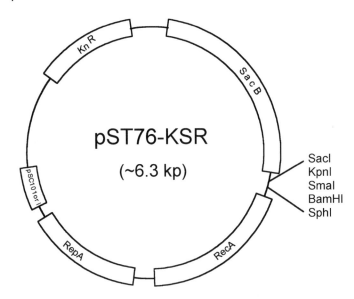

Fig. 2. Features of the shuttle plasmid pST76-KSR. The shuttle plasmid contains a kanamycin resistance marker (Kn^R) and the *E. coli recA* gene, which allows to perform mutagenesis in recombination-deficient *E. coli* strains. In addition, it harbors the negative selection marker *sacB* and the sequences responsible for the temperature-sensitive replication mode of the plasmid (pSC101ori and *repA*^ts). The unique restriction enzyme recognition sites provided by the polylinker (*Sac*I, *Kpn*I, *Sma*I, *Bam*HI, *Sph*I) are indicated.

3. Methods

3.1. Construction and Properties of the Shuttle Plasmid

1. The shuttle plasmid has to contain the mutant allele that is to be transferred onto the BAC-cloned virus genome plus flanking sequences homologous to the desired integration site on the viral genome (*see* **Fig. 1**). In addition, the shuttle vector provides several properties necessary for the mutagenesis procedure (*see* **Fig. 2**).
   a. The shuttle vector is temperature-sensitive in DNA replication. It replicates best at 30°C, but its copy number is already reduced at 37°C and it does not replicate and is rapidly lost from the bacteria at 43°C. This property is due to a modified RepA protein of plasmid pSC101 that probably undergoes a conformational change at temperatures above 37°C, which interferes with its binding to the plasmid origin of replication (*4*).
   b. The shuttle plasmid encodes the *E. coli* RecA protein that mediates recombination between the shuttle plasmid and the BAC (*see* **Fig. 2**). Therefore, mutagenesis can even be performed in *rec*A-negative bacteria like the *E. coli* strain DH10B.

   c. It contains the *sac*B gene that allows negative selection of the shuttle plasmid (*see* **Fig. 2**). *Sac*B is a gene of *Bacillus subtilis* that encodes the secreted enzyme levansucrase, which catalyzes the hydrolysis of sucrose and the formation of levans (high-molecular-weight fructose polymers). Expression of the *sac*B gene confers sucrose-sensitivity to *E. coli* because the enzyme levansucrase cannot be secreted and is mainly localized in the periplasm where it probably leads to the synthesis of toxic products *(5)*. Thus, *E. coli* bacteria expressing *sac*B cannot grow on LB agar plates containing 5% sucrose and the selection principle can be used for negative selection against *sac*B containing plasmids (*see* **Note 1**).
2. The shuttle vectors available differ by the antibiotic resistance marker encoded on the plasmid backbone (kanamycin, tetracycline, zeocin, ampicillin, or chloramphenicol resistance *(2,3,6–9)* and Messerle et al., unpublished results). In our hands, the shuttle plasmid expressing the kanamycin resistance marker works best and, therefore, we use shuttle vector pST76-KSR (*see* **Fig. 2**) now routinely for the mutagenesis procedure.
3. For construction of the recombination plasmid the mutant allele plus the flanking homologies (regions A and B, *see* **Fig. 1**) must be cloned into the shuttle plasmid. We recommend using homologous sequences of at least 2.5 kbp and of equal size (*see* **Note 2**). Cloning with the shuttle plasmids is not trivial because they are low copy and they have to be propagated at 30°C. Therefore, we routinely construct the recombination fragment in a multicopy vector like pUC or pBluescript. In the last cloning step, we transfer the insert (region A—mutant allele—region B) to the low copy shuttle vector (*see* **Note 3**). We always grow bacteria harbouring the shuttle plasmid at 30°C.

3.2. Mutagenesis of the Viral BAC by the Two-Step Replacement Procedure

   The procedure is described for using a shuttle plasmid that is based on the vector pST76-KSR encoding kanamycin resistance (*see* **Fig. 2**). When using a shuttle plasmid with another antibiotic resistance marker, substitute kanamycin with the appropriate antibiotic.

1. Transform competent bacteria already containing the BAC-cloned virus genome with the shuttle plasmid. Shake transformed bacteria for 1 h at 30°C, spread on LB agar plates containing chloramphenicol (17 µg/mL) plus kanamycin (25 µg/mL). Incubate for 1.5 d at 30°C (*see* **Note 4**).
2. Streak 10 clones from step 1 on LB agar plates containing chloramphenicol (17 µg/mL) plus kanamycin (25 µg/mL) to get single colonies the next day. Incubate at 43°C (*see* **Note 5**).
3. (Optional) Inoculate 10-mL cultures containing chloramphenicol plus kanamycin with the clones grown at 43°C. Grow at 37°C overnight. Prepare minipreps (*see* Chapter 15) and characterize the BACs by restriction enzyme digestion. Make sure that the insertion occurred at the correct position; depending on the restric-

tion enzyme used for characterization, it is sometimes possible to decide whether the insertion occurred via region A or via region B (*see* **Note 6**).

4. Streak 10 clones (one from each plate from **step 2**) on plates containing chloramphenicol only. Streak the bacteria very well in order tol get single colonies. Incubate at 37°C for 1.5 to 2 d (*see* **Note 7**).

5. Pick 10 clones (one from each plate from **step 4**) and re-streak them onto LB agar plates containing chloramphenicol (17 μg/mL) plus 5% sucrose. Incubate at 30°C for 1.5 to 2 d (*see* **Note 8**).

6. Pick about 50 clones (5 to 10 clones from each plate of **step 5**) on a plate containing kanamycin and in parallel on a plate containing chloramphenicol. Incubate at 37°C overnight (*see* **Note 9**).

7. Choose kanamycin-sensitive clones from step 6 and grow 10 mL overnight cultures with chloramphenicol (17 μg/mL). Prepare minipreps (*see* Chapter 15) and characterize the BACs by restriction enzyme digestion or PCR (*see* **Fig. 3**) (*see* **Note 10**).

After identification of the mutant clone(s), prepare Midi-Preps from 100-mL cultures (or larger cultures, compare Chapter 15). Characterize the mutant BAC by restriction enzyme analysis using several different restriction enzymes to confirm the integrity of the cloned herpevirus enome. Sequencing of relevant parts of the BAC is recommended. Reconstitute recombinant virus by transfection of the mutant BAC into permissive cells as described in Chapter 15. Save the mutant BAC clones as a glycerol stock at –70°C.

4. Notes

1. It is important to grow the bacteria containing the shuttle plasmid with the *sac*B gene in the complete absence of sucrose. Occasionally, we observed spontaneous inactivation of the *sac*B gene. In this case, selection with sucrose will not work anymore. Thus, it might be worthwhile to check the sucrose-sensitivity of the shuttle vector and of the shuttle plasmids used for recombination on a regular basis.

2. We have also constructed CMV BAC mutants using homologous sequences as short as 1.0 kb. However, recombination is favored with increasing sizes of the homologous sequences. The flanking homologies should be of equal size because recombination will occur preferentially via the longer homology arm. Please note that for generation of a mutant BAC, the first recombination step must take place via region A and the second via region B (or vice versa). If formation and resolution of the cointegrate occurs via the same region (either region A or region B only) one ends up with the parental BAC. We think that it is more likely that the successive recombination events occur via the different homology arms if they are of similar size.

3. Make sure that you save appropriate sites for this last cloning step. Sometimes it is the easiest way to clone an oligonucleotide linker into a multicopy vector that allows rapid insertion of the A- and B-fragments and of the fragment containing the mutation, and finally transfer of the complete insert to the shuttle vector.

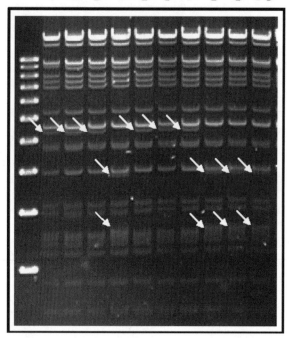

Fig. 3. Typical result of a mutagenesis experiment. Mutagenesis was performed as described in the text. Overnight cultures were grown from 10 bacterial clones selected on sucrose-containing agar plates. The isolated BACs were analyzed by *Eco*RI digestion and DNA fragments were separated by agarose gel electrophoresis. BAC plasmids carrying the mutant or wild-type allele can be identified by the DNA pattern. DNA fragments characteristic for the wild-type or the mutant HCMV BAC are indicated by arrows. Clones 1, 2, 3, 5, 6, and 7 correspond to the wild-tpye BAC plasmid, while clones 4, 8, 9, and 10 represent the mutant BAC.

4. Formation of cointegrates occurs at this step. In bacteria, single crossing-over events occur preferentially (either via region A or region B). The result is a complete insertion of the shuttle plasmid into the BAC (cointegrate, *see* **Fig. 1**). Please note that the resulting bacterial colonies after **step 1** are still a mixture of bacteria that contain cointegrates and of bacteria that contain the BAC and the shuttle plasmid separately. Incubation at 30°C can be done for up to 2 d (let the bacteria grow into big colonies) in order to augment the chance for recombination events and to increase the number of bacteria containing a cointegrate.

5. This step is to select for bacteria containing a cointegrate. Cells containing the shuttle and the BAC as separate plasmids will not grow at 43°C (or only very poorly) since replication of the shuttle plasmid is temperature-sensitive. There-

fore, the incubation temperature is critical for this step. Make sure that the shuttle plasmid does not replicate at the non-permissive temperature by incubating clones harbouring the recombination plasmid at 43°C. After a 1-d incubation period, no colonies (or only tiny ones) should be visible. If the bacteria are still growing very well, increase the temperature of the incubator stepwise by 0.5°C. Once properly adjusted, never change the temperature of the incubator. Please note that this step of the mutagenesis procedure is critical. It is important to select bacterial clones that contain a cointegrate. If bacteria that still contain the shuttle and the BAC separately are chosen for the next step, mutagenesis cannot be successful. Bacteria harboring a cointegrate can usually be identified along the second or third streak by their large colony size. Please ignore small colonies, they never contain bacteria with cointegrates. In order to make sure that the large colonies are of clonal origin, one may streak the bacteria again on agar plates with cloramphenicol and kanamycin and repeat the incubation step at 43°C.

6. We had the following experience: as long as the A- or B-fragments have a size of 2.5 to 3 kbp, insertion always occurs at the expected position. Therefore, we usually skip **step 3**.

7. The cointegrates now have the chance to resolve. Please note that the resulting clones are a mixture of bacterial cells that contain resolved plasmids and of bacteria that still contain the cointegrate. We recommend growing up the bacteria into big colonies, because the more bacteria one gets the more likely one will have bacteria in which a recombination event (resolution of the cointegrate) did occur.

8. In this step, bacteria with a resolved cointegrate are identified. Please note that selection for sucrose resistance works best at 30°C. The larger the size of the A- and B-fragments, the higher is the chance for resolution of the cointegrates. Therefore, whenever possible, provide enough homology on each side of the mutation (2–3 kbp) and a similar size of regions A and B. Nevertheless, most of the clones (about 95%) will not resolve the cointegrate. Therefore, if the sucrose selection works, one gets usually only a few colonies that grow on the sucrose-containing plates. Because DH10B are growing rather slowly, it may take up to 2 d until clones are readily visible.

9. About 80 to 100% of the clones should be kanamycin-sensitive. Sucrose-resistance and kanamycin-sensitivity are independent indications that the cointegrates did resolve. Occasionally the *sac*B gene is inactivated by mutation; in this case the clones may remain Kn-resistant and the cointegrate is not resolved.

10. If fragments A and B (the homologous sequences on both sides of the mutation) are of equal size, the chance to get the mutated or the reverted BAC should be 50:50%. It is important that the sizes of regions A and B are similar. Please keep in mind that to get a mutant BAC formation of the cointegrate has to occur via A and resolution via B (or vice versa). Sometimes it may be helpful to choose some clones from **step 3** in which insertion occurred via A and some in which insertion occurred via B, because occasionally there is a bias for resolution via one side. For identification of the mutant BAC clone it may be helpful to tag the mutation

with a restriction enzyme site that can be easily detected (for MCMV: e.g., *Hin*dIII or *Eco*RI; for HCMV: e.g., *Hin*dIII, *Eco*RI, *Eco*RV, *Bgl*II, *Xho*I).

Acknowledgment

This work was supported by the Wilhelm-Roux-Program of the Medical Faculty, Martin-Luther-University of Halle-Wittenberg and the NBL3-Program of the German Ministry for Education and Research (BMBF).

References

1. O'Connor, M., Peifer, M., and Bender, W. (1989) Construction of large DNA segments in Escherichia coli. *Science* **244**, 1307–1312.
2. Posfai, G., Koob, M. D., Kirkpatrick, H. A., and Blattner, F. R. (1997) Versatile insertion plasmids for targeted genome manipulations in bacteria: isolation, deletion, and rescue of the pathogenicity island LEE of the Escherichia coli O157:H7 genome. *J. Bacteriol.* **179**, 4426–4428.
3. Hobom, U., Brune, W., Messerle, M., Hahn, G., and Koszinowski, U. H. (2000) Fast screening procedures for random transposon libraries of cloned herpesvirus genomes: mutational analysis of human cytomegalovirus envelope glycoprotein genes. *J. Virol.* **74**, 7720–7729.
4. Hashimoto-Gotoh, T., Franklin, F. C., Nordheim, A., and Timmis, K. N. (1981) Specific-purpose plasmid cloning vectors. I. Low copy number, temperature-sensitive, mobilization-defective pSC101-derived containment vectors. *Gene* **16**, 227–235.
5. Gay, P., Le, C. D., Steinmetz, M., Ferrari, E., and Hoch, J. A. (1983) Cloning structural gene sacB, which codes for exoenzyme levansucrase of Bacillus subtilis: expression of the gene in Escherichia coli. *J. Bacteriol.* **153**, 1424–1431.
6. Link, A. J., Phillips, D., and Church, G. M. (1997) Methods for generating precise deletions and insertions in the genome of wild-type Escherichia coli: application to open reading frame characterization. *J. Bacteriol.* **179**, 6228–6237.
7. Yang, X. W., Model, P., and Heintz, N. (1997) Homologous recombination based modification in Escherichia coli and germline transmission in transgenic mice of a bacterial artificial chromosome. *Nat. Biotechnol.* **15**, 859–865.
8. Messerle, M., Crnkovic, I., Hammerschmidt, W., Ziegler, H., and Koszinowski, U. H. (1997) Cloning and mutagenesis of a herpesvirus genome as an infectious bacterial artificial chromosome. *Proc. Natl. Acad. Sci. USA* **94**, 14,759–14,763.
9. Horsburgh, B. C., Hubinette, M. M., Qiang, D., MacDonald, M. L., and Tufaro, F. (1999) Allele replacement: an application that permits rapid manipulation of herpes simplex virus type 1 genomes. *Gene Ther.* **6**, 922–930.
10. Kempkes, B., Pich, D., Zeidler, R., Sugden, B., and Hammerschmidt, W. (1995) Immortalization of human B lymphocytes by a plasmid containing 71 kilobase pairs of Epstein-Barr virus DNA. *J. Virol.* **69**, 231–238.
11. Posfai, G., Kolisnychenko, V., Bereczki, Z., and Blattner, F. R. (1999) Markerless gene replacement in Escherichia coli stimulated by a double-strand break in the chromosome. *Nucleic Acids Res.* **27**, 4409–4415.

12. Smith, G. A. and Enquist, L. W. (1999) Construction and transposon mutagenesis in Escherichia coli of a full-length infectious clone of pseudorabies virus, an alphaherpesvirus. *J. Virol.* **73,** 6405–6414.
13. Imam, A. M., Patrinos, G. P., deKrom, M., et al. (2000) Modification of human beta-globin locus PAC clones by homologous recombination in Escherichia coli. *Nucleic Acids Res.* **28,** E65.
14. Lalioti, M. and Heath, J. (2001) A new method for generating point mutations in bacterial artificial chromosomes by homologous recombination in Escherichia coli. *Nucleic Acids Res.* **29,** E14.
15. Zhang, Y., Buchholz, F., Muyrers, J. P., and Stewart, A. F. (1998) A new logic for DNA engineering using recombination in Escherichia coli. *Nat. Genet.* **20,** 123–128.
16. Murphy, K. C., Campellone, K. G., and Poteete, A. R. (2000) PCR-mediated gene replacement in Escherichia coli. *Gene* **246,** 321–330.
17. Datsenko, K. A. and Wanner, B. L. (2000) One-step inactivation of chromosomal genes in Escherichia coli K-12 using PCR products. *Proc. Natl. Acad. Sci. USA* **97,** 6640–6645.
18. Yu, D., Ellis, H. M., Lee, E. C., Jenkins, N. A., Copeland, N. G., and Court, D. L. (2000) An efficient recombination system for chromosome engineering in Escherichia coli. *Proc. Natl. Acad. Sci. USA* **97.** 5978–5983.
19. Cherepanov, P. P. and Wackernagel, W. (1995) Gene disruption in Escherichia coli: TcR and KmR cassettes with the option of Flp-catalyzed excision of the antibiotic-resistance determinant. *Gene* **158,** 9–14.
20. Muyrers, J. P., Zhang, Y., Benes, V., Testa, G., Ansorge, W., and Stewart, A. F. (2000) Point mutation of bacterial artificial chromosomes by ET recombination. *EMBO Rep.* **1,** 239–243.
21. Swaminathan, S., Ellis, H. M., Waters, L. S., et al. (2001) Rapid engineering of bacterial artificial chromosomes using oligonucleotides. *Genesis* **29,** 14–21.
22. Adler, H., Messerle, M., Wagner, M., and Koszinowski, U. H. (2000) Cloning and mutagenesis of the murine gammaherpesvirus 68 genome as an infectious bacterial artificial chromosome. *J. Virol.* **74,** 6964–6974.

19

Herpesvirus Genome Mutagenesis by Transposon-Mediated Strategies

Alistair McGregor and Mark R. Schleiss

1. Introduction

The Herpesviruses family consists of over 100 distinct viruses that are responsible for a variety of important human and veterinary diseases. These viruses are subclassified into *alpha-*, *beta-*, or *gamma-* herpesviruses, dependent upon a variety of criteria. Eight herpesviruses are human pathogens: Herpes Simplex Viruses type 1 and 2; Varicella–Zoster Virus (VZV); Cytomegalovirus (CMV); Human Herpes Viruses type 6 and 7; Epstein–Barr Virus (EBV); and Kaposi's Sarcoma Herpes Virus (KSHV), also known as HHV-8. The human herpesviruses vary in primary sites of replication and clinical manifestations of infection, but they share many molecular similarities. All are large, enveloped viruses, consisting of a linear double-stranded DNA genome, ranging from approx 120 kbp–230 kbp in size. The complete DNA sequence of each of the human herpesviruses has been elucidated, but much remains to be learned about the function of specific viral gene products in the pathogenesis of disease.

Because many of the human herpesviruses are species-specific, insights into viral mechanisms of pathogenesis often must rely on the study of related vertebrate herpesviruses. Ideally, the role of specific viral gene products in replication and pathogenesis is best elucidated by the generation of viral mutants. This is a time-consuming and labor-intensive strategy when performed using the conventional approach of homologous recombination. In some cases, the use of overlapping cosmids spanning the length of the viral genome has improved the strategy for the generation of recombinant herpesviruses. In this approach, a single mutated cosmid is transfected onto cells, along with the remainder of the cosmid set, and virus is generated as a result of recombination between

From: *Methods in Molecular Biology, vol. 256:*
Bacterial Artificial Chromosomes, Volume 2: Functional Studies
Edited by: S. Zhao and M. Stodolsky © Humana Press Inc., Totowa, NJ

cosmids. However, the recent development of technologies to clone entire viral genomes as "bacterial artificial chromosomes," or "BAC" plasmids, has made this the easiest and fastest approach for the manipulation of viral genomes, and for generation of mutant viruses *(1,2)*. Generation of viral mutants can quickly be achieved (as outlined in **Fig. 1**), using modified BAC plasmids which are transfected onto cells *(3)*. This chapter provides an overview of the general strategy employed in the construction or cloning of a herpesvirus genome as a bacterial artificial chromosome in bacteria. Also, since transposon mutagenesis has tremendous potential in the forward genetics of viruses cloned in this fashion, specific protocols are provided for the random mutation of large DNA plasmids, such as cosmids and BACs, using this technology.

2. Materials

All reagents should be of molecular biology grade. All solutions should be prepared with double-distilled or deionized water.

2.1. Generation of a BAC Plasmid Herpesvirus Clone in Bacteria

1. Herpesvirus of interest in appropriate cells that support productive infection.
2. Media with 10% fetal calf serum (FCS).

Fig. 1. *(see opposite page)* Generation of recombinant herpesvirus *via* alternative strategies. Three main approaches are used to generate recombinant virus. The use of overlapping cosmids or a BAC clone of the viral genome can produce clonal virus without purification, but homologous recombination generates a mixed population of viruses requiring further purification. **(A)** Homologous recombination using a shuttle vector to target a specific locus on the genome. The target locus/gene on the shuttle vector has been modified to insert or substitute a reporter gene marker, and is flanked by viral sequence for recombination. Progeny virus is generated by cotransfection of the shuttle vector with viral DNA onto cells, or by coinfection with wild-type virus. Recombination between viral DNA and shuttle vector produces mutant virus. Wild-type virus is also produced, and hence progeny virus needs to be purified further to obtain a clonal population. **(B)** Overlapping cosmids. A series of of cosmids spanning the length of the genome (4–8 cosmids dependent on the virus) are cotransfected together into the same cell, and virus is generated as a result of recombination between overlapping regions in the adjoining cosmids, to produce full-length viral DNA. A mutant virus can be generated by modifying the locus of one cosmid, and cotransfecting it along with the other cosmids to produce a clonal mutant virus population. This process is relatively slow and inefficient with certain herpesviruses. **(C)** A BAC plasmid clone of a herpesvirus is modified in bacteria and transfected onto cells to generate a clonal virus population of mutants. This process is relatively rapid for all herpesviruses for which BACs have been generated.

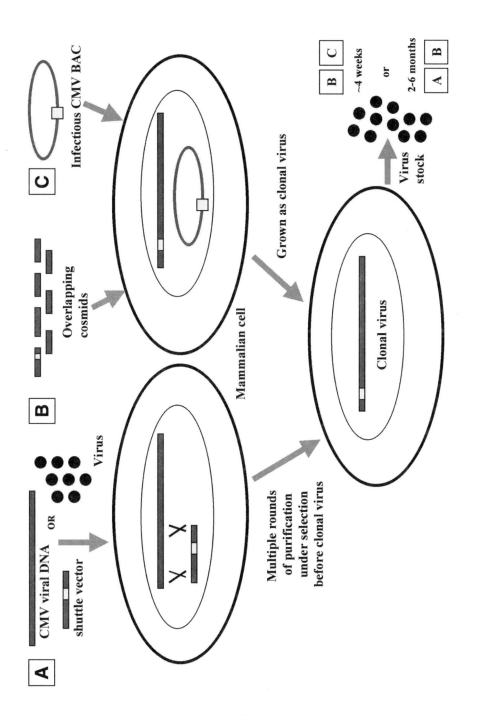

A   CMV viral DNA   OR   shuttle vector

Virus

B   Overlapping cosmids

C   Infectious CMV BAC

Mammalian cell

Multiple rounds of purification under selection before clonal virus

Grown as clonal virus

Clonal virus

Virus stock

~4 weeks

or

2-6 months

B   C

A   B

3. Low-copy number plasmid for use as shuttle vector, e.g., pACYC177 or pACYC184 (available from New England Biolabs).
4. Puromycin or mycophenolic acid/xanthine stock solutions, depending upon method of selection (see later).
5. Fluorescent microscope.

2.2. Tn5-Based In Vitro Transposon Mutagenesis of Large DNA Molecules (Plasmids, Cosmids, BAC Constructs)

1. EZ::TN kit® (kan 2) from Epicentre Technologies, Inc.
2. Target DNA (0.2 µg/µL to 0.5 µg/µL).
3. LB medium.
4. LB agar plates.
5. DH10B electrocompetent *Escherichia coli* (*E. coli*) (see section on preparation of electrocompetent cells, or use commercially available cells such as Electromax® DH10B (Invitrogen), or EC100® electrocompetent bacteria (Epicentre technology).
6. 1-mm-gapped cuvet.
7. Electroporator.

2.3. Transposon Mutagenesis of Large DNA Molecules in Bacteria

1. Electrocompetent bacterial strain DH10B carrying target BAC.
2. Donor plasmid (pTsTm8) carrying TnMax8 *(21)*.
3. 1-mm-gapped cuvet.
4. Electroporator.
5. LB medium.
6. LB agar plates.

2.4. Generation of Electrocompetent E. coli Cells and Electrotransformation of Competent Cells

1. LB agar.
2. LB medium.
3. *E. coli* strain DH10B (+/– BAC, dependent on transposon mutagenesis strategy).
4. 1-mm-gapped cuvet.
5. Electroporator.
6. 250-mL conical tubes (Falcon).
7. Ultrapure glycerol (10%).
8. Electrocompetent cells.
9. 1-mm-gapped cuvet.

2.5. Miniprep Analysis of Transposon-Mutated Large Low Copy Plasmids

1. Toothpicks or "cocktail sticks."
2. LB medium.

3. LB agar.
4. Phenol/chloroform/isoamyl alcohol (24:23:1).
5. Solution 1 (50 m*M* glucose, 25 mM Tris-HCl, pH 8.0).
6. Solution 2 (0.2 *N* NaOH, 1% SDS).
7. Solution 3 (per 100 mL: 60 mL 5 *M* potassium acetate, 11.5 mL glacial acetic acid, 28.5 mL distilled water).
8. 15 mL snap top tubes (Falcon).
9. –70°C freezer or dry ice.
10. Absolute ethanol.

3. Methods

3.1. Generation of a BAC Plasmid Herpesvirus Clone in Bacteria

A detailed summary of the stages involved in the cloning of a Herpesvirus genome as a plasmid in bacteria is outlined schematically in **Fig. 2**. However, there are a number of factors to consider prior to the construction of a recombinant virus carrying a BAC insert, and the subsequent generation of an infectious plasmid clone in bacteria (*see* **Notes 1–5**). A few general rules apply, but this is not a complete list and individual cases may have specific requirements that should also be considered. A detailed step-by-step protocol is provided in Chapter 15 by Borst et al.

3.2. Transposon Mutagenesis of a Cloned Herpesvirus Genome

The methodology of site-specific mutagenesis of BAC plasmid or cosmid clones is now relatively well established (*see* **Note 6**). Systems such as the ET/RED systems are available upon request from individual investigators (*12, 15,16*). Alternatively, the BAC plasmid, cosmid, or other large DNA molecule can be mutated randomly by use of transposon mutagenesis (*see* **Notes 7–9**). The protocol described below is for the Tn5 EZ::TN system® from Epicentre Technology, Inc. (*see* **Note 10**).

1. Thaw contents of the transposon reaction kit, as well as an aliquot of the target plasmid DNA, and place the transposase enzyme on ice.
2. Prepare the following reaction mixture in a 1.5-mL centrifuge tube:
   a. 1 µL EZ::TN 10X reaction buffer.
   b. 0.02–0.5 µg target DNA (*see* **Note 11**).
   c. x µL molar equivalent EZ::TN transposon (between 0.1–1 µL).
   d. y µL sterile water to increase reaction volume to 9 µL.
   e. 1 µL EZ::TN transposase.
   f. Total volume 10 µL.
3. Pulse-spin contents in a microcentrifuge for 10 s.
4. Heat at 37°C for 2 h.
5. Remove the sample and add 1 µL of stop solution from the kit, and mix by pipetting.

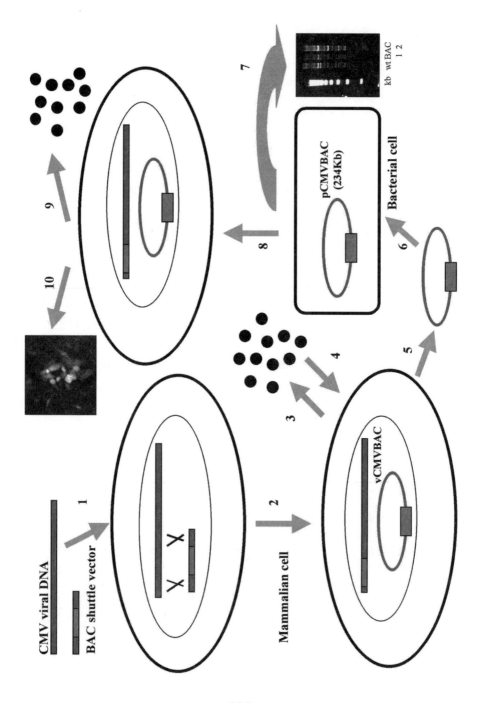

CMV viral DNA

BAC shuttle vector

Mammalian cell

vCMVBAC

pCMVBAC
(234Kb)

Bacterial cell

kb wt BAC
     1 2

6. Heat-treat the sample for 10 min at 75°C to inactivate the reaction.
7. Remove the sample and dilute to 40 µL with water and cool on ice.
8. Take 3–5 µL and use to transform *E. coli* by electroporation (see section on electrocompetent cells and electroporation) (*see* **Notes 12** and **13**). Store the remainder of the reaction at 4°C for a few days, or at –70°C if long-term storage is required. It should be noted that large plasmid reactions stored at –70°C should not be repeatedly freeze-thawed, as this increases the number of deleted plasmids in the reaction.
9. Plate out half of the transformation reaction on LB agar plates with antibiotic selection for both the target plasmid and the transposon, and incubate overnight at 37°C. Store the remainder of the transformation reaction overnight on the bench and, plate out on fresh plates if additional colonies are required. Plates with colonies can be wrapped in parafilm and stored for several months at 4°C (*see* **Note 14**).
10. To analyze colonies, pick each colony with a toothpick, and add it into 3 mL LB medium (with antibiotics) in a snap top tube, grow overnight in a shaker at 290 rpm, 37°C, and prepare miniprep DNA as described later (*see* **Notes 15** and **16**).

3.3. Transposon Mutagenesis of Large DNA Molecules in Bacteria

The protocol described is based on the use of a variant of a Tn3 transposon, Tn1721 (TnMax) that has been modified to preferentially target episomal DNA *(19,20)*. The transposon is cloned on a donor plasmid with a temperature-sensitive origin of replication *(22)* and the transposon encodes a drug resistance marker *(21)*. The machinery for transposition (transposase and resolvase) are encoded on the donor plasmid, separately from the transposon, to eliminate secondary transposition events. The transposition protocol is slightly more complicated than described for the in vitro transposon system, as the donor plasmid has to be introduced into the bacterial strain carrying the target plasmid, cosmid, or BAC prior to transposon mutagenesis. Selection of the bacterial strain carrying the BAC plasmid is obtained through selection of the ampicillin drug resistance marker encoded on the backbone of the donor plasmid. Both target DNA and donor plasmid can be maintained stably under permissive conditions (30°C) for a short period of time. The donor plasmid is rapidly lost

---

Fig. 2. *(see opposite page)* Strategy for the cloning of a herpes viral genome as infectious BAC plasmid in bacteria. Steps 1–2: Generate a recombinant virus carrying a BAC plasmid insert; steps 3–5: Infect cells with BAC virus at high multiplicity of infection, and harvest infected cells for circular replicative intermediate; steps 6–7: Transform bacteria with circular viral DNA and screen colonies for full-length BAC plasmid clones; steps 8–10: verify full-length clone is infectious by transfection of DNA onto mammalian cells.

upon switching to nonpermissive conditions during selection of mutated BACs. In the protocol described later, the target DNA is assumed to be a BAC encoding a chloramphenicol drug resistant marker and the TnMax8 (from pTsTm8) encoding the kanamycin resistance gene. Other TnMax transposons encode other markers, such as erythromycin and tetracycline *(7,21)*. However, in our experience the kanamycin marker is the best one to use when mutating BACs, as it avoids problems of leakiness or poor growth associated with the other two selection markers. A summary of the stages involved in the temperature sensitive pTsTm8 (TnMax 8) mutagenesis strategy is shown in **Fig. 3**.

1. Transform the donor plasmid into *E. coli* DH10B strain carrying the target plasmid by electroporation, and plate out the transformed bacteria on LB agar plates containing antibiotics for the target plasmid (chloramphenicol 12.5 μg/mL) and donor plasmid (ampicillin 100 μg/mL) (*see* **Note 17**).
2. The plates should be incubated for 18–24 h at 30°C and stored at 4°C until required.
3. Approximately 6–12 colonies should be picked into 14 mL snap top tubes containing 3 mL of LB medium containing ampicillin and chloramphenicol. Grow up overnight at 30°C (290 rpm) and remove 2 mL for DNA miniprep analysis as described in the miniprep DNA preparation section. Verify that the integrity of the BAC DNA is preserved by restriction digestion and analysis on a gel. Retain the remainder of the overnight culture and store at 4°C overnight.
4. Once a stable donor plasmid/target BAC colony is verified, the remainder of the overnight culture should be plated out, selecting for transposed BACs.
5. Plate out 100 μL aliquots of the culture at various dilutions (from 0.1 μL–10 μL of overnight culture) onto LB agar plates containing chloramphenicol (12.5 μg/μL) and kanamycin (20 μg/mL), and incubate overnight at 43°C to eliminate the donor plasmid, and select for transposon insertion into the BAC plasmid.

Fig. 3. *(see opposite page)* Transposon mutagenesis of a large DNA plasmid in bacteria. The following steps summarize the approach to transposon-mediated mutagenesis of a large DNA plasmid in vivo. Step 1. Temperature sensitive donor plasmid (pTsTm8) carrying the TnMax8 transposon is transformed into DH10B bacteria carrying the target DNA (BAC plasmid). Bacteria carrying both target and donor plasmids are selected by antibiotic markers: ampicillin resistance for pTsTm8, and chloramphenicol resistance for BAC plasmid, under permissive conditions for the donor plasmid (30°C). Step 2. Once stable strains are isolated, BAC transposon mutants are selected under nonpermissive conditions for the donor plasmid (43°C) and drug resistance markers kanamycin (transposon) and chloramphenicol (BAC plasmid). Steps 3–4. Mutant BAC colonies are isolated and DNA analyzed by miniprep restriction digestion and/or sequencing from the Tn into the insertion locus of the target DNA. Mutant herpesvirus BAC plasmid DNA can be transfected onto mammalian cells to regenerate mutant virus for further study.

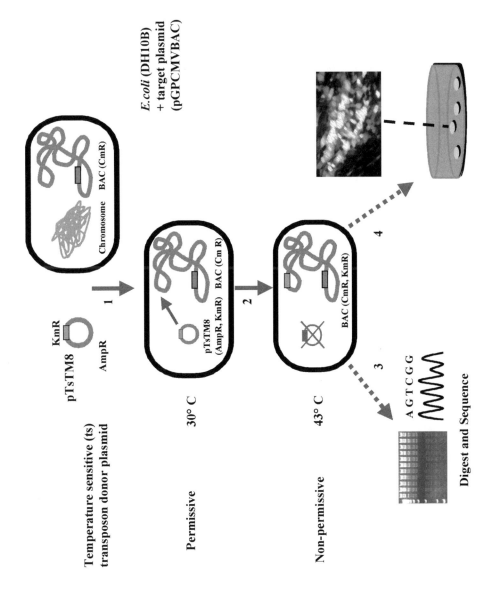

Temperature sensitive (ts) transposon donor plasmid

pTsTM8

KmR

AmpR

Permissive

30° C

Non-permissive

43° C

Chromosome   BAC (CmR)

E.coli (DH10B) + target plasmid (pGPCMVBAC)

1

pTsTM8 (AmpR, KmR)   BAC (Cm R)

2

BAC (CmR, KmR)

3

4

A G T C G G

Digest and Sequence

6. If necessary, plate out more diluted culture to get sufficient numbers of separated colonies for further screening.
7. Pick colonies into 3 mL LB medium containing kanamycin and chloramphenicol and incubate overnight at 37°C at 290 rpm (*see* **Note 18**).
8. Harvest cultures, and process as described in miniprep DNA analysis protocol section (*see* **Note 19**).

3.4. Generation of Electrocompetent E. coli Cells and Electrotransformation

This is a general protocol for the generation of electrocompetent cells with transformation efficiency up to a maximum of 10^9 per micrograms of DNA. However, the maximum efficiency will vary between bacterial strains and investigators, and should always be verified prior to transforming the transposition reaction(s). The electrocompetent cells, once generated, can be successfully stored for several months at –70°C until required. It is recommended for BAC plasmids that the *E. coli* strain DH10B be used for stability, but for other plasmids or cosmids the same constraint may not apply. It should be noted that high efficiency electrocompetent cells are commercially available (e.g., DH10B Electromax® cells from Invitrogen) as an alternative to preparing cells from scratch.

3.4.1. Generation of Electrocompetent Organisms

1. Pick a colony from a freshly plated LB agar plate into 5 mL LB medium, and incubate overnight at 37°C.
2. Transfer the 5 mL overnight culture into 500 mL LB medium in a 1-L baffled flask and incubate in shaker incubator at approx 250 rpm at 37°C, until the OD at 600 nm reaches between 0.3 and 0.4. Transfer the culture to two 250 mL conical tubes and place on ice for 1 h.
3. Pellet cells in conical tubes by centrifugation in a GSA rotor in a Sorvall centrifuge at 5000 rpm for 20 min at 4°C.
4. Decant supernatant and carefully resuspend the pellet in 200 ml of ice-cold water and pellet cells as in **step 3**.
5. Repeat **step 4** two more times, but combine the cells into one tube and reduce volume used to resuspend the bacteria to 100 mL of ice-cold water.
6. Resuspend the final pellet in 10 mL 10% glycerol and transfer all the cells to a minimum number of 2 mL microcentrifuge tubes on ice, and centrifuge for 1 min at high speed at 4°C.
7. Resuspend the pellets in 200 µL 10% glycerol per tube, and pool the cells. Aliquot cells into two 15 mL Falcon tubes.
8. Centrifuge the cells again, and resuspend in a total volume of 1–2 mL 10% glycerol. Store the cells in 200 µL aliquots and store at –70°C for up to 6 mo (*see* **Note 20**).

3.4.2. Electroporation

1. Thaw an aliquot of frozen competent cells slowly on wet ice.
2. Chill a sterile Eppendorf tube on ice and add 1–8 μL of DNA sample and 30–50 μL of competent cells. Mix by pipetting slowly up and down on ice.
3. Prepare a 15-mL snap-top tube containing 1 mL of LB medium, and place on ice.
4. Adjust the electroporator to the settings desired for electroporation. Generally, for both BTX and Bio-Rad machines, the settings are as follows: 1.3–2.2 kV, 100 Ohms, and 25 μF. The voltage setting is dependent upon the size of the plasmid DNA. Cosmids and BACs require a higher voltage setting (1.6–2.2 kV) than small plasmids (1.3–1.7 kV).
5. Take a prechilled 1-mm gapped cuvet and add the contents from **step 2**. Tap the cuvet on the bench briefly to ensure that there are no air bubbles, and that cells are evenly distributed. Next, place cuvet in the electroporator, pulse, and immediately return to ice.
6. Add 300 μL of LB medium to the cuvet, mix with cells, and transfer the contents of the cuvette to the remainder of the LB medium in the snapped top tube. Incubate for at least 90 min in a 37°C shaker at approx 200 rpm.
7. Plate out a fraction of the transformation reaction on LB agar plates containing the appropriate antibiotics, and incubate overnight at 37°C. The following day, pick colonies and screen DNA using the miniprep protocol as described later.

3.5. Miniprep Analysis of Transposon-Mutated Large Low Copy Plasmids

Prior to sequencing the transposon insertion site on the modified plasmid, cosmid or BAC, the initial strategy should be to map the insertion site by restriction digestion of miniprep DNA. This has an added advantage in that it enables the identification of any deleted plasmids. A common problem with larger target DNA constructs, such as cosmids and BACs, is obtaining an adequate amount of miniprep DNA for these purposes. Described later is a simple and rapid protocol for the isolation of miniprep BAC DNA. It should be noted that DNA isolated by this technique is suitable for restriction digestion, but not sequencing. To generate high-quality DNA for sequencing or transfection, it is recommended that the DNA be prepared by use of commercially available large-scale plasmid purification kits such as those available from Clontech (Nucleobond® 500 column large plasmid prep purification kit), and that the bacteria be grown up in LB medium (0.5 L–1 L culture) in baffled flasks. As BAC plasmids are binary copy plasmids, the low-copy plasmid protocols described in the kit should be followed.

1. Pick a bacterial colony into 3 mL of LB (plus antibiotics for transposon and BAC/plasmid/cosmid) and incubate at 290 rpm for at least 18 h at 37°C (*see* **Note 21**).

2. Number the original tubes, and transfer most of the culture to 2.5 mL tubes and number accordingly.
3. Pellet bacterial cells by spinning for 1 min at maximum speed. Decant the supernatant and add 200 µL solution 1, and resuspend the bacterial pellet by vortexing.
4. Once the pellet is evenly resuspended, add 200 µL solution 2, vortex, and allow to equilibrate for 10 min at room temperature.
5. Add 200 µL solution 3, vortex, and place on wet ice for 5 min. Centrifuge in microcentrifuge for 5 min, room temperature, at 14,000 rpm.
6. Decant supernatant into a fresh tube and mix with 150 µL phenol/chloroform. Mix by vortexing for 5 s, and microcentrifuge for 5 min.
7. Remove the top aqueous layer to a fresh tube and add 1 mL 100% ethanol. Mix by inversion and place in dry ice/ethanol bath for 20 min, or overnight at –20°C.
8. Pellet the DNA by centrifuging for 10 min at 4°C at maximum speed. A small white pellet should be visible at the bottom of the tube (*see* **Note 22**).
9. Decant the supernatant, and wash pellet with 400–600 µL of 70% ethanol.
10. Resuspend pellet in 40–50 µL of distilled water or TE. Digest 20 µL of the DNA in a single restriction enzyme reaction and analyze on a TBE agarose gel (0.6–0.9%) in the presence of ethidium bromide (*see* **Notes 22** and **23**).

4. Notes

1. Determination of the insertion locus site for the BAC plasmid. If recombinant viruses have previously been generated using non-BAC technologies for the particular herpesvirus being studied, then this may serve to indicate a likely nonessential site in the genome for BAC plasmid insertion. The absolute requirement is that the insertion site does not result in an impairment of the ability of the virus to replicate in tissue culture: the recombinant virus must replicate with similar kinetics as wild-type virus, without any disruption of essential gene function. Insertion of the BAC plasmid into the viral genome can be within a target gene(s), or in an intergenic region, and generally the choice is made based on the existing knowledge surrounding the virus in question. In general, the more is known about the virus, the easier it will be to define a suitable locus. If no recombinants have been generated in the particular virus under study, then comparable viruses should be utilized as controls to elucidate a nonessential region. In general, herpesviruses tend to have conserved colinear blocks within alpha, beta, or gamma herpesvirus groups (*4*). This strategy was employed in the case of guinea pig cytomegalovirus (GPCMV), as the genome has not yet been completely sequenced (*5*). The generation of a simple insertion mutant carrying the eucaryotic green fluorescent protein (eGFP) cassette is an advised step, as it gives the researcher an opportunity to gain valuable hands-on experience in isolating and purifying a recombinant virus prior to the generation of a BAC virus. In addition, the ability of the virus to accommodate additional sequences is an important consideration. Depending upon the reporter gene being included, as well as antibiotic or metabolic selection markers, as much as an additional 9–10 kbp of sequence may be required. In the case of certain herpesviruses, anything in excess of 4 kbp

may be problematic, because adventitious secondary deletions may occur elsewhere on the viral genome, in order for a large insertion to be tolerated. Hence, in such cases the site of the insertion locus may have to include a deletion of several genes, in order to accommodate the BAC insertion *(1)*.

2. Modifying the basic BAC plasmid to include additional useful features. The basic F plasmid is approx 6.5 kbp in size *(6)*, but it may be desirable to further modify it by introducing additional features to enable easier selection of the recombinant virus. First, as noted above, use of a visual reporter gene (eGFP) enables the easier isolation of the recombinant virus, and if available this cassette should be cloned into the basic BAC plasmid. Second, a positive selection system is generally required to enable the easier isolation of a recombinant virus, and if required, this should be introduced into the BAC plasmid as well. Examples of such systems include puromycin selection, or selection with guanosyl phosphoribosyl-transferase *(gpt)*. Third, it may necessary or desirable to remove the BAC plasmid from the recombinant virus upon manipulation at a secondary site in subsequent virus constructs. If so, the insertion site should be flanked by lox P sites *(7)*, or by unique restriction sites. Examples of such useful sites include the 8 bp cutter, *Pme*I, which is absent in all herpesviruses, and *Pac*I, which is absent in many *(8,9)*. Such strategies enable the excision of the BAC plasmid in later downstream applications. It should be noted that although it is possible to manipulate large DNA clones (<160 kbp) by restriction digests, the use of lox P- *(10)* or FRT- *(11,12)* based site-specific recombination may be simpler alternatives. A wide range of basic BAC plasmids as well as modified BAC plasmids carrying a number of additional features are available commercially, and investigators should look into these options, depending upon the needs of their particular systems.

3. Construction of the BAC transfer shuttle vector. In general, the strategy for the generation of a recombinant virus is via homologous recombination, with the BAC plasmid being targeted to a specific locus on the viral genome by the left and right arm flanking homologous sequences on a transfer vector. The intended insertion site on the transfer vector must have a suitable unique restriction site to enable cloning of the linearized BAC plasmid into the flanking sequence. Ideally, insertion should be in the middle of the flanking sequence, with at least 1 kbp present on either side, in order to allow efficient recombination with the viral genome upon cotransfection. Suitable flanking sequences with appropriate restrictions sites can easily be generated via polymerase chain reaction (PCR). As the insertion cassette is a BAC plasmid, the transfer vector effectively becomes a dimer plasmid construct. Consequently, for stability purposes, the plasmid encoding the flanking sequences for recombination cannot have the same origin of replication as the BAC plasmid. This means that vectors such as pBR322 and pUC cannot be used, as they have the *colE*1 origin of replication. There are two suitable low copy plasmid alternatives, based on the p15A origin of replication: these are pACY177 and pACYC184, available from New England Biolabs, Inc. Generation of strains of recombinant virus with engineered unique restriction sites that normally do not cut herpesvirus genomes (e.g., *Pac*I) does allow for the possibility of direct inser-

tion of the BAC plasmid into the viral genome via ligation. However, this strategy is restricted to viral genomes of the same size as HSV or smaller (<160 kbp). This approach does not work with larger herpesviruses, such as the cytomegaloviruses, because the larger DNA molecules are more susceptible to fragmentation. This strategy obviates the need for construction of a flanking sequence transfer vector, as the BAC plasmid linearized with PacI can be cloned directly into the viral genome in an in vitro PacI ligation strategy (8,9), to generate recombinant virus through transfection of the ligation reaction. However, this approach is very specialized and requires the construction of additional viruses. Hence, to save time and extra effort this approach should not be considered unless those recombinant viruses are already available.

4. Generation of a recombinant BAC herpesvirus. Once the transfer vector has been generated, it is next used in cotransfection with infectious wild-type viral DNA onto confluent monolayers of cells to generate progeny virus. Once full viral cytopathic effect is obtained throughout the monolayer, the transfection plate is harvested as a primary viral progeny stock. This stock is subsequently used to infect more cells to obtain second round virus stock. Enrichment for the recombinant virus can be achieved by the use of a positive metabolic selection technique, as mentioned above (e.g., gpt, puromycin, or neomycin selection). Individual plaques which are positive for a particular insertion marker (e.g., eGFP) may be picked to aid in the selection process. The recombinant virus is purified to a clonal population to generate an elite stock, and verified to be genetically correct by Southern blot analysis. However, this is not strictly necessary for a BAC virus, as a mixed population can still be used to transfer the viral genome into bacteria (13). However, high titer virus stock is necessary for the isolation of sufficient quantities of the circular replication intermediate in infected cells (see later).

5. Generation of a BAC plasmid in E. coli. The high titer stock is used to infect cells (multiplicity of infection of ≥4 pfu/cell). Depending upon the virus, this infection is allowed to proceed for 5–24 h postinfection (the timing is dependent upon the virus in question, e.g., 4–6 h for HSV, and 16–20 h for CMV). This time allows the virus to introduce the linear genome into the infected cell, and for that DNA to circularize as a replicative intermediate. This covalently closed genome is the basis for the BAC plasmid in bacteria, and is isolated from infected cells via a so-called "Hirt" prep. As herpesviruses have large genomes, the amount of salt used can be substantially lower than described in the original paper (14), which reduces the likelihood of arcing during high-voltage electroporation of the viral DNA into bacteria. The "Hirt" DNA prep should not be frozen at any stage, or vortexed, as both will cause fragmentation of the genome and reduce the chances of obtaining a full-length genomic clone in bacteria. The DNA is introduced into the bacteria via electroporation, and BAC-carrying bacterial colonies are selected on LB agar plates by virtue of a chloramphenicol resistance marker on the BAC plasmid, and the presence of chloramphenicol in LB agar plates. The bacterial colonies are then screened by miniprep DNA analysis technique to identify full-length viral BAC plasmid clones. Large-scale DNA preps of positive

colonies are made and can be transfected onto mammalian cells to verify infectivity of the BAC plasmid. The presence of an eGFP marker gene makes it relatively easy to identify plaque development on the transfected monolayer. The full-length BAC plasmid can then be further manipulated to generate mutants.

6. The methodology of site-specific mutagenesis of BAC plasmid or cosmid clones is now relatively well established, and systems such as the ET/RED systems are available upon request from individual investigators *(12,15,16)*. These systems allow rapid manipulation of specific sequences via the use of large oligonucleotides. The general principle is that the pair of oligonucleotides have homology to the 5′- and 3′-ends of a specific drug resistance marker (e.g., kanamycin). This allows PCR amplification of the drug resistance marker. The remainder of the oligonucleotides have homology to a specific gene sequence. This allows the investigator to target insertion of the drug marker into the viral BAC plasmid during recombination. Modified BACs are selected in *E. coli* by virtue of the introduced drug marker. This approach is described more fully in other chapters in this book.

7. The attractiveness of the transposon mutagenesis approach lies in the ability to rapidly generate a library of mutated plasmids, each carrying a random insertion of a single copy of a transposon. This technique of insertion mutagenesis is particularly useful for creating disruptions of multiple genes on a single target plasmid. It also enables the more rapid definition of a large plasmid clone, which can be accomplished by using common sequencing priming sites from the 5′- and 3′-ends of the transposon insertion site within the target DNA sequence. Over the last few years, the process of transposon mutagenesis has been improved and simplified by the development of commercially available transposon mutagenesis kits, based on versions of Tn7 or Tn5 transposons *(17,18)*. These transposons, or "transprimers," are basic elements encoding a drug resistance marker (or marker of the investigator's choice) flanked by the transposon's inverted repeats. The transposon is usually added to the reaction as a purified PCR product, or encoded on a donor plasmid incapable of replication in normal *E. coli* cells, to eliminate any background from the donor transposon source. As these transprimers do not encode any of the transposition genes, these are supplemented by the addition of purified transposon transposase/resolvase enzymes and reagents to the reaction, to enable transposition of the target DNA to occur. This approach virtually eliminates secondary transposition events on the target DNA, and the transposition reaction occurs at a relatively high efficiency in vitro over a 1–2 h time period. A library of mutant plasmids is generated in bacteria by transformation of the transposon reaction and selection for the transposon drug resistant marker. The mutated plasmids can be quickly defined by restriction enzyme digestion of miniprep DNA, and transposons precisely mapped by sequencing from the transposon into the target DNA insertion site. As both Tn5 and Tn7 systems are so similar (and in an effort to avoid redundancy), only the Tn5-based transposon system will be described later. However, for further information on the Tn7 system, readers are referred to specifications from commercial systems, such as the GPS® (Tn7) system from New England Biolabs, Inc.

8. As a contrast to these in vitro systems, an efficient in vivo transposon system is also described later. One of the major problems with an in vivo approach is that the bacterial genome, and not the plasmid DNA, is preferentially targeted for mutagenesis, as it is a substantially larger target. Fortunately, the TnMAX system is based on a modified Tn3 transposon, Tn 172, that preferentially targets episomal DNA *(19,20)*. In addition, the TnMAX system has been substantially improved by the use of temperature sensitive donor plasmids, a modification which substantially reduces background from the donor plasmid *(21)*. The genes necessary for transposition are encoded on the donor plasmid outside of the transposon inverted repeats, a feature which serves to eliminate secondary transposition events. The donor plasmid replicates at 30°C, but does not replicate at temperatures between 37°C and 42°C. Therefore, it is easily eliminated from the BAC bacterial host strain by a temperature shift from permissive temperature (30°C) to nonpermissive conditions. Similar to the Tn5 and Tn7 systems described, the TnMAX transposon only encodes a drug resistance marker (erythromycin, kanamycin, or tetracycline). This gives readers a choice in approach to transposon mutagenesis. It should be noted that this transposon strategy works well with most large target DNA molecules, but a few problems have arisen with very large BAC plasmid clones, where spontaneous deletions have also resulted during transformation of the DNA into bacteria.

9. In order to obtain maximum transposition efficiency in vitro, all necessary reagents, in particular the donor and target DNAs, must be present at optimal concentrations. When using one of the commercially available transposon kits, the most critical factor is to have the donor transposon DNA and target DNA at the correct ratio, in order to enable single event insertions to occur optimally (donor DNA should be in present in a 2 to 1 molar excess above that of the target DNA). This means that the amounts of DNA used will vary among different target DNA molecules. This is particularly true with large plasmid and cosmid clones, which are generally at lower molar concentrations. In addition, the target DNA used should be of the highest quality, and in the case of larger DNA constructs, the DNA should not be repeatedly freeze-thawed. Two possible storage procedures are suggested: store the DNA at −70°C in single-use aliquots; or refrigerate at 4°C. This avoids problems (such as partial deletion of the target DNA during transformation into *E. coli* bacteria) when establishing the transposon-mutated plasmid library. Similarly, the large DNA prep should never be vortexed, and should always gently pipeted.

10. The transposon used in this system is a modified version of the Tn5 transposon, consisting of a drug resistance marker (either kanamycin, chloramphenicol, or tetracycline) flanked by the Tn5 inverted repeats. There is also another variant available that allows the insertion of the investigator's own selectable marker of choice. The transposition event is carried out by one protein, the transposase, which is added to the in vitro reaction. The transposase used is a hyperactive variant that increases the effiency of the transposition reaction. Another similar system is available from NEB, which uses the Tn7 transposon. Both systems work very efficiently, and both

allow the possibility of modifying the supplied transposon to include additional markers, such as an eGFP marker. The Tn7 system uses a donor plasmid encoding the Tn7 transprimer, and is incapable of replication in normal *E. coli*. This offers an advantage if the transprimer has to be modified to include additional markers, as purification of the transposon away from the donor plasmid is not required to reduce background as in the case of the Tn5 system. In our experience the inclusion of extra marker genes in the transposon does not impair the transposition reaction to any significant degree using a transposon up to 3.5 kbp in size (basic size of the Tn5 transposon is about 1 kbp). However, it should be noted that in general, the smaller the transposon, the easier it is to localize the site of insertion on the target DNA. **Figure 4** provides a schematic version of the Tn5 transposition reaction on target DNA, and a scheme for selection of mutated plasmids.

11. The larger the DNA molecule, the greater the concentration of DNA required (in some cases, 1 µg of DNA may be necessary).

12. It is important to use high-efficiency competent cells for transformation. The transformation efficiency should be at least $>10^7$ per 1 µg of DNA. This is particularly important with larger target plasmids, or cosmids, as these do not transform as easily as small constructs. When testing the efficiencies of the transposition reaction and the transformation reactions, it is suggested that initially a small target control plasmid (e.g., pUC19) be used first to acquaint the novice new to transformation and transposition techniques. It should be noted that chemically competent cells can be used as an alternative to electroporation, but this protocol is not recommended for cosmids or BAC clones, because of poorer transformation efficiencies via this technique.

13. **Problem**—Competent cells transform with poor efficiency. **Solution**—Make fresh competent cells and test their efficiency by transforming a small plasmid (e.g., pUC19). A high transformation efficiency increases the yield of transposon mutants from a single reaction. Competent cells can be stored long-term at –70°C, but their efficiency will drop with time. The highest transformation efficiency is obtained with fresh competent cells.

14. **Problem**—Failure to obtain drug resistant colonies after plating out on LB agar plates. **Solution**—Two possible solutions are noted. Either the competent cells are not competent, or are not of sufficiently high competence. Alternatively, the transposon reaction failed. In order to ascertain which is the problem, the following tests are suggested. Plate out part of the transformation reaction using single antibiotics. First, plate out using the drug marker on the target plasmid to test transformation efficiency of competent cells. A large number of colonies should be obtained per plate. If only a few colonies are obtained, then the cells are of poor transformation efficiency. Second, plate out the transformation reaction using the drug marker on the transposon to determine the efficiency of the transposition reaction. Fewer colonies should be present on this plate in comparison to those obtained for target plasmids, but if a very low number or no colonies are obtained, then the transposon reaction has failed. Repeat the transposition reaction again with modified ratio of donor plasmid.

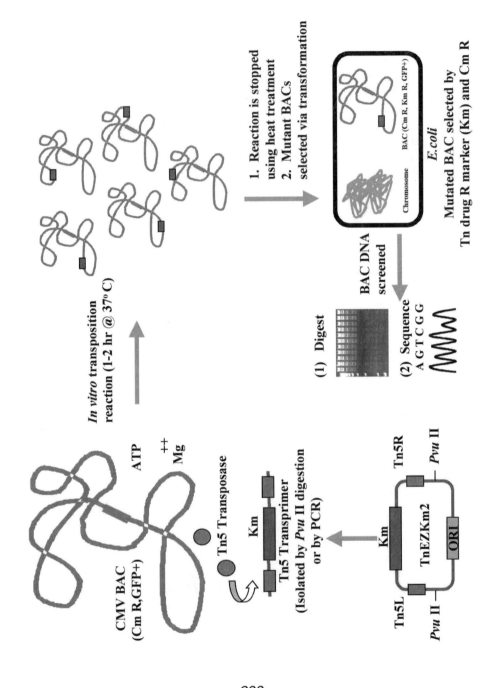

15. **Problem**—No transposon mutants are present, or only a few transposon mutants with multiple transposon inserts. **Solution**—These results indicate that the transposon/donor plasmid is present in the reaction at quantities producing a suboptimal ratio to the target plasmid DNA. Reduce the concentration of donor plasmid by 10-fold, and repeat the reaction. In addition, if possible, increase the target DNA concentration up to a maximum of fivefold in another reaction. Note that a larger plasmid requires more input target DNA than a smaller plasmid to achieve the same molarity.

16. **Problem**—Transposon-mutated plasmids carry deletions as well as a transposon insertion. **Solution**—This could indicate a variety of problems. The first possibility is that the target DNA used was frozen and thawed or vortexed. When using large plasmids, use DNA stored at 4°C to avoid this possibility. Generally, freeze-thawing of plasmids over 20 kbp in size will translate to generation of deletions during the transposition/transformation reaction. The percentage of deletions can range from 20% to 90% of the population if the plasmid has not been treated properly. The second possibility pertains to certain large BAC plasmids which, despite proper storage, are more prone to generate deletions in the transposition/transformation reaction. In this case, the size of the insert, together with the sequence cloned, would appear to present a stability problem during the transposition reaction. To improve the likelihood of obtaining full-length mutant plasmids, the use of a modified DH10B *E. coli* strain EC100® (from Epicenter technologies) is recommended as a possible solution to this problem. If deletions are still obtained, then an alternative in vivo strategy of transposon mutagenesis should be considered. The Tn5 system can itself be modified to form transposomes (a complex between transposon and transposase protein formed in the absence of magnesium), which can be used to effectively modify the BAC plasmid in bacteria *(23)*. Alternatively, the TnMax system described below can be used.

17. **Problem**—Failure to transfer the donor transposon plasmid into BAC-containing cells. **Solution**—If correct antibiotics (ampicillin and chloramphenicol) are used

---

Fig. 4. *(see opposite page)* Tn5 Transposon mutagenesis of a large DNA plasmid in vitro. Tn5 transposon DNA isolated from the donor plasmid is added to a reaction tube containing the large target DNA molecule (plasmid, cosmid or BAC). In the presence of Mg^{2+} and adenosine triphosphate (ATP), the hyperactive transposase enables transposition reactions to occur at relatively high efficiency over a 1–2 h incubation period at 37°C. The reaction is terminated by heat inactivation, and transposon-mutated plasmids are isolated by transformation of the reaction into a suitable recA⁻ strain of *E. coli*, such as DH10B. Mutant colonies are selected by the inclusion of the transposon antibiotic marker (kanamycin) in the LB agar plates. DNA from the bacterial colonies is screened by restriction digestion of minipreps to identify the approximate locus of insertion by band shift. The exact site of transposon insertion is determined by sequencing from the 5′ and 3′ ends of the transposon into the insertion site of the target plasmid DNA.

to select for the donor plasmid and BAC under permissive conditions (30°C), then undoubtedly the bacterial cells used are either not competent, or selection was not maintained for the target BAC DNA during production of the competent cells.

18. **Problem**—Kanamycin-resistant BAC mutant colonies fail to grow, or grow very poorly in LB medium overnight in the presence of both chloramphenicol and kanamycin. **Solution**—Using the TnMAX system, colonies can sometimes be growth-impaired. Try growing the mutant BAC only in the presence of one antibiotic (chloramphenicol or kanamycin). Alternatively, reduce the concentration of kanamycin in the incubation medium to 10 µg/mL.

19. Deletion mutants are common with both in vivo and in vitro systems. However, for the in vivo system and BAC target DNA, the in vivo approach should produce fewer deletion mutants. Consequently, if more than 40% of the mutant population contain deletions as well as insertions, the investigator should pick a fresh colony and repeat the selection process described above rather than continue.

20. Always test the competency of the bacteria by transforming a small plasmid (such as pUC) and plate out on LB agar plates under antibiotic selection (for pUC, use ampicillin at 100 µg/µL). In the case of transposome mutagenesis of a BAC, it is necessary to grow up the bacterial strain in the presence of an antibiotic, e.g., chloramphenicol (12.5 µg/µL), to maintain selection of the BAC in the bacterial strain which, if not included, will rapidly be lost.

21. **Problem**—Varying amounts of DNA were obtained from different colonies picked from plates. **Solution**—It is important to let the BAC minipreps grow for at least 18–24 h to get sufficient amounts of DNA for analysis. Occasionally, some colonies will grow poorly, and cultures should be visually checked prior to harvesting. If some cultures are not as turbid, these should be returned to the incubator for a further period of incubation until sufficiently high density is obtained (0.5–0.6 absorbance at 600 nm).

22. **Problem**—Cultures grow, but no mini-prep DNA was obtained. **Solution**—Lack of antibiotic in LB medium. Chloramphenicol should always be included for selection of BAC plasmid (12.5 µg/mL). Additionally, appropriate antibiotic should be included for the selection of the transposon, but the concentration used will vary between antibiotic and transposon used (kanamycin is generally used at µg/mL).

23. This protocol should provide sufficient DNA for at least two restriction digestion reactions of BAC DNA, but enough for more digestions if cosmid or other plasmid clones are used. If large amounts of DNA are required, it is suggested that duplicate tubes be set up, or alternatively a "midi" scale culture be set up, using the Clontech Nucleobond® purification kit AX100 column, or another suitable system. It is recommended that TBE gels be used for the analysis of digested BAC DNA, rather than TAE gels, as these gels tend to give better resolution for more complicated restriction patterns of DNA present at low levels. In general, the Tn insertion site can be detected by a single band shift in comparison to the wild type BAC DNA profile, if the correct restriction enzyme is selected.

Acknowledgment

The authors would like to thank Wolfram Brune for supplying the temperature sensitive TnMax plasmids.

References

1. Messerle, M., Crnkovic, I., Hammerschmidt, W., Ziegler, H., and Koszinowski, U. H. (1997) Cloning and mutagenesis of a herpesvirus genome as an infectious bacterial artificial chromosome. *Proc. Natl. Acad. Sci. USA* **94,** 14,759–14,763.
2. Borst, E. M., Hahn, G., Koszinowski, U. H., and Messerle, M. (1999) Cloning of the human cytomegalovirus (HCMV) genome as an infectious bacterial artificial chromosome in Escherichia coli: a new approach for construction of HCMV mutants. *J. Virol.* **73,** 8320–8329.
3. Brune, W., Messerle, M., and Koszinowski, U. H. (2000) Forward with BACs: new tools for herpesvirus genomics. *Trends Genet.* **16,** 254–259.
4. McGeoch, D. J., Cook, S., Dolan, A., Jamieson, F. E., and Telford, E. A. (1995) Molecular phylogeny and evolutionary timescale for the family of mammalian herpesviruses. *J. Mol. Biol.* **247,** 443–458.
5. McGregor, A. and Schleiss, M. R. (2001) Molecular cloning of the guinea pig cytomegalovirus genome as an infectious bacterial artificial chromosome in *E. coli. Mol. Genet. Metabol.* **72,** 15–26.
6. Shizuya, H., Birren, B., Kim, U. J., et al. (1992) Cloning and stable maintenance of 300-kilobase-pair fragments of human DNA in *Escherichia coli* using an F-factor-based vector. *Proc. Natl. Acad. Sci. USA* **89,** 8794–8797.
7. Hobom, U., Brune, W., Messerle, M., Hahn, G., and Koszinowski, U. H. (2000) Fast screening procedures for random transposon libraries of cloned herpesvirus genomes: mutational analysis of human cytomegalovirus envelope glycoprotein genes. *J. Virol.* **74,** 7720–7729.
8. McGregor, A., Roberts, A., Davies, R. W., Clements, J. B., and MacLean, A. R. (1999) Rapid generation of recombinant herpes simplex virus vectors expressing the bacterial *lacZ* gene under the control of neuronal promoters. *Gene Ther. Molec. Biol.* **4,** 193–202.
9. McGregor, A. and Schleiss, M. R. (2001) Recent advances in Herpesvirus genetics using bacterial artificial chromosomes. *Mol. Genet. Metabol.* **72,** 8–14.
10. Buchholz, F., Angrand, P. O., and Stewart, A. F. (1996) A simple assay to determine the functionality of Cre or FLP recombination targets in genomic manipulation constructs. *Nucl. Acids Res.* **24,** 3118–3119.
11. Smith, G. A. and Enquist, L. W. (2000) A self-recombining bacterial artificial chromosome and its application for analysis of herpesvirus pathogenesis. *Proc. Natl. Acad. Sci. USA* **97,** 4873–4878.
12. Datsenko, K. A. and Wanner, B. L. (2000) One-step inactivation of chromosomal genes in Escherichia coli K-12 using PCR products. *Proc. Natl. Acad. Sci. USA* **97,** 6640–6645.

13. Schumacher, D., Tischer, B. K., Fuchs, W., and Osterrieder, N. (2000) Reconstitution of Marek's disease virus serotype 1 (MDV-1) from DNA cloned as a bacterial artificial chromosome and characterization of a glycoprotein B-negative MDV-1 mutant. *J. Virol.* **74,** 11,088–11,098.

14. Hirt, B. (1967) Selective extraction of polyoma DNA from infected mouse cell cultures. *J. Mol. Biol.* **26,** 365–369.

15. Lee, E. C., Yu, D., Martinez de Velasco, J., et al. (2001) A highly efficient *Escherichia coli*-based chromosome engineering system adapted for recombinogenic targeting and subcloning of BAC DNA. *Genomics* **73,** 56–65.

16. Muyrers, J. P., Zhang, Y., Testa, G., and Stewart, A. F. (1999) Rapid modification of bacterial artificial chromosomes by ET-recombination. *Nucl. Acids Res.* **27,** 1555–1557.

17. Biery, M. C., Stewart, F. J., Stellwagen, A. E., Raleigh, E. A., and Craig, N. L. (2000) A simple in vitro Tn7 based transposition system with low target site selectivity for genome and gene analysis. *Nucl. Acids Res.* **28,** 1067–1077.

18. Naumann, T. A. and Reznikoff, W. S. (2000) Trans catalysis in Tn5 transposition. *Proc. Natl. Acad. Sci. USA* **97,** 8944–8949.

19. Haas, R., Kahrs, A. F., Facius, D., Allmeier, H., Schmitt, R., and Meyer, T. F. (1993) TnMax—a versatile mini-transposon for the analysis of cloned genes and shuttle mutagenesis. *Gene* **130,** 23–31.

20. Kahrs, A. F., Odenbreit, S., Schmitt, W., Heuermann, D., Meyer, T. F., and Haas, R. (1995) An improved TnMax mini-transposon system suitable for sequencing, shuttle mutagenesis and gene fusions. *Gene* **167,** 53–57.

21. Brune, W., Menard, C., Hobom, U., Odenbreit, S., Messerle, M., and Koszinowski, U. H. (1999) Rapid identification of essential and nonessential herpesvirus genes by direct transposon mutagenesis. *Nat. Biotechnol.* **17,** 360–364.

22. Posfai, G., Koob, M. D., Kirkpatrick, H. A., and Blattner, F. R. (1997) Versatile insertion plasmids for targeted genome manipulations in bacteria: isolation, deletion, and rescue of the pathogenicity island LEE of the *Escherichia coli* O157:H7 genome. *J. Bacteriol.* **179,** 4426–4428.

23. Goryshin, I. Y., Jendrisak, J., Hoffman, L. M., Meis, R., and Reznikoff, W. S. (2000) Insertional transposon mutagenesis by electroporation of released Tn5 transposition complexes. *Nat. Biotechnol.* **18,** 97–100.

20

Protective DNA Vaccination by Particle Bombardment Using BAC DNA Containing a Replication-Competent, Packaging-Defective Genome of Herpes Simplex Virus Type I

Mark Suter and Hans Peter Hefti

1. Introduction

Particle bombardment is a physical method of cell transformation that can be applied in vitro as well as in vivo *(1,2)*. The DNA- or RNA-coated gold particles that are generally used are of high density, subcellular in size, and are accelerated to high velocity to carry the genetic information into cells. Because this method does not require specific ligand–receptor interaction for application, plants, bacteria, yeast, or mammalian cells have been used as targets for DNA or RNA bombardment *(3–7)*.

Particle bombardment has been applied to mice, sheep, calves, horses, or humans to protect these mammals against infectious diseases. This procedure is generally referred to as DNA vaccination *(8)*. For this, a limited number of different genes with appropriate promoter and regulatory elements that allow the expression of the desired antigens and possibly immune enhancing cytokines have been applied to cutaneous surfaces *(9)*. From these sites, antigens are transported to local lymphnodes that are equipped with appropriate immunological elements required for the initiation of an immune response. Antigen transport from the periphery to lymph nodes is performed by specialized antigen presenting cells termed dendritic cells (DC). For successful DNA vaccination, DC need to be directly loaded with DNA. Antigens expressed by DNA transfected cells other than DC appear not to permit initiation of a detectable immune response *(10)*. By contrast, viruses with a controlled ability to replicate

From: *Methods in Molecular Biology, vol. 256:*
Bacterial Artificial Chromosomes, Volume 2: Functional Studies
Edited by: S. Zhao and M. Stodolsky © Humana Press Inc., Totowa, NJ

were shown to be able to initiate immune responses possibly by indirectly stimulating DC function *(11,12)*. Virus replication is therefore thought to be associated with immune-enhancing capabilities.

Recently, the entire genome of herpes simplex virus type I (HSV-1) lacking the cleavage/packaging signals was cloned in a bacterial artificial chromosome (BAC), termed fHSV-1Δ pac *(13)*. When transfected into mammalian cells HSV-1-like particles devoid of viral DNA are produced *(14)*. Hence, fHSV-1Δ pac DNA encodes a noninfectious replication unit of HSV-1 and appears to be safe for DNA vaccination. Moreover, multiple antigens may be encoded in BACs to broaden the immune response. The successful use of fHSV-1Δ pac for protective DNA vaccination, termed BAC-VAC, has recently been described *(15)*. The most efficient system for DNA delivery used for BAC-VAC was particle bombardment. In this chapter, procedures for loading of fHSV-1Δ pac DNA onto gold particles and particle bombardment for DNA delivery are described.

2. Materials
2.1. Coating of Purified fHSV-1Δ pac DNA to Gold Particles

All chemicals are from commercial sources and of highest possible purity. Solutions should be made with double-distilled water (ddH$_2$O). The "helios gene gun system" including gold microcarriers (Bio-Rad; Fischer Scientific, Switzerland) was used throughout.

1. 0.05 *M* spermidine (Sigma).
2. 1 *M* CaCl$_2$ (Sigma).
3. Purified fHSV-1Δ pac (*see* **Note 1**): dilute the purified DNA to 1 µg/µL in distilled water.
4. Ethanol (100%): always use a fresh bottle of ethanol. Ethanol will absorb water over time.
5. 20 mg/mL polyvinylpyrrolidone (PVP) solution: prepare a PVP stock solution of 20 mg/mL of PVP in ethanol in a small screw cap container. The solution can be stored at –20°C for several months. Before use, dilute to a working solution of 0.05 mg/mL (1:400) with 100% fresh ethanol.
6. 15 mL disposable polypropylene centrifuge tubes.
7. 1.5 mL microfuge plastic tubes.
8. Ultrasonic cleaner (e.g., Fisher FS3, Branson 1210).

2.2. Particle Bombardment

Mice are kept under specific pathogen free (SPF) conditions throughout the study.

1. Rompun (2% Xylazine, Bayer).
2. Ketamine (100 mg/mL ketamine hydrochloride, Parnell, Alexandria, NSW, Australia).

3. Sterile physiological saline (0.9% NaCl, Braun Medical AG, Switzerland).
4. Syringe (1 mL).
5. Hypodermic needle (22–30 g).
6. 6–8-wk-old C57/BL6 mice.
7. Gene Gun (Bio-Rad; Fischer Scientific, Switzerland).

3. Methods

3.1. Coating of Purified fHSV-1Δ pac DNA to Gold Particles

DNA is loaded onto gold microcarriers by a $CaCl_2$ precipitation procedure. DNA covered microcarriers are then immobilized on the inner surface of plastic tubing with a mixture of ethanol/polyvinylpyrrolidone (*see* **Notes 2** and **3**). The tubing is cut into cartridges and the DNA loaded microcarriers are propelled into cells by applying high-pressured helium gas.

1. Weigh out 12.5 mg of gold particles (0.6 μm in diameter) in a microfuge plastic tube (1.5 mL) (*see* **Note 2**).
2. Add 50 μL of 0.05 *M* spermidine solution to the measured microcarriers.
3. Vortex the microcarriers and spermidine mixture for a few seconds, then sonicate for 3–5 s using an ultrasonic cleaner to break up remaining gold clumps. You will have a brown homogenous solution without visible clumps.
4. Add 50 μL of fHSV-1Δ pac DNA (1 μg/μL) and mix by vortexing for 5 s.
5. Add 50 μL of 1 *M* $CaCl_2$ dropwise to the tube while vortexing at a moderate rate.
6. Allow the mixture to precipitate for 10 min at room temperature. The microcarriers will form a pellet and the supernatant will become relatively clear. Some of the precipitate will form along the side of the tube.
7. Spin the precipitated microcarrier solution for 10 s to pellet the gold particels, then carefully remove the supernatant.
8. Resuspend the pellet in the remaining supernatant by vortexing. Make sure you resuspend the pellet until you see no more clumps. Do not add ethanol to resuspend the pellet.
9. Wash the pellet three times with 1ml of fresh 100% ethanol (*see* **Note 4**). Spin 5 s in a microfuge between each wash step and discard the supernatant.
10. Resuspend the pellet in 200 μL ethanol containing 0.05 mg/mL PVP. Make sure the microcarrier solution is well dispersed without any clumps (*see* **Notes 5** and **6**).
11. Transfer the suspension to a 15-mL polypropylene tube containing 3 mL of 0.05 mg/mL of PVP/ethanol solution. Rinse the microfuge tube once with 200 μL of PVP/ethanol solution and add it to the 15-mL tube (*see* **Notes 5** and **6**).
12. The ethanol-PVP solution containing the DNA coated microcarriers is now ready for loading to tubes and the production of cartridges. This is very straightforward and can be done without modifications as described in the manual provided by Bio-Rad. Similarly, the application by gene gun is according to need. There is no difference between BAC DNA or other DNA-loaded microcarrier to be used for DNA vaccination (*see* **Notes 7–9**).

3.1. Particle Bombardment

1. To prepare the anesthetics, dilute 1 part of Rompun and 1 part of Ketamine in 9 parts of sterile physiological saline. Fill a syringe with 200 μL of the anesthetic (for a 25 g mouse). The Rompun/Ketamine mixture may be stored at 4°C for up to 2 wk.
2. Hold a mouse near the base of the tail and lift the animal out of the cage. Then, get the nape of the neck with the opposite hand, place the tail between two fingers to secure and control the animal. The mouse is now ready for injection. Administer the anesthetics in a steady, fluid motion into the peritoneum. The mouse will be anesthetized for the next 30–45 min.
3. The anesthetized mouse is placed under a heating lamp. Hold the mouse at the base of the tail and remove the fur by plugging the hair with two fingers; remove an area of 2–3 cm^2. Do not use any instruments, since the hair can be plugged easily when the mouse is under adequate anesthesia.
4. The mouse is now ready for DNA vaccination. The gene gun may be applied as suggested by the manufacturer. Set the pressure to 400 psi. Place the mouse on a smooth surface. The mouse will be gently pushed forward by the physical force of the gold particles propelled into the skin (*see* **Note 9**).

4. Notes

1. The purifiction of fHSV-1Δ pac DNA is described in detail in Chapter 16.
2. Gold microcarriers are available at 1.6, 1, or 0.6 μm in diameter. All seizes of carriers can successfully be used to deliver fHSV-1Δ pac DNA into the skin of mice. We prefer 0.6 μm particles because they seem to be both, easiest to handle and most efficient in DNA delivery.
3. Make sure that all reagents, in particular microcarriers and are kept dry.
4. For each coating, a fresh unopened bottle of 100% ethanol is used. Opened bottles of ethanol absorb water. This will interfere with coating of DNA loaded microcarries to the tubing.
5. PVP/ethanol solutions have to be stored in tightly closed containers to avoid evaporation of ethanol.
6. Polyvinylpyrrolidone (PVP) serves as an adhesive during the cartridge preparation procedure. It increases the amount of gold microcarriers that can be loaded on the inner surface of the tube. The immobilized particles are discharged by high gas pressure during particle bombardment.
7. For each coating, 50 μg DNA and 12.5 mg of gold microcarriers were found to be optimal. We routinely used BAC DNA at a concentration of 1 μg/μL dissolved in distilled water, rather than in TE buffer.
8. BAC DNA as any other single or double stranded plasmid DNA is supercoiled, very stable and it is thus not expected that vortexing can break up this DNA by sheer forces.
9. Successful delivery of functional BAC DNA may be assessed in vitro. Briefly, fHSV GFP DNA (*16*) is loaded onto 1.6 μm gold particles and propelled onto a

monolayer of Vero 2-2 cells in tissue culture dishes applying a reduced air pressure of 100 psi. GFP-positive cells are observed after overnight incubation and virus plaques develop gradually, indicating delivery of intact DNA, as well as proper functioning of the gene gun.

References

1. Fox, T. D., Sanford, J. C., and McMullin, T. W. (1988) Plasmids can stably transform yeast mitochondria lacking endogenous mtDNA. *Proc. Natl. Acad. Sci. USA* **85(19),** 7288–7292.
2. Ulmer, J. B., Donnelly, J. J., Parker, S. E., et al. (1993) Heterologous protection against influenza by injection of DNA encoding a viral protein. *Science* **259(5102),** 1745–1749.
3. Zelenin, A. V., Titomirov, A. V., and Kolesnikov, V. A. (1989) Genetic transformation of mouse cultured cells with the help of high-velocity mechanical DNA injection. *FEBS Lett.* **244(1),** 65–67.
4. Yang, N. S., Burkholder, J., Roberts, B., Martinell, B., and McCabe, D. (1990) In vivo and in vitro gene transfer to mammalian somatic cells by particle bombardment. *Proc. Natl. Acad. Sci. USA* **87(24),** 9568–9572.
5. Williams, R. S., Johnston, S. A., Riedy, M., DeVit, M. J., McElligott, S. G., and Sanford, J. C. (1991) Introduction of foreign genes into tissues of living mice by DNA-coated microprojectiles. *Proc. Natl. Acad. Sci. USA* **88(7),** 2726–2730.
6. Qiu, P., Ziegelhoffer, P., Sun, J., and Yang, N. S. (1996) Gene gun delivery of mRNA in situ results in efficient transgene expression and genetic immunization. *Gene Ther.* **3(3),** 262–268.
7. Ramsay, A. J., Kent, S. J., Strugnell, R. A., Suhrbier, A., Thomson, S. A., and Ramshaw, I. A. (1999) Genetic vaccination strategies for enhanced cellular, humoral and mucosal immunity. *Immunol. Rev.* **171,** 27–44.
8. Babiuk, L. A., van Drunen Littel-van den, H., and Babiuk, S. L. (1999) Immunization of animals: from DNA to the dinner plate. *Vet. Immunol. Immunopathol.* **72(1–2),** 189–202.
9. Sin, J. I., Kim, J. J., Boyer, J. D., Ciccarelli, R. B., Higgins, T. J., and Weiner, D. B. (1999) In vivo modulation of vaccine-induced immune responses toward a Th1 phenotype increases potency and vaccine effectiveness in a herpes simplex virus type 2 mouse model. *J. Virol.* **73(1),** 501–509.
10. Porgador, A., Irvine, K. R., Iwasaki, A., Barber, B. H., Restifo, N. P., and Germain, R. N. (1998) Predominant role for directly transfected dendritic cells in antigen presentation to CD8+ T cells after gene gun immunization. *J. Exp. Med.* **188(6),** 1075–1082.
11. Salio, M., Cella, M., Suter, M., and Lanzavecchia, A. (1999) Inhibition of dendritic cell maturation by herpes simplex virus. *Eur. J. Immunol.* **29(10),** 3245–3253.
12. Franchini, M., Abril, C., Schwerdel, C., Ruedl, C., Ackermann, M., and Suter, M. (2001) Protective T-cell based immunity induced in neonatal mice by a single replication cycle of herpes simplex virus. *J. Virol.* **75,** 83–89.s.

13. Fraefel, C., Song, S., Lim, F., et al. (1996) Helper virus-free transfer of herpes simplex virus type 1 plasmid vectors into neural cells. *J. Virol.* **70(10),** 7190–7197.

14. Jacobs, A., Breakefield, X. O., and Fraefel, C. (1999) HSV-1-based vectors for gene therapy of neurological diseases and brain tumors: part I. HSV-1 structure, replication and pathogenesis. *Neoplasia* **1(5),** 387–401.

15. Suter, M., Lew, A. M., Grob, P., et al. (1999) BAC-VAC, a novel generation of (DNA) vaccines: A bacterial artificial chromosome (BAC) containing a replication-competent, packaging- defective virus genome induces protective immunity against herpes simplex virus 1. *Proc. Natl. Acad. Sci. USA* **96(22),** 12,697–12,702.

16. Saeki, Y., Frafel, C., Ichikawa, T., Breakefield, X. O., and Chiocca, E. A. (2001) Improved helper virus-free packaging system for HSV amplicon vectors using an ICP27-deleted, oversized HSV-1 DNA in a bacterial artificial chromosome. *Molec. Ther.* **4(3),** 591–601.

Index